The Cultural Politics of Blood, 1500–1900

The Cultural Politics of Blood, 1500–1900

Edited by

Kimberly Anne Coles
Associate Professor, Department of English, University of Maryland, USA

Ralph Bauer
Associate Professor, Department of English, University of Maryland, USA

Zita Nunes
Associate Professor, Department of English, University of Maryland, USA

and

Carla L. Peterson
Professor Emerita, Department of English, University of Maryland, USA

Editorial matter, introduction and selection © Kimberly Anne Coles, Ralph Bauer, Zita Nunes and Carla L. Peterson 2015
Remaining chapters © Respective authors 2015
Foreword © Priscilla Wald 2105

All rights reserved. No reproduction, copy or transmission of this publication may be made without written permission.

No portion of this publication may be reproduced, copied or transmitted save with written permission or in accordance with the provisions of the Copyright, Designs and Patents Act 1988, or under the terms of any licence permitting limited copying issued by the Copyright Licensing Agency, Saffron House, 6–10 Kirby Street, London EC1N 8TS.

Any person who does any unauthorized act in relation to this publication may be liable to criminal prosecution and civil claims for damages.

The authors have asserted their rights to be identified as the authors of this work in accordance with the Copyright, Designs and Patents Act 1988.

First published 2015 by
PALGRAVE MACMILLAN

Palgrave Macmillan in the UK is an imprint of Macmillan Publishers Limited, registered in England, company number 785998, of Houndmills, Basingstoke, Hampshire RG21 6XS.

Palgrave Macmillan in the US is a division of St Martin's Press LLC, 175 Fifth Avenue, New York, NY 10010.

Palgrave Macmillan is the global academic imprint of the above companies and has companies and representatives throughout the world.

Palgrave® and Macmillan® are registered trademarks in the United States, the United Kingdom, Europe and other countries.

ISBN 978–1–137–33820–4

This book is printed on paper suitable for recycling and made from fully managed and sustained forest sources. Logging, pulping and manufacturing processes are expected to conform to the environmental regulations of the country of origin.

A catalogue record for this book is available from the British Library.

A catalog record for this book is available from the Library of Congress.

Typeset by MPS Limited, Chennai, India.

Transferred to Digital Printing in 2015

Contents

List of Illustrations	vii
Foreword by Priscilla Wald	viii
Acknowledgments	xii
Notes on Contributors	xiii

Introduction 1
Kimberly Anne Coles, Ralph Bauer, Carla L. Peterson, and Zita Nunes

Part I Race and Stock

1 Metamateriality and Blood Purity in Cervantes's Alcaná de Toledo 25
   Rachel L. Burk

2 The Blood of Others: Breeding Plants, Animals, and White People in the Spanish Atlantic 45
   Ruth Hill

3 "Rude Uncivill Blood": The Pastoral Challenge to Hereditary Race in Fletcher and Milton 65
   Jean E. Feerick

4 African Blood, Colonial Money, and Respectable Mulatto Heiresses Reforming Eighteenth-Century England 84
   Lyndon J. Dominique

Part II Moral Constitution

5 "His blood be on us and on our children": Medieval Theology and the Demise of Jewish Somatic Inferiority in Early Modern England 107
   M. Lindsay Kaplan

6 Sor Juana's Appetite: Body, Mind, and Vitality in "First Dream" 127
   Anna More

7 Blood and Character in Early African American Literature 146
   Hannah Spahn

Part III Medicalizing the Political Body

8 Flowing or Pumping? The Blood of the Body Politic in Burton, Harvey, and Hobbes   171
   Robert Appelbaum

9 Linnaeus and the Four Corners of the World   191
   Staffan Müller-Wille

10 "Who Got Bloody?": The Cultural Meanings of Blood during the Civil War and Reconstruction   210
    James Downs

11 Colonial Transfusions: Cuban Bodies and Spanish Loyalty in the Nineteenth Century   229
    David Sartorius

Further Reading   251

Index   258

List of Illustrations

Cover *Commiphora gileadensis* (L.) C. Chr. (Balm of Gilead, Opobalsam): finished drawing of tree's habit (with roots). Courtesy of the Yale Center for British Art

1.1 Title page of *El ingenioso hidalgo Don Qvixote de la Mancha* (editio princeps 1605). Courtesy of the Rosenbach Museum and Library, Philadelphia   38

9.1 "Man's" place in the system of nature. Detail from Carl Linnaeus, *Systema Naturae, sive Regna Tria Naturae Systematice Proposita per Classes, Ordines, Genera, & Species* (Leiden: Theodor Haak, 1735)   192

9.2 Frontispiece of Carl Linnaeus, *Hortus Cliffortianus* (Amsterdam, 1738). Courtesy of WikiCommons   200

9.3 Entry for "Homo Sapiens" in Linnaeus's personal, annotated copy of the tenth edition of *Systema Naturae* (1758). Linnean Society London, Library and Archives, Linnaean Collections, *Systema Naturae*, 10th edn (1758), Sign. BL.16. By permission of the Linnean Society of London   201

9.4 Page with descriptions of human varieties in Linnaeus's personal, annotated copy of the twelfth edition of *Systema Naturae* (1766). Linnean Society London, Library and Archives, Linnaean Collections, *Systema Naturae*, 12th edn (1758), Sign. BL.21. By permission of the Linnean Society of London   202

Foreword

Follow blood back to its source, and we arrive at the beginning of all life: the sea. This is an origin story that the science writer Bernard Seeman tells in his 1961 story of blood, *The River of Life: The Story of Man's Blood from Magic to Science*. His story begins terrestrially—with an "earth ... hot with the violence of its birth"—but moves quickly to the sea as the source of life, and blood.[1] In his narrative, the slow course of evolution over millennia leads quickly to a "form" that "closed itself around the sea" (11). But when it left its primal home, it carried its memory materially with it—literally, in the blood: "The watery environment that was essential to life and which primitive life forms could not leave, became the blood" (11). The sea-turned-blood allowed these organisms to venture onto the primordial earth, "but, wherever they went, they could never fully forsake the mother sea" (11–12).

As surely as human beings need blood to survive, we need stories to make sense of our lives. Not all stories are created equal, of course. Rudyard Kipling does not offer his *Just So* stories as science, for example; he is not concerned with the empirical evidence of how the leopard got his spots or the camel got his hump. Science has its own stories about the world. Theories, discoveries, and explanations all rely on language, and, typically, the conventions of narrative as they circulate through the scientific community and beyond. Based as they may be on state of the art technologies and rigorously controlled experiments, these stories are nonetheless subject to human limitations. Technologies depend on human analysts, and they obscure as well as reveal. Scientists' cultural assumptions influence everything from the choices that underpin scientific experimentation to the interpretation of the results and the language in which they are reported. Those limitations do not detract from the valuable understanding of the world that scientific inquiry makes possible. Rather, they set the terms for productive collaborations between scientists and literary critics. Literary critical analysis of the stories of science can—like any technology—elucidate the assumptions and open them up for inspection.

Seeman's story of blood offers an excellent example. Writing for a general young adult audience, he works to capture the drama of creation in the language of a just so story: "The earth was hot with the violence of its birth. There was no life. There was not even the sea" (9).

The language of creation moves quickly to the description of a birth: "This new form closed itself around the sea. The sea no longer flowed through this creature, carrying sustenance to its cells. Instead, food and oxygen were removed from the sea outside, passed through certain openings or membranes into the fluids of a closed circulatory system and thence to each of the cells ... Thus the river of life was born" (11). In the process of evolution that Seeman describes, blood, the life force, contains this memory of the sea: "The organism out of which man ultimately evolved, encompassed and enclosed a portion of the sea. The watery environment that was essential to life and which primitive life forms could not leave, became the blood—an internal environment which, being portable, allowed this higher organism a mobility that had previously been impossible ... But, wherever [such organisms and their descendants] went, they could never fully forsake the mother sea" (11).

I have begun with, and dwelled on, Seeman's account because of how richly it exemplifies the themes of this volume. This scientific creation story conveys the centrality of blood to human existence and to social relations. In Seeman's account, the sea becomes "mother" to us all. This is an inclusive creation story, extending back to a time—and medium— that predates humanity's divisive social distinctions. But the choice of "mother" also illustrates the inextricability of blood from kinship, which distinguishes as it conjoins. "Blood relations," as the following essays show, names the socially foundational metaphor of kinship and the paradox of biology. "Blood ties" are both the metaphorical glue of any kinship system and the material conditions that make one's nearest biological kinfolk the best candidates for a life-sustaining blood or organ donation in the case of a life-threatening event. The power of blood is so difficult to decipher because it is at once the foundational social metaphor and the most basic necessity for life.

Although *The Cultural Politics of Blood* is not about HIV/AIDS, the pandemic dramatized the themes it elucidates; its impact is evident in the work of critics whose imaginations could not help but be formed by it. The early years of the pandemic powerfully illustrated the dangers inherent in the inextricable circulation of blood as both substance and metaphor. In both iterations, the complexity of blood manifested a world grappling with its unprecedented connectedness. The topics considered in the essays in this volume are evident in the medical science, socioeconomics, and cultural implications of HIV/AIDS. Paula Treichler christens HIV/AIDS "simultaneously an epidemic of a transmissible lethal disease and an epidemic of meanings or signification," by which she means the pandemic is "cultural and linguistic as well as biological and biomedical."[2] Drawing on the work of Ferdinand de Saussure, she explains, "the term *signification* ... calls attention to the way in which a language (or any other 'signifying system') organizes rather than labels experience (or the world)" (331, n2). As Douglas Crimp similarly insists, "AIDS does not exist apart from the practices that conceptualize it, represent it, and respond to it. We know AIDS only in and through those practices."[3] The pandemic crystallized the inextricability—and irreducibility—of biology and semiotics. The considerable documentation in the mainstream media of scientific medicine's coming to terms with the mysterious ailments made the connections increasingly difficult for a lay readership as well as medical professionals to ignore. The semiotics of HIV/AIDS shaped the socioeconomic and cultural aspects of the pandemic, but also significantly affected diagnoses, treatments, and health outcomes.

It is a tragic irony that HIV/AIDS was not initially recognized as transmissible through blood—evidence of the power of story in the workings of medical science. The length of time the story went bloodless, moreover, is especially surprising considering how it was shaped by the history of blood: specifically, the metaphors of circulation as simultaneously life-sustaining and life-threatening. HIV/AIDS infused a changing story about a global body politic that interleaved intensifying senses of connectedness and estrangement. The terror of contagion spoke volubly of the danger of strange bodies in intimate contact, but there was a perverse reassurance when nations gave way to networks, and the anonymous contacts of an atomistic world became the unexpected intimacies of a global village.

Coursing through the circulatory systems of individuals and illuminating their contacts and routes, HIV/AIDS bore witness to the signifying relations of an emerging world system: the bloodlines of the theosocial. In popular culture, it has given rise to an obsession with viruses that turn the infected into the undead. In the form of zombies, the conversion depicts the fear of the crassly material—the body without its animating soul. But in the vampire, the complexity of the blood metaphor finds its fullest expression. Some still bear the marks of Bram Stoker's putrescent anti-hero, but the trend in vampires, at least since the 1980s, is toward the witty, urbane, and of course deadly sexy. The transcendent bloodlessness of these beings is tempered by an appetite that signals the recursive pull of the human in all of its lusty barbarities. Blood, after all, will out.

Although the following essays are not specifically concerned with vampires or HIV/AIDS, they bring to the surface the terms of a cultural preoccupation as they trace the bloodline back more than five centuries to show the co-evolution of science and stories in the articulation of human relations. The historical breadth of *The Cultural Politics of Blood* shows how the metaphor of blood, and the bodily distinctions it produced, accompanied the long history of colonialism and left its legacy in the mutual entanglements of scientific medicine and cultural geopolitics. The volume chronicles how the biologization of human distinctions expressed as blood, kinship, race, or some other marker, have facilitated the naturalization of power relations and, in turn, how those relations come back to life in the laboratories, clothed in the science of blood. This long history of human difference "written in blood" is compulsory reading for anyone interested in the myths of our contemporary moment and the bloodless futures we might imagine.

Priscilla Wald
Professor of English and Women's Studies
Duke University, USA

Notes

1. Bernard Seeman, *The River of Life: The Story of Man's Blood from Magic to Science* (New York: W.W. Norton, 1961), 9.
2. Paula Treichler, *How to Have Theory in an Epidemic: Cultural Chronicles of AIDS* (Durham, NC: Duke University Press, 1999), 11, 1.
3. Douglas Crimp, "AIDS: Cultural Analysis/Cultural Activism," in *AIDS: Cultural Analysis/Cultural Activism*, ed. Douglas Crimp (Cambridge, MA: MIT Press, 1988), 3–16, 3. Originally an issue of *October* 43 (Winter 1987).

Acknowledgments

We thank Robert Levine, Karen Nelson, and the Center for Literary and Comparative Studies in the Department of English at the University of Maryland for their support of the conference "Bloodwork: The Politics of the Body, 1500–1900," which took place 6–7 May 2011 at the University of Maryland and provided the starting point for this volume. The conference was made possible through generous funding on the part of the Office of the Vice President of Research and the College of Arts and Humanities. We are grateful to these funding bodies and to the department itself for hosting the event; particular gratitude is due to Kent Cartwright, who as chair of the department, shepherded the conference to its fruition.

We dedicate this volume to our students—past, present, and future.

Notes on Contributors

Robert Appelbaum is Professor of English Literature, Uppsala University, Sweden, and Fellow of the Stellenbosch Institute for Advanced Study, South Africa. He is the author of many articles and several books on the interconnections between politics, literature, culture, and the body, including *Aguecheek's Beef, Belch's Hiccup, and Other Gastronomic Interjections: Literature, Culture, and Food Among the Early Moderns*, winner of the 2007 Roland H. Bainton Award. His latest book is *Working the Aisles: A Life in Consumption* (2014).

Ralph Bauer is an Associate Professor of English and Comparative Literature at the University of Maryland, College Park, USA. His research interests include the literatures and cultures of the colonial Americas, early modern studies, hemispheric studies, and the history of science. His publications include *The Cultural Geography of Colonial American Literatures: Empire, Travel, Modernity* (2003; 2008), *An Inca Account of the Conquest of Peru* (2005), and (co-edited with José Antonio Mazzotti) *Creole Subjects in the Colonial Americas: Empires, Texts, Identities* (2009). He is currently at work on a monograph provisionally titled "The Alchemy of Conquest: Prophecy, Discovery, and the Secrets of the New World."

Rachel L. Burk's book project, *Con sangre entra: Blood and Purity in Early Modern Iberia*, is a study of literary and visual works that marshal distinct understandings of blood to resist the racializing doctrine of blood purity in Spain and Portugal during the sixteenth and seventeenth centuries. Currently an Assistant Professor at Notre Dame of Maryland University, USA, she has taught at Tulane, the University of Pennsylvania, and the Universidade de Lisboa, the last as a Fulbright scholar. She holds a PhD in Hispanic Studies from the University of Pennsylvania and an AB in English from Columbia.

Kimberly Anne Coles is an Associate Professor at the University of Maryland, USA. Her book *Religion, Reform, and Women's Writing in Early Modern England* (2008; 2010) examines the influence of women writers on religious identity and its cultural expression in the sixteenth century. She has published articles in *ELR*, *Modern Philology*, and *RES* on the topics of women's writing, gender, and religious ideology. Her current book project, *"A fault of humour": The Constitution of Belief in Early Modern*

England, contends with the physiological and philosophical context that makes moral constitution (the framework within which religious ideology is understood) a feature of the blood.

Lyndon J. Dominique is Assistant Professor of English at Lehigh University, USA, specializing in eighteenth-century literature and issues related to critical race studies, colonialism and transatlanticism, gender, and social justice. He received his BA with honors from the University of Warwick and his PhD from Princeton University. He is the editor of the anonymously published 1808 novel *The Woman of Colour* (2007) and the author of a monograph, *Imoinda's Shade: Marriage and the African Woman in Eighteenth-Century British Literature, 1759–1808* (2012). Currently, he is working on a book about the concept and narrative forms of social justice peculiar to literature written in eighteenth-century Britain.

James Downs is an Associate Professor of History at Connecticut College, USA. He earned his BA from the University of Pennsylvania and his PhD from Columbia University. He is the author of *Sick from Freedom: African American Illness and Suffering during the Civil War and Reconstruction* (2012). He has published articles in *The New York Times*, *The Lancet*, *Time Magazine*, and *The Chronicle of Higher Education*, and he blogs for the *Huffington Post*. He co-edited *Taking Back the Academy* (2004) with Jennifer Manion and edited *Why We Write: The Politics and Practice of Writing for Social Change* (2006).

Jean E. Feerick teaches in the Department of English at John Carroll University, USA. She is the author of *Strangers in Blood: Relocating Race in the Renaissance* (2010) and co-editor, with Vin Nardizzi, of *The Indistinct Human in Renaissance Literature* (Palgrave, 2012). Her interests include theories of race and nation in sixteenth- and seventeenth-century literature, literature and science, and early modern ecology. Her essays have appeared in *Shakespeare Studies*, *Early Modern Literary Studies*, *English Literary Renaissance*, *Renaissance Drama*, and *South Central Review*, and a forthcoming essay on race and grafting will appear in *A Handbook on Shakespeare and Embodiment: Gender, Sexuality, and Race*. She is currently at work on a project on Shakespeare and the four elements.

Ruth Hill is Professor of Spanish and Andrew W. Mellon Chair of the Humanities at Vanderbilt University, USA, where she teaches courses in Transatlantic Studies, Latin American Studies, and Hemispheric American Studies. She is the author of *Sceptres and Sciences in the Spains: Four Humanists and the New Philosophy* (2000), which explores the reception of Bacon, Gassendi, Newton, and Vico in Spain, Portugal, Mexico, and Peru, and of *Hierarchy, Commerce, and Fraud in Bourbon Spanish America* (2005), which addresses the reforms and failures of Enlightenment modernization efforts. She is currently working on a history of Aryanism in the Americas entitled *Incas, Aztecs, and Other White Men: A Hemispheric History of Hate*, and on a study of whitening and racialization in South America and the American South called *Race and the Hemispheric South from Simón Bolívar to Hinton Rowan Helper*.

M. Lindsay Kaplan teaches in the English Department of Georgetown University, USA. She authored *The Culture of Slander in Early Modern England*, co-edited, with Valerie Traub and Dympna Callaghan, *Feminist Readings of Early Modern Culture: Emerging Subjects*, edited a contextual volume of *The Merchant of Venice*, and has written essays on gender, Jews, and race, most recently "The Jewish Body in Black and White in Medieval and Early Modern England," in *Modern Philology*. Her current book project demonstrates how the concept of Jewish perpetual servitude, developed in medieval Christian theology, contributes to the construction of early modern race.

Anna More is Professor of Hispanic Literatures at the University of Brasília, Brazil. Her research focuses on the colonial period of Latin America, particularly on seventeenth-century Mexico. She is the author of *Baroque Sovereignty: Carlos de Sigüenza y Góngora and the Creole Archive of Colonial Mexico* (2013) which received an honorable mention for the LASA Mexican Studies Humanities Award. She has also written articles on seventeenth-century science, poetics, and historiography in Mexico and is currently editing the forthcoming Norton Critical Edition of Sor Juana Inés de la Cruz.

Staffan Müller-Wille is Senior Lecturer and Co-director of Egenis, the Centre for the Study of the Life Sciences, at the University of Exeter, UK. He has published extensively on the history of natural history, heredity, and genetics. Among the more recent publications is a book co-authored with Hans-Jörg Rheinberger on *A Cultural History of Heredity* (2012).

Zita Nunes is Associate Professor of English and Comparative Literature at the University of Maryland, USA. The author of *Cannibal Democracy: Race and Representation in the Literature of the Americas* (2008), she has published widely on African Diaspora literature and politics. Her current book project is *The Hue of the Dove: The Peace Movement and Black Internationalism*. She is also working on a reevaluation of the Third

Pan-African Congress (1923) from the perspective of extensive journalistic commentary in black student newspapers in Portugal, France, Belgium, the Netherlands, as well as a number of African countries.

Carla L. Peterson is Professor Emerita in the Department of English at the University of Maryland, USA. She specializes in nineteenth-century African American literature, culture, and history and has published numerous essays in this field. She is the author of *"Doers of the Word": African-American Women Speakers and Writers in the North (1830–1880)* (1995), and more recently, *Black Gotham: A Family History of African Americans in Nineteenth-Century New York City* (2011), a social and cultural history of black life in nineteenth-century New York City as seen through the lens of family history.

David Sartorius is Associate Professor of History at the University of Maryland. His research focuses on race, empire, and the African diaspora in the Caribbean. His publications include *Ever Faithful: Race, Loyalty, and the Ends of Empire in Spanish Cuba* (2013) and "Race in Retrospect: Thinking with History in Nineteenth-Century Cuba," in *Race and Blood in the Iberian World*, ed. Max S. Hering Torres, María Elena Martínez, and David Nirenberg (2012). He has also co-edited two special issues of journals dedicated to transnational history: "Dislocations across the Americas" (*Social Text*, 2010) and "Revolutions and Heterotopias" (*Journal of Transnational American Studies*, 2012).

Hannah Spahn is Wissenschaftliche Mitarbeiterin at the Department of English and American Studies, University of Potsdam, Germany. She is author of *Thomas Jefferson, Time, and History* (2011) and *Thomas Jefferson und die Sklaverei: Verrat an der Aufklärung?* (2002) and co-editor, with Peter Nicolaisen, of *Cosmopolitanism and Nationhood in the Age of Jefferson* (2013). Her current book project focuses on African American conceptions of character and cosmopolitanism in the long nineteenth century.

Introduction

Kimberly Anne Coles, Ralph Bauer, Carla L. Peterson, and Zita Nunes

Social scientists, historians, and literary historians have struggled to explain why group identity and group difference were so often mapped onto the human body in pre-modern and early modern times, when biology was not yet available as an explanatory model; nonetheless, it is clear that the articulation of systems of social discrimination through a language of the body predates the emergence of biology as a science. *The Cultural Politics of Blood* surveys how conceptions of the blood—one of the four bodily fluids known as humors in the early modern period—permeate discourses of human difference from 1500 to 1900. In gathering this collection of essays, we explore how medical theory, at different points in Western history, has supported fantasies of human embodiment and human difference that serve to naturalize hierarchies already in place. We begin with the assumption that one of the most enduring and controversial signifiers of difference, namely that of "race," is still under construction today and that our understanding of the term would profit through an engagement with its long, evolving, history. The essays here interrogate how fluid transactions of the body have been used to justify existing social arrangements over four hundred years in England and Spain and in the Anglo- and Ibero-Americas. We choose this expansive geopolitical area as our field of investigation because of both the shared, if adversarial, colonial interests of England and Spain, and the complex political relationships that their respective colonial activities produced. Our volume examines how medical theory concerning the nature of blood shaped cultural and political agendas, and vice versa, over an extended history and within various contexts.

The deconstruction of nineteenth-century "scientific" racism may have challenged claims to a biological explanation of cultural, economic, and social difference, but this critique does not account fully for the enduring sociopolitical power of its logic, rooted in naturalized conceptions of difference, which predate the nineteenth century.[1] The consensus among literary and cultural historians of the pre-modern and early modern periods has been that this terminology of the body is acquitted of race because it frequently gestures toward what we define as culture: "while the language of race [in early sources]—gens, natio, 'blood,' 'stock,' etc.—is biological," the medievalist Robert Bartlett writes, "its ... reality was almost entirely cultural."[2] Bartlett's assertion reflects the widespread opinion among scholars that the discriminations of pre-modern and early modern peoples were based upon cultural, as opposed to physiological, difference—precisely because these discriminations were so often directed against cultural and religious others. There are two significant problems with this view: first, to say that somatic vocabularies underwrite a cultural difference and not a physiological reality is to describe any racial ideology; second, these opinions rest upon modern apprehensions of both culture and physiology.[3]

The essays collected here reopen the question of the pre-modern origins of the modern concept of race. A central theme is the recognition that the concept of race as used in the modern Anglophone world is not a trans-historical category by which we can explain the history of colonialism across time and space. Rather, it is the product of a quite distinct colonial history as it unfolded in the British settler and plantation colonies from the seventeenth century, as age-old ideas of hereditary blood interacted and fused with classically informed scientific theories of the human body, in particular humoral theory, whose constellation depended on factors that were understood to be contingent. If we broaden both the geographical and historical scope to include the Iberian world and the fifteenth through the nineteenth centuries, as we do in this collection, we find that the interaction between notions of blood in terms of heredity and notions of blood in terms of science took on myriad forms that were locally, historically, and culturally specific. Our volume is not organized according to period or discipline, but according to themes connected by the overarching concept of blood. It is trans-historical and transnational and draws together literary critics and historians of cultures on both sides of the Atlantic world. In place of a chronological, developmental history of the relationship between race and blood, this organization emphasizes the complex and interwoven nature of these concepts. It focuses upon the materials of manufacture: how race is sewn together from the fabric of available social, political, and medical discourses. But if we suggest that race is a political fiction, we nevertheless insist that it has historical consequences. The tangled relationship of blood and race as it is understood at different historical moments, with evolving frames of reference and altered rationale, is itself a history of the race concept.

Our strategy, therefore, parallels the production of the cultural scripts themselves: episodic, opportunistic, and wide-ranging. It is precisely because the narratives of literature and history concerning race are mutually constitutive that they are impossible to tease apart. Our volume explores how the fiction of race is told, reshaped, and retold over a fraught colonial history and terrain. Since race is a discourse that serves specific political needs or answers economic agendas, shifting the focus of both time and place affords the opportunity to view, each time, the circumstance of its creation and the materials of its making. There is value in studying the individual occasion when the discourse is produced—one discrete chapter at a time—both in terms of what it tells us about the cultural moment of its formation, and in terms of what it tells us about its strategies of production.

In the early modern period, race is a concept at the junction of a set of intersecting concerns of lineage, religion, and nation.[4] This volume draws attention to humoral theory as the discourse that underwrites the ideologies that cut across all of these categories. The humors are certainly not the only mode of explanation for bodily difference, but they prove to be the most durable model from classical antiquity up through the eighteenth century. (Staffan Müller-Wille, for example, argues in this volume that Carl Linnaeus's 1735 classification of humankind into races—or what Linnaeus termed "varieties"—was organized according to his own method of humoral taxonomy.) While only one of the humors, blood nonetheless came to serve as metonymy for a humoral disposition internal to the body that was thought to determine the very nature of human beings. The malleable quality of humoral theory allowed it to license a range of political arrangements and discriminations—it proved wax-soft to the stamp of ideology. But such malleable discourse does not retain its shape for long. We do not claim that the circumstances of sixteenth-century racial logic convey forward; rather, we suggest that emergent and residual discourses occupy the same cultural moment. Materials of manufacture get recycled, and echoes of a former thought can be heard in later discourse. The essays collected here tell a (hi)story throughout which blood remains a consistent, although not entirely stable, signifier.

In early modern England, the term "race" commonly referred to family lineage, or bloodline, and relied upon pervasive notions of what were believed to be the properties of blood.[5] Current research suggests that many of the terms and relations of animal husbandry were retained and applied to human physiology through the eighteenth century—precisely because the term "race" initially stemmed from animal husbandry before it became attached to human (noble) stock. As Charles de Miramon explains, the transformation of heredity is not confined to breeding books, or limited to discourses about races of dogs. Rather, "the revival of hereditary blood early in the fourteenth century," and its attribution to human beings of noble birth, "is ... [an] example of the cultural and political evolutions that explain the birth of race."[6] Of course, Miramon's terms are exaggerated: there is no "birth," invention, or point of origin to race. There is rather a process by which concepts for rationalizing human difference appear and adapt, fuse and fade away, relocate and are repurposed.[7] Older conceptions of race are dismantled and their vocabularies appropriated to serve different political objectives and economic ends.

The constitution of noble subjects was thought to be in equilibrium, a privilege that not only endowed them with superior physical, intellectual, and moral capacity, but also conferred their license to rule. This early modern fiction of race, indebted to a fourteenth-century formulation, reified social hierarchy in the body and reproduction. In his treatise concerning humoral complexion, Levinus Lemnius describes the moment of conception as the moment when the soul enters the body and mixes with "the Parentes Seed, which is ... of the purest and best concocted bloude."[8] While, in Lemnius's Aristotelian conceit, the soul spreads its influence throughout the body as its substantial form, it is clear from his description that the humoral substance of the body proper is a composite of parents' blood. "Parentes Seed," however, was not the only "concocted bloude" introduced into a child's body.

Like semen, breast milk was considered the concocted blood of the parent.[9] But the parent was not the only possible source of this blood; numerous regimens, in England and on the continent, voiced considerable concern over the choice of wet-nurse for noble children.[10] Jean Feerick, one of the contributors to this volume, surveys these claims about "uncouth milk" in her own book concerning hereditary blood and its degeneration: "in giving her milk to a suckling child, the wet nurse is thought to convey through it ... the precise balance of humoral fluids characteristic of her own body."[11] But, as Feerick notes, the fact that this advice is justified through farming practices, not only underscores its commonplace logic, but also how much this logic of heredity—and the fears concerning the possible decline of the disposition of blood—is indebted to animal husbandry.[12]

Both physical and moral virtue were understood to convey through bloodlines. Programs for rearing the offspring of English nobles, such as Thomas Elyot's *Boke named the Governour* (1531), concentrate upon retaining their inherited constitution. Elyot pays particular attention to the selection of the wet-nurse, "[f]or as some auncient writers do suppose often times the childe souketh the vice of his nourise, with the milk of her pappe."[13] Like many English medical theorists, Elyot focuses upon the "complection" of the nurse, insisting that she not be "seruile," but be "of the right and pure sanguine."[14] That Elyot associates the virtue of the nurse with her status is crucial here: since noble women could not be counted upon to nurse their infant children, Elyot's intention was to guard against the contamination of noble bloodlines through their exposure to the blood of servants. He therefore recommends a woman of status as a nurse. But this is an inherently political gesture. The humoral disposition of noble blood determined the intellectual and moral character of the ruling class. The pollution of noble blood could result in the rank corruption of government. Therefore, Elyot sees the "utter destruction of a realme" in the wrong selection of a nurse.[15]

The mild hysteria evident in Elyot's treatise underscores the instability of a social hierarchy premised upon notions of blood—since blood is itself inherently unstable. Humoral complexion is vulnerable to a host of "non-naturals" (food and drink, intake and expulsion, sleep, motion, air, and passions), and numerous regimens instructed the nobility in how to order their diet (since diet attends to all of the "non-naturals" that affect disposition) to guard against the degeneracy of their nature. As Lemn[ie] observes: "we see the common sorte and multitude, in behauiour and maners grosse and vunnurtured whereas the Nobles and Gentlemen (altering theyr order & diet, and digressing from the common fashion of their pezantly countreyme[n]) frame themselues & theirs, to a verye commendable order, and ciuill behauiour."[16] Of course, such declarations reiterate the fantasy of physiological superiority that upholds the social hierarchy already in place. But they also speak to the extent to which "ciuill behauiour"—moral and social virtue—were the understood effects of humoral disposition. In this ideological framework, the nobility inherited the right to rule due to superior physiology, but they could be alienated from that entitlement through bodily degeneration.

* * *

Part I of this volume, "Race and Stock," examines these early discourses concerning rank—and their disintegration—in both Old and New World contexts. In her contribution to this collection, Rachel Burk explores the impulse to maintain blood purity at the precise moment when discourses concerning the integrity of noble blood are being dismantled in Spain. After the expulsion of the Jews from Spain in 1492, and the forcible conversion of those who remained, Spanish society consisted nominally of Christians and *mudéjars*—Moors who lived under Christian rule but who did not convert. Those Jews and Moors who converted to Christianity (*marranos* or *conversos* and *moriscos* respectively) were considered to be "New Christians," an identity that was passed from generation to generation through family lineage; that was associated with social shame; and that barred stigmatized individuals from exercising certain professions that were reserved for "Old Christians." After the *mudéjars* were expelled in 1610, the entire population was nominally Christian. In this context, social prestige traditionally attributed to the nobility came to be predicated upon the distinction of being an "Old Christian." While this ideology of *limpieza de sangre* did much to lend Spain's national culture the cohesiveness it had formerly lacked, it came at the expense not only of *conversos*, *moriscos*, and religious dissidents but also of some sectors of the traditional aristocracy—whose well-documented family trees were rarely spotless in regard to *limpieza de sangre*.

Burk's chapter, "Metamateriality and Blood Purity in Cervantes's Alcaná de Toledo," attends to literary significations of *limpieza de sangre* in early modern Spain. The famous metafiction of Miguel de Cervantes's masterpiece *Don Quixote* invites the reader to reflect on its own material and intercultural production, Burk argues, because it characterizes the Arab-Iberian presence as a physical trace inscribed in the printed work; it thus exposes the cultural and ideological fiction of *limpieza de sangre*, which was sustained in large part by a much-falsified paper record defining bodily identity. The metamaterial commentary of the novel, which constantly undermines its own claim to the purity of its textual genealogy, thus serves as a parody of the obsession with blood purity that girded Spanish national identity during the Counter-Reformation.

The virtue of nobility—in decline and under scrutiny—is refused in favor of a revised category of "Old Christian." This shift in the source of virtue also marks a shift in the systems of power that virtue guaranteed, since government posts were barred to all but "Old Christians." Older modes of thinking concerning the virtue of the ruling class are reworked rather than abandoned. *Don Quixote* provides profound insights into the transforming cultural politics of blood during the early modern period not only in the Old World but also in the New. Thus, Cervantes's tragic-comic aristocratic hero is modeled on the chivalric ideals that had become an utter anachronism by the early seventeenth century in Counter-Reformation Spain but attained a second life in the conquest of the New World. Conversely, the low-born squire, Sancho Panza, can dream of becoming governor of some overseas island populated by people who are not of Spanish "blood."

Ruth Hill, in her chapter, "The Blood of Others: Breeding Plants, Animals, and White People in the Spanish Atlantic," shows how discourses of animal husbandry, formerly used to support and affirm the inherited quality of noble blood, are appropriated to a New World essentialism. What she terms "folk biology"—found in farmers' manuals and veterinary handbooks—is a little-known source of early modern thinking linking skin color with blood in the Hispanic world. Analyzing dictionary entries (beginning with Covarrubias's 1611 *Tesoro de la lengua castellana o española*) on key terms such as "blood," "milk," and "caste," as well as common-sense popular idioms in early modern Spain and Spanish America, Hill sheds new light on well-known debates about theories of degeneration and "whitening" (*blanqueamiento*) in the Spanish Atlantic. Her previous book, *Hierarchy, Commerce, and Fraud in Bourbon Spanish America*, disputed that "casta" in eighteenth-century viceregal Spanish America can be equated to race; rather, she emphasized the "cluster of somatic, economic, linguistic [and] geographic ... circumstances" that produced social hierarchy.[17] But folk biology, Hill argues here, became harnessed to the ideology of Euro-centrism with enduring consequences for the history of race in the Hispanic world. It provided an early modern precedent of essentialist thinking in terms of chromatic categories and geo-global affiliation.

In "'Rude Uncivill Blood': The Pastoral Challenge to Hereditary Race in Fletcher and Milton," Jean E. Feerick examines critiques of noble blood in a British colonial context. While Wales had long been under English rule, it was a location, like Ireland and Scotland, that placed particular pressures (diet, climate, foreign nurses) on the stability of noble blood. Milton uses the colonial landscape, Feerick argues, as a place to dismantle cultural narratives concerning hereditary blood, depicting it as "variable, immoderate, and inflamed" (contrary to the scripts that record it as stable, balanced, and the font of "ciuill behauiour"). She argues that a satirical representation of declined bloodlines in John Fletcher's *The Faithfull Shepherdess*, a known source for John Milton's *Comus*, prompted Milton to portray embodied differences in rank as sociopolitical fiction in his masque. Milton places cultural assumptions of the correspondence between low birth and vice in the mouths of the noble sons of the Lord President of the Council in the marches of Wales (for whom the masque was staged). But the masquing figures who perform depravity are evidently elite. Further, the Lady of the masque does not remain untouched by corruption; rather, she overcomes it. Through these strategies, Feerick argues, "Milton enacts a reordering of the ... moral universe," depicted in the masque and available in its cultural context, "in which divinity attaches to elite bodies and evil to laboring ones." In *Comus*, noble blood is not a stable source of virtue; rather, virtue is the product of hard labor on a terrestrial plane.

The issues of noble blood and laboring bodies (and of residual and emergent notions of race) came into collision in Aphra Behn's novella *Oroonoko* (1688). Considerations of blood and race in studies of eighteenth-century British fiction began several decades ago with critical appraisals of Behn's work, and its royal African protagonists, Oroonoko and Imoinda. Since then, scholars have broadened their field of inquiry to include a series of lesser-known novels of the period. Lyndon J. Dominique has been at the forefront of this work with his reprint of the anonymously authored *The Woman of Colour* (1808) and his critical book *Imoinda's Shade*. In his essay for this volume, "African Blood, Colonial Money, and Respectable Mulatto Heiresses Reforming Eighteenth-Century England," Dominique reaffirms the critical centrality of women of African descent in contemporary discussions about race in eighteenth-century British studies. The essay's focus on African blood, colonial money, and their connections to wealthy women of color in British novels set in England allows him to consider the "specific responses to money when the specter of African blood" attaches to it in a marriage plot. Dominique's chapter throws into relief the confused ideas concerning status when the category of class is emergent, and rank is not yet dormant. But his essay also highlights the cultural crosscurrents when social status is not yet completely detached from notions of hereditary blood, but nation is becoming a more important category of social belonging and social exclusion.

Invoking Roxann Wheeler's conception of complexion as a sign of respectability, Dominique suggests how fictional wealthy women of all colors attempt to pass in England, not as white women but as women of African descent with complexions of English respectability. In the less progressive novels, the female protagonists cannot pass because the moral and physical flaws associated with their mother's African blood—which are sometimes reinforced by their father's vicious or foreign blood—exclude them from respectable national recognition. As such, they offer a direct literary link to the origins of the nineteenth-century tragic mulatto as a flawed national and racial figure whose blood cannot nourish the roots of the nation's family tree.

The notion of blood as constitutive of virtue, respectability, and morality persists at least until the eighteenth century, evidenced by works such as Daniel Defoe's, "A True-Born Englishman" (1701). Defoe's satire bears the traces of a history of racial thinking (however ironically deployed): "But if our virtues must in lines descend, / The merit with the families would end, / And intermixtures would most fatal grow; / For vice would be hereditary too."[18] In Defoe's utterance, as Jennifer Brody has observed, "it is the viscous substance of blood itself that conveys both virtue and vice."[19] The language of Defoe's poem not only insinuates the crossover of the term "race" from family lines to national groups, but also supplies evidence that both kinds of racial ideology—one that supports social hierarchy, another that affirms national superiority—rest upon the invisible qualities of blood. In the more progressive novels that Dominique analyzes, mulatta protagonists are portrayed as women of African descent with African complexions that are as respectable as English ones. The function of these respectable heroines is to force white Britons to fix their own domestic prejudices about, and behavior toward, people of African descent, and allow their already mixed blood to commingle with white blood at the root of the nation's family tree.

* * *

Humoralism, and the medical discourse that applied it, claimed that blood determined a host of values within individual bodies: health, physical endurance, skin color, temperament, intelligence, and morals. The essays in Part II, "Moral Constitution," consider the interconnectedness of body and soul in early modern medical/humoral discourse. Such interconnectedness might well explain why religion is often such a crucial term in the rationalization of bodily difference, and further, why raced subjects are often morally encoded. While both physicians and metaphysicians had debated the relationship of soul to body, and the precise habitation of the soul within the body, since antiquity, the theories of Pietro Pomponazzi in the early part of the sixteenth century forced a reinterrogation of the definition of soul. Pomponazzi argued that since the soul "acted materially in sense-perception and immaterially in intellection, it must partake of both ontological realms." He therefore assigned the soul to the material realm, positing the soul as "the highest material form" of human being.[20] Of course he was correct: it had proved beyond the means of philosophy to show how sense-perception and imagination could be part of intellection—an activity emanating from the higher, rational soul—and to not implicate the soul as material form.

The rupture between natural and Christian philosophy that Pomponazzi initiated was considerable. In his influential *Liber de anima*, Philip Melanchthon simply cedes the field to Pomponazzi: he declares the soul is immortal, but admits that his assertion cannot be proven by philosophy; rather, it is affirmed by scripture.[21] But Melanchthon further argues that a full description of the whole human body is required in order to know the powers of the soul, since the soul can only be known through its actions.[22] Here, Melanchthon is following Luther who maintained that the whole human being, body and soul, was subject to grace. In trying to provide a clearer conception of what the soul might be, however, Melanchthon returns it to the purview of natural philosophy, and follows Galen's definition of the human soul, grounded in the body.[23] In so doing, Melanchthon "demonstrate[s] that the manifestations of ... dysfunction" attributable to reason (mens/anima) "were physically pathological and spiritually sinful," and that sinfulness could also be expressed in physical manifestations.[24] Indeed, Galenic medical commentary consistently frames Melanchthon's published work throughout the sixteenth and seventeenth centuries.[25] One can perceive the reason for this framing in Melanchthon's own directive that physicians ought to be the guardians of morals and the custodians of men's natures.[26] But the situation of Melanchthon's *De Anima* within the frameworks of medical theory also begins to make sense of the assertion by Timothy Bright, the chief physician to the Royal Hospital of St Bartholomew (1585–91), that by the latter part of the sixteenth century, many had considered the behavior of the soul "to be subiect to the phisicians hand" and had "esteemed the vertues themselves, yea religion, no other thing but as the body hath ben tempered."[27]

By the mid–late sixteenth century, psychology, or the philosophical study of the soul, became, in many ways, the legitimate concern of Christian physicians. Almost all Christian philosophers conceded (with Melanchthon) that the immortality of the soul was beyond philosophical demonstration. But what could be known about the soul was manifest in the body. In an early modern world, where moral constitution is a condition of the blood, a fact of humoral complexion, what we would term "culture"—religious affiliation—is read as nature.[28] The reason why this matters is that the discourse of natural philosophy composed numerous arguments against religious others throughout the sixteenth century. Luther, Melanchthon, Zwingli, and Calvin had all offered arguments depicting religious opponents as humorally distempered and deluded, subject to the madness that an excess of melancholy produces.[29] As James Hankins has observed, there was a "virtuous mean of 'true religion'," depending, of course, upon which religion was counted as "true."[30]

As Jeffrey Jerome Cohen has pointed out, the humoral construction of religion and other cultural practices that get "written on and produced through the body" was not confined to the early modern period; what is particular to it is the extent to which this construction was methodized.[31] The consequence of the Pomponazzi affair, Eckhard Kessler claims, was that "philosophy [was] no longer ... identical with Aristotle, nor Aristotle with St. Thomas and the teaching of the church." Christian philosophy was reinvigorated, but Christian and natural philosophy were no longer married.[32] Whether or not clerics (future and former) theorized on the relationship of soul to body, and many did, Thomas Wright asserts in his treatise, *The Passions of the Minde* (1601), that it was the work of *"naturall Philosophers*, to explicate ... how an operation that lodgeth in the soule," which is to say, the working of the rational mind, "can alter the body, and mooue the humors from one place to another."[33]

Prior to this divorce, however, the matter of belief could be fashioned on different terms. In her contribution to this collection, M. Lindsay Kaplan demonstrates how discourses of theology and natural philosophy are deployed in combination to subordinate the Jewish body at a time when the political subordination of Jews in medieval Europe was proving unsuccessful. In "'His blood be on us and on our children': Medieval Theology and the Demise of Jewish Somatic Inferiority in Early Modern England," Kaplan points out that none of the discourses in classical or early medieval scientific texts depict the Jewish body, in particular, as susceptible to bleeding; but when united with religious formulations that mark the Jewish body with the shame of the crucifixion, Jewish bodies are effeminized and essentialized. Kaplan shows how discourses that claim the periodic bleeding of Jewish bodies proceed from theological assertions—but they are nonetheless subsequently "proved" through natural philosophy. In articles such as "Jessica's Mother: Medieval Constructions of Jewish Race and Gender in *The Merchant of Venice*," Kaplan has appraised the extent to which the racial representation of Jews in early modern English culture was a medieval inheritance. Here, she suggests that the demise of the theological-scientific discourse that she traces is concurrent with the disempowerment of Jews throughout Europe. Her argument therefore shows how "the shape of scientific knowledge" responds to "cultural assumptions" and power relations.

Similar to Kaplan, Anna More demonstrates how the materials of scientific discourse are shaped according to cultural or political agendas; in this case, the poem that she analyzes responds to an internal struggle concerning gender and church politics. This orientation is very different from the one that Kaplan describes in which scientific discourse serves the needs of both church and public politics. In "Sor Juana's Appetite: Body, Mind, and Vitality in 'First Dream'," More finds the late seventeenth-century poem of Sor Juana Inés de la Cruz firmly situated within the debates concerning faculty psychology and the state of the soul. Sor Juana's poem, as More points out, draws heavily from Luis de Granada's *Introduction of the Symbol of Faith*, a late sixteenth-century treatise that relied upon a Galenic account of physiology. In an attempt to free the rational soul from the corruption offered by the body, Sor Juana depicts a night journey, when sleep depresses the effects of sensory apprehension on cognition. The soul is then "unburdened of" the "external rule" of the body and able to float free of the flesh and fluids that inevitably hamper, and potentially obstruct, its operations. Indeed, in the imaginative world of the poem, sleep purifies the images introduced to the intellect (through memory, not sensory apprehension), and the mind receives only "most clear vapors / of the four tempered humors." Through a dream vision, Sor Juana releases the rational mind from the constraints of humoral pathology—freeing cognition from bodily corruption in order to achieve clear inspiration. By claiming a purity of apprehension for the vision of her poem, cleared of intemperate humors, Sor Juana invokes an authority outside of the church—and above the church—in order to license her own cultural production.

Unlike most of the essays in this volume, More's does not depict a wider sociopolitical contest. Rather, it shows at a local level how a cultural practice—belief—can be understood within a set of physiological operations. It also demonstrates how the discourse of natural philosophy can be brought to the service of the individual to negotiate moral, and even sacred, authority. Early modern arguments concerning the vexed issue of mind–body interaction bequeathed to later periods not only Cartesan philosophy, but also a persistent suspicion that the qualities of soul and body were sympathetic and reinforcing. These two legacies would seem in contradiction since René Descartes famously refused the existence of both vegetative and sensitive soul in order "to isolate ... both thought and soul from bodily praxis."[34] Nevertheless, the motto "nihil est in intellectu quod non prius est in sensu" is widely available through the eighteenth century.[35] We tend to exaggerate the extent to which new theories displace old: discourses compete and contest, braid and abrade for some time.[36]

In their efforts to overcome the mind–body dualism and theorize the role of sympathy (as opposed to religious belief) in character formation, earlier Scottish thinkers often downplayed the particular relevance of blood. This is due, in part, to the role that blood assumed in Descartes's theory.[37] For example, the lines of inquiry evidenced in David Hume's "Of National Characters" express what has been described as the peculiarly bloodless and disembodied nature of liberal notions of character. Yet, these notions coexist with continued references to blood as both subject matter and metaphor. In her essay, "Blood and Character in Early African American Literature" (Chapter 7), Hannah Spahn notes that the discourse of blood and nation-formation present in the American Declaration of Independence left an asymmetry that African American writers later sought to remedy. Spahn suggests that, on the one hand, these writers held blood to be "essentially" responsible for particular personal or collective characteristics; on the other hand, however, they also invoked the biblical topos "of one blood" to emphasize from a political and philosophical perspective the universal aspects of blood as a fluid uniting all humanity from a political and philosophical perspective. Spahn argues that the tension between these two positions of particularity and universality stem from a conundrum at the heart of sympathy-based accounts of character that were widespread during the revolutionary period. In contrast to those scholars who understand race and blood to be conjoined as indicators of racial *identity*, Spahn takes a fundamentally new approach by invoking the concept of *character* and examining the ambivalent role of blood in African American conceptions of it.

Spahn begins her essay with a discussion of how the ambiguous role of blood in sentimental discourse surfaces in the Scottish-inflected conception of character in the Declaration of Independence, especially in Thomas Jefferson's draft version which invokes words and phrases such as "blood," "common blood," "common kindred," and "consanguinity." In this document, Jefferson recounts the violent disruption of *both* African and English kinship ties—that is, those of Africans brutally kidnapped from their homeland and those of Americans who severed bonds with their English parent during the revolutionary war. As Spahn points out, however, these disruptions give rise to an incomplete parallelism between two potential "nations" of enforced exiles since only one nation—the United States of America—becomes a political reality, and in so doing excludes the other from national belonging. Hence, Jefferson's many references to blood apply *exclusively* to the destruction of the English family. The blood at the root of Thomas Jefferson's "tree of liberty" is English, as is the family that grows from it.

As a response to this exclusivity, African American writers proposed alternative conceptions of blood and character by refuting the liberal racism of the Declaration and inverting Jefferson's model in three distinct ways. First, they insist that African blood was also spilled in the American war for liberty, thus mixing it with that of white Americans in fostering the growth of the nation. Second, they point to the ways in which the spilling of African blood caused by the brutality of the Atlantic slave trade challenges the assumed linkage of white American blood to good moral character. Finally, by taking Jefferson's own personal life (and his vexed relationship with Sally Hemmings) as exemplar, they undermine not only assertions of white American moral character but also the purity of the classification itself.

* * *

As notions of lineage ceased to have value in terms of systems of government, the concept of family, as political model and metaphor, became increasingly compelling. What were once ontological connections of family relationships—ones that licensed political authority—become connections in the imaginary of social relationships that license political participation. In the emerging fiction of a national family, blood is an index of shared identity. As Spahn's essay shows, whether spilled in war or labor, blood was a point of entry—a means of authorizing membership within the nation. Groups that the nation sought to exclude (due to insufficient family resemblance) invoked blood shed in the building or protection of the nation as proof of their national character.

Defoe's description of "The well-extracted brood of Englishmen" shows the extent to which concepts concerning rank and lineage had been conflated with fantasies of a national family of one blood. Defoe self-consciously interrogates these cultural scripts—but we can assume, from the language of the poem and its satiric form, that the ideologies were not visible to some, or even to most, of his countrymen. Older fictions of blood still appear to be influential. Defoe's diction in the ironic pronouncement that "virtue gives nobility" (and not the other way around) supplies some evidence: "virtue" is still the concept upon which social authority is premised.[38] Notions of national character are clearly persuasive at the time of his writing. As Defoe explains in his preface, "A true Englishman is one that deserves [the] character ... but as for a true-*born* Englishman, I confess I do not understand him."[39] He does not imagine character to adhere to the blood. But it seems clear from his assertion that he has been "taxed with bewraying my own nest," since the first publication in 1701, "and abusing our nation by discovering the meanness of our original" that ideas of national purity, and attendant fictions of national virtue, still had currency.[40]

Indeed, Defoe wrote his poem in defense of a foreign prince. Evidently, ideas of rank and lineage become entangled with notions of national character when the race of ruling families had been undermined, but the race of nations (or of "true-born-English nationality"[41]) was still under construction as a political fiction. But we should notice the extent to which blood is used as a key term in a political vocabulary in Defoe's poem. Political arrangements from the early modern period through the eighteenth century were founded on a host of assumptions—but not least was the assumed inner constitution of peoples. Medical readings of the disposition of blood and humors justified many of the political arrangements both within and between Western nations. Since the rubric of the final section, Part III, "Medicalizing the Political Body," implies both the idea of the physical body in political space and that of the body politic, the essays assembled here consider both how medical theory adds weight to the political power that presses on the human subject, and how it affects the peculiar organization of the political body as a whole. These essays excavate the underpinning logic that allowed blood to supply the political vocabulary that supported hierarchical structures of governing status.

Only two chapters in this volume, that of Anna More and of Robert Appelbaum, do not engage a specific sociopolitical arrangement of difference; but both model how a particular discourse is produced. William Harvey's discovery of the circulation of blood had a decisive impact on political theory through Harvey's friend, Thomas Hobbes.[42] The coming of a mechanical model of statecraft through Hobbes and his contemporaries has already received attention. But, as Appelbaum notes, the coupling of the theory of the modern body and that of the modern state still calls for further elucidation. Hobbes applied Harvey's ideas of circulation directly to his theory of human physiology and indirectly to his theory of the state. In his essay for the collection, "Flowing or Pumping? The Blood of the Body Politic in Burton, Harvey, and Hobbes," Appelbaum examines the joining of science and statecraft by way of a contrast between the theories of the body and the state expressed in the various versions of Burton's *Anatomy of Melancholy* and Hobbes's major political writings. In Burton, the flowing of wealth and justice through the state replicates the blood in the body, and supplies the commonwealth with the requisite sustenance to keep it in health. But Appelbaum's essay describes how "the switch to an understanding of blood as a mechanical functioning could also effect a switch to an understanding of political life." In Hobbes's conception, blood circulates, and wealth and justice circulate too. But while Burton and Hobbes have similar apprehensions of political concepts such as monarchism and law, their very different models of the body politic produced different ideas of how social transactions took place—whether flowing as an expression of social affinities and relationships, or pumping through the commonwealth as a mechanism for its maintenance.

While Appelbaum demonstrates how discourses of blood map onto, and even help to formulate, political theory, Staffan Müller-Wille (Chapter 9) shows how blood and humors provide a map of the wider world. Indeed, the work of Müller-Wille demonstrates the limits of mechanical theory, and the extent to which (while embraced by political theorists such as Hobbes) it failed to circulate.[43] His own volume, *Heredity Produced: At the Crossroads of Biology, Politics, and Culture, 1500–1870*, edited with Hans-Jörg Rheinberger, traces some of the same cultural bloodlines as this present collection. In his chapter, "Linnaeus and the Four Corners of the World," Müller-Wille takes on an important early text at the crossroads of the decline of neoclassical natural history and the rise of modern anthropology, with its distinct modern conception of race—*Systema Naturae* by the Swedish naturalist Carl Linnaeus. While Linnaeus has often been regarded as the father of modern racial thinking for inventing four "varieties" (as Linnaeus called it) of humans, aligning skin color with one of the various continents—reddish (America), whitish (Europe), blackish (Africa), and tawny (Asia)—Müller-Wille provides an important and much needed corrective of this view by showing that for Linnaeus, these characteristics were grounded in humoral theory. These human "varieties" were therefore malleable, not stable, produced by climate and lifestyle. In other words, Linnaeus did not so much look forward to the nineteenth century as he looked backward in an attempt to systematize the geo-cultural connections posited by neoclassical natural philosophy in which all variation within a species was caused by local, environmental factors. As Müller-Wille writes, "[h]eredity, environment, and culture remained inextricably entangled in Linnaeus's conception of human variation."

The evolution of mechanical philosophy and of the language of nation happens, to some extent, in cooperation: while Galenic materialism and mechanical philosophy coexist and contest each other for some considerable time, as ontological values of blood decline, fictions of national group and national identity borrow from their discourses in order to gather strength. The final two essays of the volume map the political state—and political participation in it—in terms of blood. In his essay, "'Who Got Bloody?': The Cultural Meanings of Blood during the Civil War and Reconstruction," James Downs turns away from commonplace discussions of Civil War medical practices and developments to instead analyze blood as *representation*. He argues that writers throughout the Civil War and Reconstruction periods imbued references to blood with metaphorical power and cultural meaning that existed apart from medical conditions. Metaphors of blood took on a range of meanings from representations of nationhood, patriotism, and honor to critiques that undermined the mythology and romance that many white nineteenth-century Americans associated with war. Bloodshed became visible only when it could be processed through one of these cultural registers.

Situated within print culture, white officials, and many white soldiers as well, referred to bloodshed with great frequency in military, federal and personal correspondences, diaries, journals, and newspapers. Yet their references are almost exclusively to the spilling of white blood and are filtered through notions of nationhood, patriotism, honor, and sacrifice. References to black blood are strikingly absent since the shedding of black blood lacked such cultural resonances. In turn, due to the difficulty of access to literary outlets, documentation by black soldiers and freed slaves of black bloodshed in battle or through disease remains rare. It is only toward the end of the nineteenth century, when a greater number of African American writers gained hold of publishing power, that accounts of how black people's bloodshed contributed to the nation begin to proliferate. Downs concludes by arguing in favor of an expanded archive that would include literary sources so that historians might further investigate cultural representations of blood in the historical record. Until then, they risk buying into the official logic proffered in the nineteenth century of the cultural meaning of blood.

David Sartorius opens his essay with a saying popular among Cuban planters during the nineteenth century: "Con sangre se hace azúcar" ("Sugar is made with blood"). This prized commodity involves producers and consumers in a relationship to the violent institution of slavery, and Sartorius traces references to blood in the words of those mutually entangled through the experience of slavery in the Caribbean: planters, abolitionists, and the enslaved. In each case, Sartorius acknowledges, references to blood gave meaning to various forms of allegiance and to the motivations for inclusion and exclusion in racial, social, and political communities. Sartorius, however, is not satisfied with this conventional understanding. He questions whether, in fact, the ties between blood, race, and political allegiance are consistent and enduring over the colonial period. Sartorius demonstrates that, although blood continued to inform discussions about slavery, it declined in importance in debates over political allegiance as the century progressed. The introduction of anthropology as well as evolutionary biology and psychology shifted the paradigms for explaining humanity, politics, culture, society, and intellectual capacity. Sartorius suggests that blood informed but did not fully explain the power of concepts such as native and foreigner, ancestry and fictive kinship, and, most importantly, loyalty. Sartorius notes that late nineteenth-century scientific experiments helped reinforce the conceptual shift. He calls attention to the discovery that disease could be borne through the blood without transmitting racial characteristics. This finding combined with evolutionary biology's assertion of a common origin for a single human species to change the metaphorics of blood. Rather than immediately suggesting human separation according to inherent qualities, blood, according to Sartorius, took a back seat in explanations of commonality based on religion or loyalty as a foundation for education. By the end of the nineteenth century, racialized understandings of blood were challenged by the sense that all humans, having demonstrated loyalty to Spain, could evolve through education into ideal subjects and citizens.

* * *

Collectively, the essays in this volume emphasize the material and metaphoric significance of blood. Blood is both vital to our lives and obscured from our sight, visible only at times of injury or death; yet, its unseen course under the skin allows it to transport a host of meanings that we attach to it. Our volume charts the traffic of blood in the circulation of the race concept and suggests a compelling history of race written in blood.

Notes

1. In his lecture at the Collège de France dated 28 January 1976, Michel Foucault assumes a "mobile ... polyvalent discourse" of race that appears "at a very early stage"—certainly in the sixteenth and seventeenth centuries; but, he goes on to observe, "the word 'race' itself is not pinned to a stable biological meaning" (see *"Society Must be Defended" Lectures at the Collège de France, 1975–76*, ed. Mauro Bertani and Alessandro Fontana [New York:

Picador, 1997], 77). He further argues that the discourse of *racism* is premised on a shift to a "biological and medical sense" of the term in the first half of the nineteenth century (80). But while modern racism, what we now refer to as "scientific" racism, is not possible prior to the development of the biological terms used to describe it, it is clear that both the term "race" and its medical sense had sixteenth-century existence. Foucault acknowledges that "race" is not "free-floating" in pre-modern and early modern histories, and that it is a "relatively stable" designation of a "historico-political divide" (77). This collection explores the extent to which medical theory and natural philosophy—relating to Aristotelian principles and Galenic materialism—underwrite difference at certain historical points in colonial history. The project of race inscribes cultural difference in the body and reproduction. Most, although not all, of the essays assembled here explore this particular nexus. Galenic materialism certainly lacks the stability of biology, but it nonetheless supplies the political force of a received, and widely apprehended, knowledge of the body. The description of difference is always metaphoric: our volume investigates when markers of "blood" are deployed *as* metaphors, and when they describe a medical or philosophical discourse that informs ideology.
2. Bartlett, *The Making of Europe: Conquest, Colonization, and Cultural Change, 950–1350* (Princeton University Press, 1993), 197. For the counter-argument, see Jeffrey Jerome Cohen, "On Saracen Enjoyment: Some Fantasies of Race in Late Medieval France and England," *Journal of Medieval and Early Modern Studies* 31 (2001), 113–46; 115–16.
3. Much of the argument put forward in this section has already been rehearsed by David Nirenberg in his excellent essay, "Was there Race before Modernity? The Example of 'Jewish' Blood in Late Medieval Spain," in *The Origins of Racism in the West*, ed. Miriam Eliav-Feldon, Benjamin Isaac, and Joseph Ziegler (Cambridge University Press, 2009), 232–64; particularly 233–6. See also, Nirenberg, "Race and the Middle Ages: The Case of Spain and Its Jews," in *Rereading the Black Legend: The Discourses of Religious and Racial Difference in the Renaissance Empires*, ed. Margaret Greer, Walter D. Mignolo, and Maureen Quilligan (University of Chicago Press, 2007), 71–87; and Ania Loomba, "Race and the Possibilities of Comparative Critique," *New Literary History* 40 (2009), 501–22.
4. The introductions to a number of collections published in the 1990s importantly stressed the diversity of meanings in early modern race, as well as its cross-section with understandings of region, nationality, religion, and gender. Most of these, however, emphasized race as a cultural understanding rather than an embodied term. See Margo Hendricks and Patricia Parker (eds), *Women, "Race," and Writing in the Early Modern Period* (London: Routledge, 1994); Joyce Green MacDonald (ed.), *Race, Ethnicity, and Power in the Renaissance* (Madison, NJ: Fairleigh Dickinson University Press, 1997); *The William and Mary Quarterly* 54.1 (Winter 1997), special edition, "Constructing Race: Differentiating Peoples in the Early Modern World."
5. There have been three recent (or recent enough) monographs that consider the humoral basis of early modern race in England: see Roxann Wheeler, *The Complexion of Race: Categories of Difference in Eighteenth-Century British Culture* (Philadelphia: University of Pennsylvania Press, 2000); Mary Floyd-Wilson,

20 Introduction

> *English Ethnicity and Race in Early Modern Drama* (Cambridge University Press, 2003); and Jean Feerick, *Strangers in Blood: Relocating Race in the Renaissance* (University of Toronto Press, 2010).

6. "The Invention of the Concept of Race," *The Origins of Racism in the West*, 200–16; 208.
7. Ania Loomba and Jonathan Burton make this point at the start of their documentary companion, *Race in Early Modern England: A Documentary Companion* (Basingstoke: Palgrave Macmillan, 2007), 1.
8. *The touchstone of complexions generallye appliable, expedient and profitable for all such, as be desirous & carefull of their bodylye health* (London, 1576), 62v.
9. See Gail Kern Paster, *The Body Embarrassed: Drama and the Disciplines of Shame* (Ithaca, NY: Cornell University Press, 1993), 194.
10. Ibid., 215–80. See also, Rachel Trubowitz, "'But Blood Whitened': Nursing Mothers and Others in Early Modern Britain," in *Maternal Measures: Figuring Caregiving in the Early Modern Period*, ed. Naomi J. Miller and Naomi Yavneh (Aldershot: Ashgate, 2000).
11. See Feerick, *Strangers in Blood*, 55–77; 202.
12. See, as an example, Jaques Guillemeau, *Child-birth, or The happy deliuerie of vvomen* (London, 1612), "The nursing of children," Ii4v. Laurent Joubert asserts in his *Popular Errors* the extreme view that the milk of the wet-nurse exceeds the influence of parents' sperm in its influence on the infant's humoral complexion; he therefore likens the selection of a wet-nurse to that of a wife in that her "ancestry, [her] blood, and [her] conduct" must be researched in order "to have the best lineage possible" (see *Popular Errors* [c. 1578], trans. Gregory David de Rocher [Tuscaloosa: University of Alabama Press, 1989], 193).
13. *The Boke named the Governour* (London, 1531), 16v.
14. Feerick, *Strangers in Blood*, 60.
15. *Boke named the Governour*, 16r.
16. *The touchstone of complexions*, 16v.
17. *Hierarchy, Commerce, and Fraud in Bourbon Spanish America: A Postal Inspector's Exposé* (Nashville: Vanderbilt University Press, 2005), 200.
18. *A True Collection of the Writings of the Author of the True-Born Englishman* (London, 1703), 218.
19. Jennifer DeVere Brody, *Impossible Purities: Blackness, Femininity, and Victorian Culture* (Durham, NC: Duke University Press, 1998), 3.
20. Eckhard Kessler, "The Intellective Soul," in *The Cambridge History of Renaissance Philosophy*, ed. Charles B. Schmitt et al. (Cambridge University Press, 1988), 503.
21. Philip Melanchthon, *Liber de anima*, in *Opera quae supersunt omnia*, ed. Carolus Gottlieb Bretschneider and H. E. Bindeil, 28 vols (Halle-Braunschweig, 1834–60), XIII. col. 16: "Anima rationalis est spiritus intelligens, qui est altera pars substantiae hominis, nec extinguitur, cum a corpore discessit, sed immortalis est. Haec definitio non habet physicas rationes, sed sumpta est ex Sacris literis" (quoted in Kessler, "The Intellective Soul," 517).
22. See Sachiko Kusukawa, *The Transformation of Natural Philosophy: The Case of Philip Melanchthon* (Oxford University Press, 1995), 91–2.
23. Ibid., 91. Melanchthon writes: "Ergo Galeneo anima praesertim sensitiva et vegetativa, aut temperamentum est, aut spiritus vitalis ac naturalis in

animantibus, hoc est, aut temperamentum, aut spiritus vitalis est principium vitae ac motus in animantibus, seu est res movens corpus. Ita rem aliquam moventem Galenus digito monstrare conatus est, ut quae res esset anima suspicari possemus" (*Liber de anima*, XIII, col. 10; Kusukawa, *Transformation*, 90, n. 73).

24. Angus Gowland, *The Worlds of Renaissance Melancholy: Robert Burton in Context* (Cambridge University Press, 2006), 42. See also, Kusukawa, *Transformation*, 89; and Emily Michael, "Renaissance Theories of Body, Soul, and Mind," in *Psyche and Soma: Physicians and Metaphysicians on the Mind–Body Problem from Antiquity to Enlightenment* (Oxford University Press, 2000), 163–5.

25. One such edition provides the source for Robert Burton in his 1621 *Anatomy of Melancholy*: *Joannis Magiri, doctoris medici et philosophi ... commentarius ... In aureum Philipp Melanchthonis labellum de Anima* (Frankfurt, 1603). The point to be made here is that, with access to the library at Christ Church, Oxford, Burton did not have access to a published free-standing edition of Melanchthon's work.

26. "Ad medici necessariam esse considerationem harum dissimilitudinum, constat. prodest autem in communi vita ad tuendam valetudinem, ad regendos mores, ad circumspectionem in familiaritatibus. Omnino necessaria diligenia est, vitare monstrosas, superbas, malevolas & perfidiosas naturas" (ibid., 360).

27. Timothy Bright, *A Treatise of melancholie* (London, 1586), iiir. That Bright seeks to counter this apprehension situates his work within the wider cultural debate that was taking place not only in England but also in continental Europe. Bright's treatise is fairly unique in terms of its Platonic situation of these arguments, driven by his own inclination toward Calvinism. While William Perkins and other English Calvinists devote a good deal of time and ink to trying to drive a wedge between soul and body (in an attempt to recuperate spiritual crisis as an instrument of God in no way connected to the body or to the humoral excess of melancholy), Bright is isolated as a medical theorist who advances the position. (Even in his reliance upon Bright in his later *Anatomy of Melancholy*, Robert Burton does not reproduce this argument; his section on "religious melancholy" is a case-in-point.) The rehearsal of all of the arguments and actors in this debate is impossible, but Dennis Des Chene provides an excellent overview. See *Life's Form: Late Aristotelian Conceptions of the Soul* (Ithaca, NY: Cornell University Press, 2000), particularly chapter 4; see also, Kessler, "The Intellective Soul," in *The Cambridge History of Renaissance Philosophy*, 485–534.

28. Intellectual historians such as Angus Gowland and James Hankins have recently argued that belief itself—the excess, defect, or lack of religion—was apprehended and understood largely in terms of temperament in the latter part of the sixteenth century, although the initiation of such thinking, and the theory upon which sixteenth-century Christian physicians directly draw, is derived in the fifteenth. See Gowland, *The Worlds of Renaissance Melancholy*, and "The Problem of Early Modern Melancholy," *Past and Present* 191 (2006); and Hankins, "Monstrous Melancholy: Ficino and the Physiological Causes of Atheism," *Laus Platonici philosophi: Marsilio Ficino and his influence*, ed. Stephen Clucas, Peter J. Forshaw, and Valery Rees (Leiden: Brill, 2011), 25–43.

29. Gowland, "The Problem of Early Modern Melancholy," 105–6.
30. Hankins, "Monstrous Melancholy," 29.
31. Cohen, "On Saracen Enjoyment," 116.
32. Kessler, "The Intellective Soul," in *The Cambridge History of Renaissance Philosophy*, 507.
33. Thomas Wright, *The Passions of the Minde* (London, 1601), 6; emphasis added.
34. Garrett Sullivan, *Sleep, Romance and Human Embodiment: Vitality from Spenser to Milton* (Cambridge University Press, 2012), 137.
35. Attributed to Aristotle, Aquinas provides this formulation. The phrase becomes a central claim of Empiricism, available in the works of John Locke, George Berkeley, David Hume, and John Stuart Mill.
36. While Descartes's mechanical philosophy had an undeniable scope of influence, the progress of its acceptance was, as is the usual course of theories, slow and contested. Indeed, his situation of the contact point between body and soul in the pineal gland failed to settle the question of mind–body interaction at all. Stephen Fallon has wittily remarked that in this assertion, "the question of mind–body interaction [had] been miniaturized rather than resolved" (see *Milton Among the Philosophers: Poetry and Materialism in Seventeenth-Century England* [Ithaca, NY: Cornell University Press, 1991], 28). This was, in fact, the particular point on which he was assailed through the eighteenth century (in large measure because animals shared this anatomical feature, but none would credit them with rational thought or soul).
37. While Descartes accepted William Harvey's theory that blood circulated, he did not accept Harvey's account of the heartbeat, forceful systole. Whether this misunderstanding on the part of Descartes was an actual or willful one is an open question. But Harvey's account of circulation appealed to Descartes because it allowed for a single motion of the heart to be the source of all motions in the body by particle to particle transfer—that of muscle, veins, glands, and spirits. There was no need, then, for secondary causes or other faculties. For an account of this, and how Harvey's theories concerning blood influence Descartes, and are circulated by him, see Roger French, "Harvey in Holland: Circulation and the Calvinists," in *The Medical Revolution of the Seventeenth Century*, ed. Roger French and Andrew Wear (Cambridge University Press, 1989), 46–86, 50.
38. *A True Collection of the Writings of the Author of the True-Born Englishman*, 196.
39. Ibid., 178, emphasis added.
40. Ibid., 177.
41. Ibid., 193.
42. In fact, the principal instrument of dissemination for Harvey's theories was through Hobbes and through the philosophy of Descartes (see, French, "Harvey in Holland"). The obscurity of Harvey's theories was partly responsible for its very slow acceptance. But even well after Harvey's theories had disproved Galen, humoral theory continued to prevail as the principal medical discourse through the eighteenth century.
43. The charge of atheism laid against both Hobbes and Descartes might go some distance in explaining this: both were assailed for not accounting for the operation of the soul within the body (Descartes's quibble on the pineal gland was apparently insufficient; see n. 36, above).

Part I
Race and Stock

1
Metamateriality and Blood Purity in Cervantes's Alcaná de Toledo

Rachel L. Burk

> One day I was in the Alcaná market in Toledo, a boy came by to sell some notebooks and old papers to a silk merchant; as I am very fond of reading, even torn papers in the streets, I was moved by my natural inclination to pick up one of the volumes the boy was selling, and I saw that it was written in characters that I knew to be Arabic. And since I recognized but could not read it, I looked around to see if some Morisco who knew Castilian, and could read it for me, was in the vicinity, and it was not very difficult to find this kind of interpreter, for even if I had sought a speaker of an older and better language I would have found him.
>
> In short, fortune provided me with one, and when I told him what I wanted and placed the book in his hands, he opened it in the middle, read for a short while, and begun to laugh.
>
> I asked him why he was laughing, and he replied that it was because of something written in the margin of the book as an annotation. I told him to tell me what it was, and he, still laughing, said:
>
> "As I have said, here written in the margin is written: 'This Dulcinea of Toboso, referred to so often in this history, they say she had the best hand for salting pork of any woman in all of La Mancha.'"
>
> —*Don Quixote* 1.9[1]

Early in the First Part of *Don Quixote* (1605), the narrator intervenes in the story of the knight's exploits to explain that his source for the text up to this point, originally found in the archives of La Mancha, has run out. Fortuitously, he tells us, he has discovered Cide Hamete Benenjeli's continuation of Quijote's adventures at a stall in Toledo's Alcaná marketplace. It is this manuscript—in a language he cannot read, with amusing marginalia, and found in a pile of junk—that forms the supposed basis for what becomes a central masterpiece of European literature.

A defining beginning to the proliferation of fictional levels that make up the complex narrative strategy of *Don Quixote*, the Alcaná episode sets up central metafictional premises of the work by introducing the narrator, a *morisco* translator, and a second manuscript from the Arab-Manchegan historian, fictional author of the novel's supposed source texts. For all the merited study this chapter has received, critics have paid little attention to a complementary discourse to that of the work's supposed origins: an elaborate portrayal of textual materiality that alludes to concerns about embodiment in early modern Spanish society, that is, to blood purity. The Alcaná passage creates a fiction of the material creation of the novel, the concrete pre-texts of *Don Quixote*, in the form of soiled papers for sale cheaply in an open-air market in the less-known Jewish neighborhood of Toledo. That is to say that not only is the novel acutely metafictional, it is also metamaterial. A hallmark of the text is its extended discussion of books and manuscripts as objects inscribed, imagined, manufactured, collected, burned, saved, spoiled, found, and sold. These include the material texts that serve as ancestors to the book in the reader's hand. What's fundamental to the novel's fictional universe is not only the metafictional discussion of the printed word, but also the pervasive representation of the manuscript, including its discrete social purposes, ways of signifying, and forms of circulation.

This depiction of the materiality of the text is also one among the multiple representations of Arab-Iberian culture in the novel, one that comments on this community's singular relationship to manuscripts and writing. The narrator's visit to the Toledo marketplace brings to the fore Cervantes's claims about the textual artifacts ("notebooks" and "old papers" to be sold as pulp) at the same time as it underlines the status of the novel as a translation from prohibited Arabic to imperial Castilian. The text presents its own creation as a series of exchanges—material, linguistic, economic—that associate the book as a physical object with the contemporaneous cultural politics of the *moriscos*, the population of formerly Muslim Iberians forcibly converted to Christianity, translated— if you will—from one belief to another. At the same time, the text persistently equates human bodies with bodies of texts. At stake in the discussion of the physical character of *Don Quixote*, I argue, is blood purity, the emerging system for classifying bodies by blood and belief. The "Prologue" to Part One suggests the novel itself as an imperfect, even illegitimate, textual body, the ugly stepchild of its author, while in the "Examination of the Books," the priest and barber put Quijote's library on trial, burning heretical books in place of heretical bodies. Seen in this light, the Alcaná passage, which represents the source of the *Quixote* as a messy, incomplete collection of cheap paper and illicit script, takes on further meaning. Read together, these three passages create a fictive "genealogy" of the materiality of the novel that is heterodox, an embodied text with a hidden, illegitimating past. This, too, speaks to the *morisco* question. Cervantes makes the ostensible pedigree of his novel as a physical object comment on larger cultural issues of embodiment related to exclusion, lineage, and race. Specifically, the metamaterial discourse addresses the ways in which *morisco* bodies were newly seen as distinct in substance from Old Christian bodies, thanks to doctrine of blood purity.

From its most basic formulation, the relationship between characters and the book in *Don Quixote* is both discursive and material; it has to do with meaning as well as with the physical form in which meaning is expressed. At the turn of the seventeenth century in Spain and Portugal, the Inquisition, charged with policing Catholic orthodoxy, had a related goal: to distinguish the bodies of multi-ethnic Iberians, to make them legible like title pages. Blood purity (Sp. *limpieza de sangre*; Pt. *limpeza de sangue*) was a discriminatory legal designation of religio-ethnic identity that classified converts to Christianity like the *moriscos* and their descendants as bodily impure. Through a discourse of metamateriality, which draws attention to objects that signify and emphasizes both their material existence and semantic content, Cervantes speaks to the premises of racialization that underlie the blood purity doctrine and questions the fiction that a bloodline marked bodies with hidden difference.

Critical contexts

Early modern Iberia insistently attempted to transform the conceptual and the metaphorical into concrete realities. *Limpieza de sangre* was the prime, but in no way singular, example of what Margaret Greer calls the period's materializing instinct. Gil Anidjar has argued that the emergence of blood purity signaled an epistemological shift in the European understanding of community, an absolute inversion of an earlier understanding of Christendom in which the spiritual community of Christian believers was transformed into race and the body politic.[2] Examining how Cervantes approaches textual materiality in light of these observations, my reading of *Don Quixote* stands at the intersection of two seemingly disparate currents of critical investigation about early modernity, one rooted in the history of the book and the other centered on the emergence of race in Europe's first modern empire.

In discussing the celebrated metaliterary aspect of *Don Quixote*, critics have called our attention to the role that the multiplicity of genres (romance, novella, picaresque) played with respect to the emergence of the novel and to the multiplication of authors (prologuist, narrator or second author, *morisco* translator, Arab historian) that created a satire of singular textual authority. Cervantes's unrivaled use of metalepsis, to which I will return later, is central to the text's metaliterary discourse.[3] In the past decade *Quixote* scholarship has taken a turn toward materiality studies, underscoring that Cervantes does not ignore the textual artifact itself, the concrete link between the worlds of the reader and of the author. Cervantes incorporates various kinds of writing and a range of writing technologies in both the First and Second Parts, elaborating a discourse about textual materiality as a sociocultural phenomenon.[4]

Carroll B. Johnson and Juan Carlos Rodríguez address in detail the "moorishness" of the Arabic manuscript in Chapter Eight, understood previously as another part of the satire of authority.[5] Mercedes Alcalá Galán devotes a chapter of her enlightening *Escritura desatada* to what she calls the "clandestine *Quixote*" of Cide Hamete, a concept that is vital to my reading. She brings to the fore the historical context: the persecution of *moriscos* in Cervantes's lifetime and the legal status of any manuscript in Arabic in Phillip II's Spain.[6]

The contemporary study of blood purity in Hispanism began in the last century with historians (Sicroff, Nirenberg, Kamen)[7] who detail *limpieza* as a legal, social, and demographic phenomenon as it bred a complex ideological reality from its fifteenth-century beginnings. Like-minded critics of race have sought to include *pureza de sangre* and the Iberian Peninsula within a broader evolution of European racial ideas (Balibar and Wallerstein, Nirenberg, Anidjar).[8] Literary critics have explored the idiosyncratic representational history of Iberian blood (Mariscal, Dopico Black, Beursterien, Greene),[9] in particular its relationship to questions of honor and as a displaced discourse. The study of Arab-Iberian culture under Christian rule (García Arenal, Burshatin, Caro Baroja, Domínguez Ortiz and Vincent, Barletta, Dadson, Fuchs, García Arenal and Rodríguez Mediano)[10] and its presence within

Cervantes's works (Graf, Johnson, Armas)[11] has countered the ideology of Spanish national identity, premised on religious-cum-racial "purity," by exploring its foundational hybridity. Likewise historians have begun to productively investigate the relationship between Iberian blood purity and the development of the New World caste system.[12]

The interest in this chapter is in demonstrating how historio-cultural work on blood purity can broaden and deepen our understanding of Cervantes's discussion of materiality in *Don Quixote*. In bringing together the self-conscious impulses of the novel and its engagement with textual materiality with broader cultural examination of the *moriscos* and the rise of racializing discourses of embodiment, Cervantes ironizes the urge to make the metaphysical physical. He uses the play of fictionality of his novel to underline that blood purity is a rhetorical rather than fleshly reality, one produced and maintained by documents, not bodies.

Scrap paper and an indiscriminate reader

In the Alcaná, the boy selling messy piles of crumbling, dog-eared papers is not a bookseller; his goods are more humble. Neither codices nor vellum manuscripts, but only "notebooks and old papers." The vendor does not consider them objects of study, translation, and eventual publication, as indicated by their haphazard display and low market valuation. The narrator purchases them altogether for a half *real*, the price for an inexpensive lunch. Nonetheless, the narrator browses the junk dealer's wares with interest. He admits to being an indiscriminate reader, a voracious consumer of the written word in every form to the point that he will read "torn papers in the street." Indifferent to differences in the quality and authority of texts, he does not favor documents from the official archives of La Mancha over Arabic scribbled on scraps from the Alcaná. He disregards any hierarchy of materials, the way that form dictates the nobility of the content; likewise he feels no compunction to hide his appreciation for all languages, irrespective of his knowledge of them, social taboos, or legal prohibition. The narrator suffers from a minor form of text-inspired madness, much like Quijote himself: he reads any discursive object, disregarding material form and what that form is supposed to tell the reader about content. Thus, he serves as another double for the protagonist, given that his "natural inclination" is to read indiscriminately, which leads him to transgress customary behavior. His would-be madness originates in the textual objects themselves. The narrator's idiosyncratic relationship to material texts deepens the satire of textual origins first iterated in the "Prologue" to the First Part in which *Don Quixote* is pegged as inherently flawed, "without a single grace." By iterating a (fictional) material dimension, the Alcaná episode makes concrete the text's inadequacy. Not only is the content of the history at hand dubious, but it is even of uncertain quality as an object.

Sebastián de Covarrubias's 1611 dictionary sheds light on the second author's reading materials:

> PAPER. There is one called white that is the ordinary one in which one writes and another crude that serves to wrap goods, that is called estraça paper, from the Italian verb *straziare*, which is to break; because this paper gets ripped to wrap goods.[13]

The "torn paper" is of the second type described, crude stock that could be gray or brown in contrast to finer-grade paper, made of new white linen and employed for writing. *Carta straccia*, in the parlance of the Italian book trade, was waste paper, either made for purposes other than writing or spoiled by misprinting. This often meant it was made of recycled materials, leaving the end product less desirable as a writing surface. As Covarrubias clarifies, these broken or marred papers were to contain goods sold at market, employed for their material rather than discursive properties.

The wares actually for sale in the Alcaná, eventual source for the book in hand, are similarly ignominious. Covarrubias describes the *cartapacio* as a small, portable notebook, or collection of smaller notebooks, for students of an ordinary and common kind.[14] Equally, the "papeles" available are "viejos" ("old" rather than antique or "antiguas"), implying their poor quality as well as their age.

Cervantes hints at a possible end for the parcels of papers (since they were clearly not intended for reading). The boy means to sell his stock to a silk dealer: "moving more quickly than the silk merchant, I bought from the boy all the papers and notebooks for a half real." Silk has its own connection to the early modern book trade: it was used in creating luxurious bindings, particularly in Moorish styles popular in Europe during the fifteenth and sixteenth centuries. Book-binders regularly employed scrap paper, odd pieces of leftovers from book printing, to make pasteboards or book covers, which would then be covered in lavish textiles or leather. A possibility broached by Mercedes Alcalá Galán is even less exemplary: Cide Hamete's manuscript was destined for worm food.[15]

What piques the narrator's interest in the Alcaná among the manuscripts in the merchant's display is the script, the Arabic characters that he recognizes as such but cannot read. Documents like the ones for sale in the Alcaná were prohibited by Philip II's 1566 Royal Decree (called the *Pragmática*), which barred Moorish cultural practices of all kinds, and paid special heed to banning Arabic, both spoken and written. Thus the narrator's uninhibited approach to reading and feckless willingness to accept all source texts as equal is more than a harmless eccentricity: it is a strategic overlooking, the stakes of which would be apparent to an early modern reader. The fictional source for *Don Quixote* would have been illegal in Philip II's Spain, a clandestine manuscript in a prohibited language belonging to a suspect minority. Thus, the narrator's insouciance toward Arabic and the prohibitions on *morisco* culture naturalizes what would have been, outside the world of the text, a very dangerous exchange, of illegal goods for cash and an illegal language for the language of record.

The Royal Decree of 1566 went well beyond the discouragement of heterodox religious belief, aiming to suppress the Moorish way of life, even forcing converted Muslims to abandon their names. Central to its repressive program was its language policy: *moriscos* were required to learn Castilian in three years. At the end of the term, speaking, reading, and writing Arabic became illegal, leading to the ridiculous prospect that those unable or unwilling to use Spanish were legally mute.[16]

The possession of Arabic texts frequently prompted Inquisitorial investigation. Many manuscripts of Islamic Spain either left the Peninsula with the diaspora of the Moorish nobility by the end of the fifteenth century or were burned by Cardinal Cisneros in his efforts to Christianize and hispanize Granadan Moors at the beginning of the sixteenth century. The most valuable of those remaining were appropriated into private library collections including Cisneros's own. Some, however, were maintained in secret by *morisco* communities despite the risks involved. Within these communities, often closely knit but distant from one another, manuscripts in Arabic and *aljamiado* (Romance written in Arabic script) were of singular importance for the maintenance of a cultural identity under threat. Despite the immediate danger posed by the Inquisition, *moriscos* kept and coveted prohibited documents, some of which were religious, some folkloric, others legal and familial records. According to Vincent Barletta, the manuscripts themselves, irrespective of their subject matter, constituted guarded tokens of a better past; they were tokens in part because most *moriscos*, long cut off from centers of Islamic learning, were unable to read them.

Official Spanish efforts to eradicate Arabic were rooted in the status of the language in Islam. Holy among its believers, Arabic is the language of the Qur'an and thus of revelation. The majority of Muslims read scripture in the same language in which it was written, avoiding the exigencies of translation. In Castilian propaganda, Arabic, as graphia or written sign, was viewed as a foundation for religious and cultural difference. Besides being virtually indecipherable to a reader of Romance, it was read "backwards" (right to left), which was understood as symbolic of its essential incorrectness. Anti-*morisco* tracts regularly cited the Arabic language as demonstrative of the irreconcilable differences between communities that made the assimilation of *moriscos* impossible. According to Alcalá Galán: "The language [Arabic] becomes, in its unintelligibility, a metaphor for the lack of communication and the failure to come together. The curious thing is this apparent difficulty in translateability was interpreted as hostility and essential intrinsic enmity [of Arabic writers for Romance readers]."[17] Written Arabic and the manuscript in Arabic took on an unusual stature in sixteenth-century Spain, according to records of Inquisition trials. "Among the Old Christians paper written in Arabic provoked true fear, while for the *moriscos*, although they could not understand it, it had an almost magical quality."[18] Possession of books or prayers in Arabic were among the most severely punished crimes by the Inquisition, regardless of whether the owner was literate, because they believed that they were treated as religious talismans as well as communicative texts.[19]

Don Quixote comments on *morisco* manuscript circulation and the transmission of material texts in light of what writing in Arabic came to represent for New Christians and Old during the sixteenth century. Vincent Barletta explains: "[*morisco*] literature [is] at its core a secret literature composed and recopied by and for small, intricately connected communities of readers."[20] Thus, by buying, translating, and publishing the discovered text, the narrator interrupts the closed circuit of *morisco* manuscript circulation, usurping the silk merchant for whom the papers were intended. Indeed, the text may suggest that the seller and intended buyer could be *moriscos*, as silk weaving was traditionally a Moorish trade as practiced in Toledo.[21] When viewing the Alcaná episode in dialogue with its socio-historical context, we must conclude that Cervantes's narrator is anxious to have, understand, and make public Moorish knowledge that is not his own. (Admittedly this knowledge is about a baleful Christian knight.) Similarly, the *morisco* obliges him without compunction, although the translation described—of Arabic script into the vernacular—was an especially perilous endeavor for him during the reign of Philip II. The second author and *morisco* translator together transform the private and encoded form of a minority into the public and "decoded" form of the dominant idiom. The translation they perform is not only linguistic, but also cultural, from marginal to majority, and material, from manuscript to printed text. Had this transaction taken place in the historical Toledo of 1600, it would have been enough to land both in an Inquisition cell.

Prologue, stained genealogies, and the textual body

In the "Prologue" to the First Part of *Don Quixote*, the prologuist insists—satirically and in contrast to custom—that the reader take note of the imperfections in the text he introduces, initiating a series of ludic genealogies and problematic embodiments in his explanation of how we, as readers, should understand *Don Quixote*. The prologuist frequently ambiguates his references, confusing Don Quijote, the character, and *Don Quixote*, the novel, and likewise conflating anatomical and textual bodies. This "brain child" of the writer reflects his maker, as "[e]verything engenders its like." Don Quijote is "ugly son," fruit of his "sterile and poorly cultivated wit."[22] Evoking and undermining the trope of the work as child of the writer's intellect, the prologuist draws attention to this rhetorical commonplace, perhaps best known in Ovid, taking it to an absurd extreme. At the same time, he underlines the troubled "inheritance" of his book-son on multiple fronts: genealogical, literary, corporeal.

The genealogy of the novel invoked here alludes to pressing concerns about bodies and inheritance in early modern Spain in relation to *limpieza de sangre*.[23] By the 1520s, the Castilian monarchy had managed the conversion to Christianity of Iberian Jewish and Muslim populations through coercion and threat. However, over the course of the sixteenth century, legal and social sanctions contravened efforts to assimilate the new converts, citing with increasing conviction a bodily difference—their impure or stained blood—that marked them as distinct and lesser. For the first time in European thought, difference between "peoples" or "nations," which had been flexible categories, based on religion, history, culture, language, and geography as well as the body, were essentialized through blood purity. It naturalized social and cultural diversity, making it absolute in a way previously inconceivable, given that the multifold differences of what we would now call ethnicity were inexact, open to conversion, chance, and change. The sin of the father, in this case, "mistaken" religious belief, was visited upon the son and grandson through the familial bloodline.

One impact of the *limpieza* doctrine was a culture-wide obsession with genealogical-cum-physical purity, among Old Christians as well as New. It became an issue of honor, fundamental to social standing. While a *morisco* such as our translator, culturally hybrid in dress and language, would be an obvious victim of social stigma and blood purity law, the threat to the majority of Spaniards was present but less overt. By the sixteenth century the ancestral presence of a *converso* relative, however distant, "stained" the family tree although intermarriage had been unexceptional before the fifteenth century. Blood purity was a hidden register of racial identity and a troublingly ambiguous one. There was no visible trait attributable to all New Christians; blackness of skin was not a tell-tale sign of impurity. There was no blood test; neither the color nor viscosity of let blood revealed its degree of purity, according to contemporaneous medical understanding. The ambiguity of blood purity as a register created a culture of intense social anxiety. Cervantes himself explicitly satirizes the mortal fear that an unknown familial transgression could come to light in *El retablo de las maravillas*, pointing out the instability of social position within a hierarchy based on the purportedly immutable quality of one's blood.

A Spaniard seeking to establish his blood purity did so institutionally through the production of documents called *probanzas de limpieza de sangre*, or proofs of blood purity. Issued by the Inquisition, these documents attested to Old Christian ancestry and good repute. These were based on Inquisitorial investigations of archival records as well as the soliciting of affidavits, which created a legitimizing legal system to substantiate a supposedly physical truth. For example, the 1664 application of D. Juan de Alfaro of Córdoba for a position as a notary in the Inquisition is a dossier of close to 30 documents attesting to the applicant's *limpieza*.[24] Along with letters, it includes a printed form with blanks as well as a guide, also printed, for providing information on blood purity by those asked to evaluate the job candidate. Not only does the dossier establish the importance of witnesses in establishing purity, but the use of a printed formula is remarkable for the degree of bureaucratic institutionalization it suggests. Both "fama" (reputation) and "honor" are treated as if they were easily quantifiable, like a driving record.

Many Spaniards attempted to cover up their lineage in order to "pass" as Old Christian in the present, papering over their familial past, legally and symbolically. Cervantes explored this phenomenon in his exemplary novellas, as Barbara Fuchs elaborates.[25] They excised family histories and genealogical lists, removing traces of New Christian relatives for the purpose of producing blood purity *pruebas* (or proofs) that conformed to legal formulas for determining Old Christian lineage. The accuracy of the *probanzas* to report blood purity was questionable. It was not uncommon to be falsely impugned as a *converso*; on the other hand, it was also common practice to fake documents with paid-for testimonies.

In the "Prologue," after condemning his issue as illegitimate and deformed, the implied author proposes a second relation to *Don Quixote*, seemingly distancing himself from the character-text's insufficiencies: "although I appear to be the father I am the stepfather of don Quijote."[26] While reinscribing the author as stepfather may sever the genealogical link to the writer, it amplifies concerns about the insufficiency of the text's lineage, now suggesting its illegitimacy. As Dopico Black points out, a stepfather implies either that the biological father has died or that our author has been cuckolded in a society without divorce.[27] Nonetheless, the prologuist returns to a strong claim of responsibility for "his child." The confused bloodline of the book-man takes on another dimension in light of possible illegitimacy: the "son" is "feo" or ignoble, not "lindo," or noble. The mapping of anatomy onto the material text was a cultural commonplace, according to Dopico Black. Book copies were often called bodies, particularly in the aggregate: "cuerpos de libros." The Royal Censor granted permission for the printing of a certain number of "bodies" in his license. The vocabulary of the codex itself makes frequent reference to the body: a bound book has a face (*carátula*), spine (*lomo*), fingers (*índices*), and foot (*el pie de la página*).[28]

Early in the First Part, body-book discourse comes to the fore when the barber and priest meet to "dismember" Don Quijote's collection of books. Cervantes actively politicizes the trope of the textual body, son of the author's *ingenio* (wit). In the "Examination of the Books" (*Escrutinio de los Libros*) in Chapter Six, the knight's friends pass judgment on the contents of his library in a mock auto-da-fé in which books act as substitute for actual bodies. Dopico Black points out that, according to the criteria of the Priest and the Barber, books not only have bodies, but those bodies can be good or bad. Romances of chivalry are damned wholesale for having intemperate bodies, never complete in all their parts or proportional following Aristotelian norms. Chivalric authors, in the estimation of the priest, seek to produce monsters rather than wholesome figures. Both the physiological disproportion of the romances, as well as their spiritual error, are cause for them to be thrown out of the library and Christian society, anticipating the expulsion of *moriscos* to take place in 1609, after the publication of the First Part of the novel in 1605.[29]

The discovery of Cide Hamete's manuscript in Chapter Nine happens at the culminating moment of the joust between Quijote and the Basque as they charge one another with makeshift weapons raised:

> [W]e left the brave Basque and the famous Don Quixote with their swords raised and unsheathed, about to deliver two downstrokes so furious that if they had entirely hit the mark, the combatants would have been cut and split in half from top to bottom and opened like pomegranates.[30]

The reader expects the knight to get the worst of the encounter, as he has recently been unhorsed by a windmill. Instead, as we know, the source for the adventure runs out: "at this extremely uncertain point, the delectable history stopped and was interrupted, without the author giving us any information as to where the missing parts could be found."[31] The episode (and the supposed material text that transcribes the episode) "ended up disfigured" and is left both "maimed and crippled."[32] Embodying the archival source, Cervantes substitutes the textual body for Don Quijote's, a metalepsis that is first rhetorical, replacing character with text, and then narrative, once the second author intervenes into the fictional plane.

Having articulated the purported imperfection of *Don Quixote* in the "Prologue" and elaborated on the interchangeability of books and bodies in the "Examination," Cervantes moves on, in the Alcaná episode, to give tangible (albeit metafictional) proof of the physical flaws of his work. Cervantes depicts the novel as originating in a stitched-together assemblage of fragments of dubious quality, a composite textual body made up of heterogeneous parts: a truncated archival record, notebooks filled with illegible characters, decaying papers, and trash from the streets. Moreover, the amalgamated collection of pre-texts reveals the palimpsestic trace of Arabic, leaving it marked—stained—with otherness. A patchwork, it is open and composite, suggesting mutability. In a word, it is fraught with menacing connotations for Baroque Spain where the origin and composition of the individual bodies of subjects as well as the political body were very much at issue.

Metalepsis and the editio princeps

The elaboration of a metamaterial discourse in *Don Quixote* mirrors the often explored metafictional one created through the embedding of narrative levels. Significant to this strategy is not only multiple narrators (couched as ersatz authors), but also the figure of metalepsis involving books and writers. In narratology, metalepsis describes the contamination of the diegetic with the extradiegetic, the violating of fictional planes to various effects—comical, epistemological, even ontological. This transgressive metafiction creates a façade of historicity. In our case Cervantes affirms the "reality" of his central story through metalepsis. During the "Examination of the Books," for example, the priest picks up a copy of Cervantes's *Galatea* and tells the barber that the author is an old friend, who is "better versed in misfortune than poetry."[33] As a fictional device, metalepsis borrows the reality of the extra-textual world, creating a reality effect within fiction and, resultantly, a pleasurable disorientation for the reader. In this case, characters discuss the historical Miguel de Cervantes, author of the 1585 pastoral romance *La Galatea* as well as the book in the reader's hand. The author appears not as a seen-and-heard character but, analogically, as his book, which has crossed from the real world into the fictive. In both parts of the *Don Quixote*, books do just this kind of metaleptical traveling between planes.[34]

But metalepsis can work in the other direction, as well. Brian McHale sees this deliberate violating of levels as also creating a kind of short circuit that allows fiction to intrude on the ontological plane of the author.[35] The fiction of the material text, as I have described it here, acts in this fashion by incorporating the concrete world of the book into the illusory world of the text. It creates a metaleptical relationship between the real-world artifact of the book and the fictive genealogy of its creation described in the text. That is, the text acts to preemptively include printed copies of *Don Quixote* into its fictional world—first editions, for example, published by Juan de la Cuesta in Madrid in the first decade of the seventeenth century.

Considering *El ingenioso hidalgo don Quixote de la Mancha* as a material object in the form Cervantes knew it provides insights into his sophisticated joke. The first edition of 1605 was printed by the editorial house of Juan de la Cuesta in Madrid and sold by bookseller Francisco Robles in the same city (see Figure 1.1). Its first print run was 1500 copies (or "cuerpos"), an amount that suggested Robles's confidence in the work having at least marginal success in sales. It was not a masterful work of typography but an adequate one, more or less typical of the early modern Spanish printing industry.[36]

In addition to the text we now know as Cervantes's novel, the edition included ancillary pages, mostly before the main text as in a contemporary edition. This front matter denoted the authorizing legal and commercial apparatuses of printer, bookseller, and censor, situating the

Figure 1.1 Title page of *El ingenioso hidalgo Don Qvixote de la Mancha* (editio princeps 1605)

printed work of *Don Quixote* within the structures that authorized its commercial production and distribution. While the title page reveals the bookseller's rudimentary marketing scheme, the "Tasa" (Tax or Valuation) and "Privilegio" (Privilege), standard legal texts, evidence the Habsburg monarchy's interest in controlling the rapidly expanding printing industry through price fixing, licensing, and censorship. The Royal Council established the price that could be asked and granted the authors permission to publish for a length of time, a predecessor of copyright, as articulated in the pages devoted to Valuation and Privilege. Missing from the *princeps* was the "Aprobación" (Endorsement or Approval) from the Inquisition, certifying the review and sanction of a censor, although this approval had been granted for the edition. The "List of Errata" ("Fe de erratas") affirmed that the printed text matched, in great part, the manuscript approved for printing. Typical of 1600, the Robles first edition also includes a conventional dedication to a patron, the Duque de Bejar, lifted wholesale from *The Works of Garcilaso de la Vega with Annotations* by Fernando de Herrera (1580).

In laudatory preface poems other authors lent their fame to a new poetic work. Something like contemporary readers' blurbs, they represented the insertion of the work into circles of cultural repute as the other front matter located the text within social institutions. As to the *Quixote*, Lope de Vega, playwright and rival to Cervantes, reported in a letter of 1604 that Cervantes had been asking around for poets willing to write for him. Finding no one, Cervantes wrote them himself under assumed names as the prologuist famously explains, the first sally in the sustained parody of textual authority that the text is known for. Ruefully falsifying the dedicatory poems, attributing them to characters in others' works, and insisting on their part in the play of narrative levels, Cervantes makes clear that his fiction operates at the level of the textual apparatus as well, even though the historical author had little real control over it.

The exact details of the *editio princeps* and what it tells us about early readership make for fascinating study. For our purposes, however, my description of the 1605 first edition serves to demonstrate how Cervantes fictionalizes this and all printed editions of the *Quixote* through the metamaterial discourse. The discussion of the material life of *Don Quixote* in *Don Quixote* encourages us to read the ancillary material of the printed book metaleptically, as part of the body of the text and thus of the fiction of text's origins. The book in the reader's hand, whatever the edition, takes on discursive significance. Cervantes has created a metaleptical machine: the metamaterial discourse incorporates every future edition of the text into its discussion of textual materiality.

This artful narrative gesture should be understood in dialogue with the biopolitics of the early modern Spanish state and the equating of real and textual bodies throughout *Don Quixote*. The tidy printed *Don Quixote* in its first instantiation (and, by extension, later ones) shows the sanction of multiple authorities, a work fully integrated into public culture of its moment. Yet the story of its origins told in the narratorial intervention in Chapter Nine suggests just the opposite: its marginality and illegitimacy, even its dissimulated impurity.

As an object, the mannered façade and machine-made textual body of the edition belies its (supposed) heterogeneous origins, its diverse, polyglot materiality. In substance, it contrasts to the (fictive) materiality of the pre-texts, underlining the distinction between what it is and what it (reportedly) began as. Unlike the textual body of the supposed source text that is left maimed during the battle with the Basque, the real book is amply complete at almost 700 pages ending with the adventuring knight back at home. Printed in two columns of tidy gothic script on white paper (*papel blanco*), made of good linen and cotton from the paper mills of the Monasterio de El Paular outside Madrid, measured and cut into uniform quarto sheets, the edition is at a notable distance from the scrap paper of the Alcaná merchant. Equally, the front matter of the first edition asserts the book's insertion into extradiegetic structures of legitimation (printing industry, book trade, and governmental regulation). This public—and particularly governmental sanction—is very much at odds with the off-market, illegal transaction between narrator, merchant, and translator portrayed in the Alcaná chapter.

If we read the ancillary material of the printed book metaleptically, as part of the body of the text and thus of the fiction of the text's origins, the front matter appears as evidence of an attempt to whitewash, to paper over, the "real" origins of *Don Quixote* as a physical entity. Understanding the printed text in light of the (fictional) genealogy of its own making, the object itself looks to dissimulate its own unruly material past. Like a Spaniard who has shored up his imperfect pedigree with testimonies and affidavits, *Don Quixote, impreso y aprobado*, passes for *limpio*. The novel, printed and sanctioned, passes for pure.

Conclusion

In important respects *Don Quixote* is a rumination on the power of the written sign as discourse and object. To that end, the novel offers us various narratives, humorously contested and overwritten, of its origins: authorial, linguistic, and, in the narratorial intervention treated here, material. This series of anterior versions and artifacts that the text depicts through the play of metafiction and metalepsis has implications beyond the hermeneutic. It characterizes the Arab-Iberian presence as a physical trace inscribed in the printed work, a bloodline, purported evidence of its distant but still residual heterogeneity and alterity. To what end, one may ask, is the backstory of *Don Quixote*'s untidy, diverse pre-texts. Reading the backstory of *Don Quixote* within the ideological matrix of Counter-Reformation Spain, in particular within the pervading climate of genealogical-cum-racial anxiety, allows us to see how Cervantes employs metamateriality to put the lie to the premises of blood purity. Via this insidious doctrine, the body itself came to be seen as recalcitrant: it passed on the mistaken belief of ancestors irrespective of the sincerity of one's conversion. The distinction between the unworthy pedigree of the pre-texts of the novel and the actual, real-world edition reveals how supposed impurity depends on tropes of presence and absence, on rhetoric alone. Ultimately the pre-texts leave no trace of inherent and unwilled wrongness. In the final analysis, Cervantes shows how the materialization of difference, the doctrine of blood purity, is itself a fiction, based in part on a much-falsified paper record to support and define bodily identity. The supposed genealogy of the embodied material text in *Don Quixote* suggests that what Spanish bodies, textual, individual and national, ultimately had in common was the attempt to mask the character assigned them.

Notes

I completed this research remembering the kindness and insight of Álvaro García de Zúñiga, to whom I dedicate the chapter.

1. All translations are from Edith Grossman, trans., *Don Quixote* by Miguel de Cervantes (New York: Ecco Books, 2005). Cervantes scholars writing in English often prefer "j" rather than "x" in the spelling of Don Quijote for the <jota> sound to reflect orthographic changes in Spanish spelling. This essay acknowledges both the older Spanish printed form and its contemporary version. I use the orthographic difference as a means to distinguish the novel's title, *Don Quixote*, from the character, Don Quijote.
2. Gil Anidjar, "Lines of Blood: *Limpieza de sangre* as Political Theology," in *Blood in History and Blood Histories*, ed. Mariacarla Gadebusch Bondio (Florence: Sismel, Galluzzo, 2005). Published as this collection was going to print was Anidjar's *Blood: A Cultural Critique of Christianity* (New York: Columbia University Press, 2014).
3. Twentieth-century narratologists like Gérard Genette understand Cervantes as an early cultivator of the trope.
4. See Caroll B. Johnson, *Cervantes and the Material World* (Urbana-Champaign: University of Illinois Press, 2000); Roger Chartier, "Don McKenzie and Don Quixote," in *Books and Bibliography: Essays in Commemoration of Don McKenzie*, ed. John Thompson (Wellington: Victoria University Press, 2002), 19–35; Francisco Rico, *El texto de "Quixote". Preliminares a una ecdótica del Siglo de Oro* (Barcelona/Valladolid: Ediciones Destinos/Centro para la Edición de los Clásicos Españoles y Universidad de Valldolid, 2005); Fernando Bouza and Francisco Rico, "'Digo que yo he compuesto un libro intitulado *El ingenioso hidalgo de la Mancha*'," *Cervantes* 29: 1 (2009), 12–30.
5. Carroll B. Johnson, "The Virtual *Don Quixote*: Cide Hamete Benenjeli's Manuscript and Aljamiado Literature," in *Essays on Golden Age Literature in Honor of James A. Parr*, ed. Barbara Simerka and Amy R. Williamsen

(Lewisburg: Bucknell University Press, 2006), 172–88; Juan Carlos Rodríguez, *El escritor que compró su propio libro. Para leer el Quijote* (Barcelona: Random House Mondadori, 2003).

6. Mercedes Alcalá Galán, *Escritura desatada: poéticas de la representación en Cervantes* (Alcalá de Henares: Centro de Estudios Cervantinos, 2009).

7. Albert Sicroff, *Los estatutos de limpieza de sangre: Contraversias entre los siglos XV y XVIII* (Madrid: Taurus, 1985); David Nirenberg, *Communities of Violence: Persecution of Minorities in the Middle Ages* (Princeton University Press, 1998); Henry Kamen, *The Spanish Inquisition: A Historical Revision* (New Haven: Yale University Press, 1998).

8. Etienne Balibar and Immanuel Wallerstein, *Race, Nation, Class: Ambiguous Identities* (New York: Verso, 1998); David Nirenberg, "Race and the Middle Ages: The Case of Spain and its Jews," in *Rereading the Black Legend: The Discourses of Religious and Racial Difference in the Renaissance Empires*, ed. Margaret R. Greer, Walter Mignolo, and Maureen Quilligan (University of Chicago Press, 2007), 71–87; David Nirenberg, "Was there Race Before Modernity? The Example of Jewish Blood in Late Medieval Spain," in *The Origins of Racism in the West*, ed. Miriam Eliav-Feldon, Benjamin Isaac, and Joseph Zeigler (Cambridge University Press, 2009), 232–64; Gil Anidjar, "Lines of Blood: *Limpieza de sangre* as Political Theology," in *Blood in History and Blood Histories*, ed. Mariacarla Gadebusch Bondio (Florence: Sismel, Galluzoo, 2005).

9. George Mariscal, *Contradictory Subjects: Quevedo, Cervantes, and Seventeenth-Century Culture* (Ithaca, NY: Cornell University Press, 1991); George Mariscal, "The Role of Spain in Contemporary Race Theory," *Arizona Journal of Hispanic Cultural Studies* 2 (1998), 7–22; Georgina Dopico Black, *Perfect Wives, Other Women: Adultery and Inquisition in Early Modern Spain* (Durham, NC: Duke University Press, 2001); John Beursterien, *An Eye on Race: Perspectives from the Theater in Imperial Spain* (Lewisburg: Bucknell University Press, 2006); Roland Greene, *Five Words: Critical Semantics in the Age of Shakespeare and Cervantes* (University of Chicago Press, 2013).

10. Mercedes García Arenal, *Inquisición y moriscos. Los procesos del Tribunal de Cuenca* (Madrid: Siglo XXI de España Editores, 1978); Israel Burshatin, "The Moor in the Text: Metaphor, Emblem, and Silence," in *Race, Writing, and Difference*, ed. Henry Louis Gates, Jr (University of Chicago Press, 1986); Julio Caro Baroja, *Los moriscos en el reino de Granada*, 4th edn (Madrid: LISTMO, 1995); Vincent Barletta, *Covert Gestures: Crypto-Islamic Literature as Cultural Practice in Early Modern Spain* (Minneapolis: University of Minnesota Press, 2005); Trevor Dadson, "Official Rhetoric versus Local Reality: Propaganda and the Expulsion of the *Moriscos*," in *Rhetoric and Reality in Early Modern Spain*, ed. Richard J. Pym (London: Tamesis, 2006), 1–24; Antonio Domínguez Ortiz and Bernard Vincent, *Historia de los moriscos: Vida y tragedia de una minoría* (Madrid: Allianza, 2007); Barbara Fuchs, *Exotic Nation: Maurophilia and the Construction of Early Modern Spain* (Philadelphia: University of Pennsylvania Press, 2008); Mercedes García Arenal and Fernando Rodríguez Mediano, *Un Oriente español: los moriscos y el Sacromonte en tiempos de Contrarreforma* (Madrid: Marcel Pons, 2010).

11. Eric Graf, "When an Arab Laughs in Toledo: Cervantes's Interpellation of Early Modern Spanish Orientalism," *Diacritics* 29.2 (1999), 68–85; Carroll B. Johnson, "The Virtual *Don Quixote*: Cide Hamete Benenjeli's Manuscript and

Aljamiado Literature," in *Essays on Golden Age Literature in Honor of James A. Parr*, ed. Barbara Simerka and Amy R. Williamsen (Lewisburg: Bucknell University Press, 2006), 172–88; Carroll B. Johnson, *Translating a Culture: Cervantes and the Moriscos* (Newark: Juan de la Cuesta, 2010); Frederick De Armas, *Don Quijote Among the Saracens: A Clash of Civilizations and Literary Genres* (University of Toronto Press, 2011).

12. María Elena Martínez, *Genealogical Fictions: Limpieza de sangre, Religion, and Gender in Colonial Mexico* (Palo Alto, CA: Stanford University Press, 2008); Lúcia Helena Costigan, *Through Cracks in the Wall: Modern Inquisitions and New Christian Letrados in the Iberian Atlantic World* (Leiden: Brill, 2010); Max S. Herring Torres, Martínez, and Nirenberg, eds, *Race and Blood in the Iberian World* (Berlin: Lit Verlag, 2012).
13. Sebastián de Covarrubias, *Tesoro de la lengua castellana o española*, ed. Ignacio Arellano and Rafael Zafra (Madrid: Iberoamericana-Veuvert, 2006), 851; translations of works in Spanish other than the *Quixote* are mine throughout.
14. Covarrubias, *Tesoro*, 313.
15. Alcalá Galán, *Escritura desatada*, 133.
16. Repressive even by sixteenth-century standards, the Pragmática of 1566 directly provoked Granadan *moriscos* into taking up arms against the Spanish government in a two-year conflict known as the Second Granadan War, which ended in the forced relocation of the largest remaining population of *moriscos* in the Iberian Peninsula. For further study of the persecution of converted Muslims, see García Arenal and Rodríguez Mediano.
17. Alcalá Galán, *Escritura desatada*, 118.
18. Ibid.
19. García Arenal, *Inquisición y moriscos*, 55.
20. Vincent Barletta, *Covert Gestures: Crypto-Islamic Literature as Cultural Practice in Early Modern Spain* (Minneapolis: University of Minnesota Press, 2005), 6.
21. Alcalá Galán, *Escritura desatada*, 134.
22. Grossman, *Quixote* by Miguel de Cervantes, 3.
23. The most comprehensive historical study of blood purity law remains Albert Sicroff's *Los estatutos*, although other historians have made lucid contributions in more general works, in particular Henry Kamen's chapter "Racialism and its Critics" in *The Spanish Inquisition: A Historical Revision* (1999). Deborah Root's "Speaking Christian" (1988) treats the evolving relationship between *nobleza de sangre* and *pureza de sangre* in the sixteenth century. An excellent introduction to the relationship between blood purity, the Inquisition, and the body can be found in Georgina Dopico Black's *Perfect Wives*, particularly the Preface and Chapter One.
24. "Informaciones de limpieza de sangre de D. Juan de Alfaro, natural y vecino de Córdoba, como notario del Santo Oficio," Biblioteca Nacional de España, Mss 12564/7.
25. Barbara Fuchs, *Passing for Spain: Cervantes and the Fictions of Identity* (Urbana-Champaign: University of Illinois Press, 2003).
26. Grossman, *Don Quixote* by Miguel de Cervantes, 12.
27. Georgina Dopico Black, "Canons Afire: Libraries and Life in Don Quixote's Spain," in *Don Quixote: A Casebook*, ed. Roberto González Echevarría (Oxford University Press), 110.
28. Dopico Black, "Canons Afire," 108.

29. Ibid., 107–9.
30. Grossman, *Don Quixote* by Miguel de Cervantes, 65.
31. Ibid.
32. Ibid., 66.
33. Ibid., 52.
34. The chivalric romances inventoried and critiqued by the priest and barber in Quijote's library are published works that would have been available to a middle-aged *hidalgo* from the provinces.
35. Brian McHale, *Postmodernist Fiction* (New York: Routledge, 1987).
36. Studies of the early modern Spanish book trade and the early editions of Cervantes in particular are many. See María Marsá's *La imprenta en los Siglos de Oro* (Madrid: Ediciones del Laberinto, 2001); Juan Carlos Rodríguez's *El escritor que compró su propio libro* (Barcelona: Random House Mondadori, 2003); and Francisco Rico's *El texto del "Quixote"* (Barcelona/Valladolid: Ediciones Destinos/Centro para la Edición de los Clásicos Españoles y Universidad de Valladolid, 2005).

2
The Blood of Others: Breeding Plants, Animals, and White People in the Spanish Atlantic

Ruth Hill

Cognitive scientists Scott Atran and Douglas Medin argue that empiricism and essentialism characterize folk constructions of nature just as they characterized Aristotelian, pre-theoretical (or pre-scientific) thinking about nature. "People ordinarily assume that the various members of each generic species share a unique underlying nature, or [biological] essence ... People the world over assume that the initially imperceptible essential properties of a generic species are responsible for the surface similarities they perceive."[1] One of the most urgent debates in the cognitive sciences today centers on the rapport between this folkbiological, commonsense, or pre-theoretical essentialism and racial or ethnic categories. Is it innate in humans to perceive humans as natural kinds, that is, to perceive and sort them as we do different plants and brutes in nature? Are humans hardwired to sort humans into groups according to appearance and assumed biological essence? Or, could it be that racial and ethnic ideologies frequently make operational beliefs about blood due to domain transfer from ANIMAL and PLANT to HUMAN?[2]

This controversy rages on without resolution, but it has already established the urgency of widening our academic lens to accommodate human diversity within the frame of natural history. Toward achieving that accommodation, I propose here that folkbiology be our critical *entrée* into the origins of whitening or *blanqueamiento* equations in colonial Spanish America. Folkbiology allows us to look anew at miscegenation (hybridization) and whitening—clearing the human body of non-European blood—in humans. Equations for whitening in the Spanish Atlantic were inextricably linked to myriad notions and equations of *degeneration* (L. *degeneratio*—i.e., deviating or departing from [*de*] one's blood, origins, kind, or "generation" [*generatio*]), in animals including humans.[3] This degeneration often occurred through selective or unintentional crossbreeding that produced hybrids, whether the latter were constituted, ideologically, as improvements or deteriorations of the original blood.

The entry on blood (*sangre*) in the first Spanish dictionary of hard words, Sebastián Covarrubias's *Tesoro de la lengua castellana o española* (1611), reveals the cognitive and linguistic duality of the noun that is addressed from a variety of perspectives in this volume: "Blood ... Strictly speaking, blood is that which is inside our veins, that is, according to the Romans; however, we do not distinguish *inter sanguinem et cruorem*. Sometimes blood signifies relation."[4] Two types of usage—the literal and the figurative—are voiced here. Moreover, Covarrubias emphasizes that Spaniards do not distinguish between *blood* (what flows through the veins) and gore (L. *cruor*), which is the coloring substance, that is, the matter that gives blood its color. Several other entries from Covarrubias's *Tesoro* are contingent upon or overlap with *blood*. For example, *milk* (*leche*) clearly participates in a literal sense of *blood*: "Milk. The juice from cooked blood that nature sends to the female's breasts, in animated substances, so that she can nourish her offspring."[5] While this definition entails a literal understanding of blood (specifically, the blood cooked by the human or brute body to produce milk), it is well known that a raft of colonial discourses deployed a mother's (or wet-nurse's) milk as the essence of a child's or group's morphology, intellect, and morality.[6]

Nature (*naturaleza* or *natura*) in that 1611 *milk* entry presents the same cognitive and semantic flexibility as *blood* does in the early modern Spanish world: it denotes the natural world, but it also "[s]ometimes means status and essence." Further, in humans, "it is taken to mean *casta* and birthplace or native country."[7] The term *casta* meant lineage, breed, stock, blood, or nature, and it was used with plants, brutes, and humans alike. Like *casta*, *raza* meant a breed of animals, but it also referred to lineage or blood in animals and humans. Witness the Spanish proverb, "El can de buena raza, si oy no caza mañana caza [Sooner or later, a purebred dog hunts]," as explicated in a 1626 collection: "Proverb which expresses that though a well-born man ... might make a little mischief, blood eventually leads him to act with honor."[8] Here a lustrous genealogy or pedigree is translated out as blood, which becomes the essence of noble behavior. Blood is the unnamed essence in another old saw that personifies canines: "Can que madre tiene en villa, nunca da buena ladrida" ("The dog with a country mother never really learns to bark").[9] This anthropomorphic saying jocosely warns against sexual unions and marriages between respectable people and peasant youths—"young men and women, native sons and daughters, who are not on the level of those from outside [that is, the city]."[10] The wild or uncouth mother transmits her inferiority to her offspring. Again, there was little separation between the ANIMAL and HUMAN domains, in both of which blood, milk, nature, *casta*, and *raza* were essentialized.

Indeed, what all of the foregoing entries have in common, from the critical perspective of the cognitive sciences (psychology and anthropology in particular), is their function: they are essences, or *essence placeholders*. As Atran and Medin explain, "even when people do not have specific ideas about essences, they may nonetheless have a commitment to the idea that there is an underlying nature (i.e., they may have an 'essence placeholder')."[11] This psychological essentialism is universal, although the concept "of an essence placeholder allows that people may come up with different mechanisms for conveying or modifying causal essence."[12] Folkbiological essentialism (shared by Aristotle and his followers on both sides of the Atlantic) differs crucially from the assumptions about biological essentialism held by scholars of race. It is a *constructionist essentialism* that allows for perceptible physical change over time and place, in plants, brutes, and humans alike.[13]

Folkbiological essentialism was especially visible in natural histories, books of secrets, veterinary handbooks, and farmers' manuals in Spain when Thomist Aristotelianism held sway over the schools. In *Libro de los secretos de agricultura, casa de campo y pastoril* (*Book of Secrets of Agriculture, Country Life, and Husbandry*), a wildly successful handbook first published in 1617, the Catalonian Friar Miguel Agustín writes of the fields best suited for growing wheat: "Just as we weigh the variety of complexions in men from different provinces and regions according to the air and disposition of the heavens, so we do likewise in agricultural fields, which have different complexions, because some are strong and others weak; some dry, others humid; the soil is heavy in some, and thin in others; some rocky, others even; and, finally, some sandy and others fat."[14] Whereas the essence invoked in the foregoing example was complexion, in other, equally vivid instances of constructionist essentialism the essence was explicitly blood, breed, or pedigree. Witness the Renaissance Jesuit Father Mariana's *Historia de España*, which is cited in the entry on *raza* in the first dictionary published by the Spanish Royal Academy of Language: "Just as in crops and brutes, the family lines and pedigree of men [*la raza de los hombres y casta*] is transformed and bastardized due to the properties of the stars and earth, especially over time."[15] Similarly, even mating a pure and numerous breed (*raza*) of stallions with different breeds of thoroughbred mares (*razas de yeguas castizas*) might result in defective offspring. Pedro García Conde, royal horse and mule doctor and shoer, notes in *Verdadera albeytería* (*True Veterinary Science*) (1707) that Andalusian horses are the official thoroughbreds of Spain, but they nonetheless come to ruin in Old Castile, New Castile, and the Kingdoms of Leon, Galicia, and Portugal, because of the vegetation and cold air in those areas.[16]

Works related to natural history such as books of secrets and husbandry reflect not only a constructionist essentialism but also commonsense assumptions about hybridization and degeneration in the organic world. In the PLANT domain, hybridity was fraught with tensions that probably produced some of the contradictory equations for human whitening that appeared later. On one hand, it was considered far better to cross a domesticated variety of tree with a savage or wild variety, for such a union would make the domesticated variety more robust, fertile, and durable, and the savage variety domestic, according to Friar Agustín.[17] On the other hand, one could not breed plants by crossing different kinds because their incompatible essences, though they be hidden for a time, would eventually manifest themselves: "it is always better to graft each kind onto its peer, with which the tree and graft may become a union and brotherhood [*hermandad*], otherwise they never come to a good end, for it is impossible that these two [different] natures will not be found out someday."[18] The Catalonian Agustín insisted on the irreconcilable fate of mixed-offspring in the PLANT domain, reiterating that a domesticated plant or tree "in the end takes after the nature from which it issues." One could cross two different kinds of trees, or even a good and bad vine, "even when they were incommensurate." However, "the truth is, these things from grafting trees [and vines] onto their opposites are not lasting." And yet, he conceded that there was a way to make "one and the same fruit take a half from one species and the other half from another."[19] Manuel Ramírez de Carrión, for his part, exhibited in his 1629 book of secrets, *Maravillas de naturaleza* (*Wonders of Nature*), an essentialist stance toward color in fruit trees: "Red apples took their color from the mulberry bush where they were first grafted." The mulberry was its essence or virtue, and "[n]ature diffuses the virtue of the contributing soul from one generation to the next." That did not, however, mean that different varieties of plants or trees (fruit-bearing and non-fruit-bearing, for example) could not be productively crossed: "a mulberry bush grafted onto a white poplar makes white mulberries."[20]

It must be underscored that the core concept of the human hybrid, half-breed, or mongrel actually arose in the domains PLANT and ANIMAL. The entry *hybrid* (L. *hybrida*), in Antonio de Nebrija's 1492 Latin–Spanish dictionary, states that it means "progeny of wild and tame" (*hijo de fiero y manso*) and it is associated with bastardization.[21] The Spanish *bastardo* was "said also of fowl and brutes when they are engendered by two different kinds or breeds, and because such fowl and brutes deviate from their origins [*degeneran de su natural*], they are called *nothos* in Greek, or *degeneratio*."[22] The term *mestizo* (L. *mixtio*), which was to designate first the human mixed-blood in general and then the Indian "half-breed" or "mongrel" in particular, appears repeatedly in Gabriel Alonso de Herrera's 1513 *Agricultura*. In Covarrubias's *Tesoro*, the entry *mestizo* reads, "mestizo. That which is engendered from different kinds of animals." In his *Trésor des deux langues françoise et espagnolle* (1607), after which Covarrubias modeled his own *Tesoro*, César Oudin records *mestizo/mestif/mestis* as the offspring of a Moor and a White, or the offspring of different kinds of dogs or other animals.[23] The Spanish physician Nicolás Monardes mentions "mestizo trees [*árboles mestizos*] that are neither pines nor cypresses" in *La historia medicinal de las cosas que se traen de nuestras Indias Occidentales*.[24] In the Spanish Royal Academy of Language's *Diccionario* it is clear that *mestizo* was used with animals: "MESTIZO, ZA. Adjective that applies to the animal with a father and mother of different *castas*. It comes from Latin *Mixtus*. Lat. *Hybris, idis. Hybridus, a, um*."[25] It is also clear, from the usage example provided immediately thereafter, a passage taken from the Spanish Crown's corpus of colonial laws protecting the *mestizo* child born to Spanish and Indian parents, that the domain ANIMAL overlapped with the domain HUMAN.

Like *mestizo*, the term *mesta* (the name for the annual meeting of sheep breeders) was directly associated with ancient notions of crossbreeding: it was derived from the Latin *mixta* ("commingled") because different breeds of sheep were brought there to be sorted and marked before they could be sold.[26] This adds another layer to the essentialization of mother's milk that came up in my earlier discussion of *blood*. The contributions of Aristotelian natural history and folkbiology to that essentialization, in the Spanish colonial context, have gone unremarked, but they were fundamental. Consider, for example, a "natural wonder" recorded by Ramírez de Carrión that oscillates between the ANIMAL and HUMAN domains: "The wool of a kid (goat) that is weaned on sheep's milk becomes finer, whereas goat's milk makes a lamb's wool coarser. From which one can infer that choosing the right wet-nurses is of grave consequence to one's children."²⁷ Because milk was defined as the juice produced by cooked blood, cross-breastfeeding—the weaning of lambs by goats, or of kid goats by ewes (then believed to be different varieties or kinds of the same animal family)—was believed to alter a defining characteristic of each of these animals: the coat, hair, or skin. By analogy, the breastfeeding of children by human females of a different variety or kind would cause perceptible alterations, the latter of course every bit as susceptible to hierarchization as alterations in sheep wool (the softer or finer, the more valuable). Examples such as this one encourage us to speculate further about how the early modern mind analogically approached human hybridity and how nature was instrumentalized by local or distant ideologies to pronounce clearing the blood easier for some human hybrids than for others.

In the early modern Spanish world, the horse and its hybrids commanded a respect and scholarly attention that we can scarcely imagine today. The thoroughbred horses originating in Andalusia, Spain, were prized since at least the sixteenth century. Indeed, hundreds of years before the Belle Meade plantation in Nashville and the Monticello plantation ouside of Charlottesville, Spain and its imitators in England and the Lowlands were selectively breeding the most coveted stallions in the world. The most widely accepted etymology for the term *mulato*, found in Spanish dictionaries and myriad discussions of human diversity, points to the equine hybrid *mulo* ("mule"), the offspring of a mare and an ass. Similarly, the Spanish term for illegitimate child, *borde*, and for a type of mule (*burdégano*), was derived from the Latin *burdus* (offspring of a stallion and an ass). Surprisingly, scholars of eighteenth-century *casta* painting in the Spanish Atlantic do not comment on equine hybrids and horses that appear alongside human hybrids in the genre. It made perfect sense for artists to include horses and equine hybrids in *casta* paintings commissioned by Europeans: throughout early modern Spain and Spanish America, a horse or equine hybrid signified one's social status. In Spain, verisimilitude demanded that Sancho Panza ride a mule and that Don Quijote convince himself that he was a riding a thoroughbred, not a nag. In the Spanish Atlantic, the possession of a horse suggested a proximity to Spanish blood, I am convinced, as in the painting in which the motto tells us that a Spanish mother and a *mestizo* father pass on their love for horses to their *castizo* son from an early age ("Mestizo y española dan al castizo la afición al caballo desde bien niño").²⁸ But there was a lot more to it than that.

According to Aristotelians, including many Spanish *albéitares*, horses and humans were noble animals because animals with blood were noble,[29] and horses were the most noble of irrational animals.[30] The circulation of blood was discovered and published by the Spanish *albéitar* Francisco de la Reina, in 1547, many decades before it was discovered in humans.[31] Shoer and *albéitar* Francisco García Cabero ardently defended his profession in *Templador veterinario de la furia vulgar en defensa de la facultad veterinaria o medicina de bestias* (*Veterinary Sedative for Common Fury in Defense of the Veterinary Faculty, or Animal Medicine*) (1727), against the medical establishment's opinion that its methods and its patients were inferior to medicine.[32] In *Curación racional de irracionales* (*Rational Healing of Irrational Animals*) (1728), he argued that physicians and horse doctors used the same tools and methods for curing their patients, because both disciplines dealt with the physiological, or sensitive, dimension of their patients.[33] He analogically reasoned as follows: on the same day and time that a physician treats a White slave owner, he may also treat a Black slave, because medicine applies equally to them both as humans, or members of the same kind: sensitive and intellectual beings. They are different, but medicine is indifferent. Likewise, horses and humans are different animals, but the *albéitar* treats a horse or mule, and the physician treats a human, using the same arts. Although the intellectual and sensitive animal (human) is superior to the merely sensitive animal (equine), they both present the illnesses of sensitive animals. Likewise, White masters are superior to Black slaves, but both are humans (intellectual and sensitive animals) when they fall ill.[34]

Albéitares trained in Thomistic Aristotelianism, even when they embraced the "new albeyteria," as Domingo Royo termed it in *Llave de albeytería* (*Key to Albeitería*) (1734),[35] knew that horses were not men. Still, horses "almost had reason,"[36] and they acted as if they were reasoning.[37] In the 1729 revised edition of his *Compendio de albeitería* (*Compendium of Albeitería*), Fernando de Sande y Lago informed readers that a soldier from Switzerland once told him that his captain had a horse so smart that when they put a clock in front of it, the stallion would give as many kicks as the hours struck, "without being a second off." It would also lie down and act as a mattress for its master when the ground was cold, "and it did this with utter tranquility, as if it were capable of rational understanding."[38] The same horse and mule doctor glowed that purebred stallions abhorred mating with their mothers, supporting his proposition with an *exemplum* from a much-consulted Spanish marriage treatise: "[the author] states that he really heard from a lord worthy of belief ... that upon discovering that they had tricked [the stallion] into mating with his own mother, he bent his head down and ripped off his testicles with his teeth."[39]

Discourses of human nobility and horse nobility surfaced, intermingled, even in debates about religion and blood in Spain. In the sixteenth century, the Archbishop of Toledo compared himself and other church officials to horse traders in order to legitimate the blood purity statutes that excluded from crown and church posts persons with Moorish, heretic, Jewish, or out-of-wedlock ancestors. If you offer a horse trader an imperfect horse as a gift, the Archbishop asserted, he will not accept it because what matters most to him is its lineage or breed. Even if the horse is a thoroughbred, determining its pedigree is his principal concern.[40] Folk, in contrast, often resorted to the continuum of human and equine to lambast pedigree claims made by some humans, as in the old saw, "Pariente de parte del rocín del baile" (literally, "related to the lord's horse"). As the seventeenth-century compiler explained, "it labels [them] as being of such lowly origins that the closest they could get to that relationship would be as servants or horse grooms for the nobleman to whom they claim to be related, or for his ancestors." Then there was the equally venerable saying, "No hay linaje sin putas ni muladar sin pulgas" ("Every dungheap has its fleas and every bloodline its whores"), which did not require exegesis.[41]

It is quite surprising that scholars continue to overlook certain taxa for human mixed-bloods in colonial Spanish America originating in *albeitería*. In the Peruvian Inca Garcilaso's *Comentarios reales de los Incas* (*Royal Commentaries of the Incas*), he examines new human kinds in viceregal Peru to conclude his lengthy discussion of plants and animals in the New World. The chapter is called "New names to call different breeds or kinds [*generaciones*]." It begins thusly: "We were forgetting the best of what went over to the Indies, which is Spaniards and Blacks ... From these two groups [*naciones*] others have been made over there, mixed in every sort of way, and in order to differentiate them they give them different names." After defining *mestizos* and *mulatos*, the Inca informs his readers that persons who are one-quarter Indian and three-quarters Spanish are called *cuatralbos*, and persons three-quarters Indian and one-quarter Spanish, *tresalbos*. To horse doctors and horse traders alike, the *cuatralbo* stallion was a superior breed of racer: "the horse that has four [*cuatro*] white [*albo*] feet, which they call *cuatralbo*, is noble and fast, though it is not strong."[42] The *tresalbo* breed had three white feet and was highly prized, though less so than the *cuatralbo*. However, if one of the three white extremities feet was the right foot, it was considered a *tresalbo argel*. The folk taxon *argel* (literally, "Algerian") contained the horse with a white right foot, which was rejected as unlucky by nobles and commoners alike.[43] As the Spanish proverb puts it, "Del hombre malo y del caballo argel, quien fuere cuerdo guárdese de él" ("Whether it's a bad man or an Algerian stallion, avoid him if you're smart").[44]

It would only take two crossings of Spanish and Indian blood to produce a *cuatralbo*: the first crossing, between an Indian woman and a Spaniard, yields the mestizo whose blood is half Indian and half Spanish; and the second crossing, of a mestizo woman and a Spaniard, yields the *cuatralbo* whose blood is quarter Indian and three-quarters Spanish. Note the simplification wrought by folkbiology of blood equations used for selective horse breeding: reasoning anthropomorphically, the *cuatralbo* person should have four parts of Spanish or White blood (analogous to white feet on a non-white horse), not three. However, I suspect that in late sixteenth-century Peru, the human *cuatralbo* was considered Spanish on all four sides (or feet), that is to say, that Indians could clear their bodies of native blood after two crossings with Spaniards. The human *tresalbo* would be produced after two crossings, as well: the first crossing, between an Indian woman and a Spaniard, yields the half-breed (*mestizo*) whose blood is half Indian and half Spanish; the second crossing, of an Indian woman and a mestizo, is a *tresalbo* whose blood is three-quarters Indian and quarter Spanish. However, in this scenario the second crossing does not raise the social status of the offspring because it reverses course, ideologically: in the sixteenth century, persons in Peru ranked the *tresalbo* lower than the *cuatralbo*, if my interpretation of both taxa is correct. The *tresalbo*, an interruption in the Europeanization of indigenous ancestry or blood, was to be designated *salta-atrás* (literally, a "jump-back") as the Spanish colonization of South America wore on.

I now turn to the blood equations found in El Orinoco ilustrado (1741–45), a natural history of the Orinoco region that once included what is today Colombia and part of Venezuela. This account, which would later be translated into French and consulted by the likes of Buffon, Humboldt, and Thomas Jefferson, was penned by the Valencian Jesuit José Gumilla (1686–1750), a missionary who spent more than two decades preaching to Indians and Blacks in the region. His heuristic itineraries for whitening members of both groups appear in chapter 5 of the first part of the revised and expanded edition from 1745, where he explains that *four* crossings are required to whiten Blacks and he mocks the supposedly eradicable nature of Blackness. The popular myth that the Black or mulatto in a person does not come out, "la falsa opinión de que la especie de mulatos *no sale*,"[45] was proverbially expressed in Spanish as "Sobre negro no hay tintura" ("There's no dye that covers up black"). This old saw must have originated in husbandry and textiles, for it is omnipresent in agricultural discourses. Ramírez de Carrión, for instance, observes that white rams may produce black or multicolored sheep, but black rams never produce white sheep.[46] The Jesuit preacher proceeds to chart the whitening of Blackness for his readers:

> In sum, you can be certain that a mulatta by the fourth generation also becomes White by the very same steps that a *mestiza* does, through the following series of unions:
> I. From European male and Black female issues the mulatta (two quarters from each side).
> II. From European male and mulatta issues the *cuarterona* (one-quarter mulatta [quadroon]).
> III. From European male and *cuarterona* issues the *ochavona* (one-eighth mulatto [octoroon]).
> IV. From European male and *ochavona* issues the *puchuela* (completely White female).[47]

Here the first crossing yields the *mulata*; the second, the *cuarterona*; the third, the *ochavona*; and the fourth, the *puchuela*, or purely white woman. It is somewhat perplexing that this chart, in which Whiteness is the perceptible effect of an assumed cause or essence (European blood), does not use the term *blood*, and yet, we are clearly dealing with blood, algebraically expressed or not. Why do Whiteness, Blackness, and Mulattoness displace blood?

Perhaps a caveat that we read in the first Spanish dictionary definition of *blood* can help us make sense of this absence: "however, we [Spaniards] do not distinguish *inter sanguinem et cruorem*." This compounding of blood and color-causing agent (gore, or *cruor*) convinces me that the Spanish *negrura* (blackness), *negro* (the color black), *blancura* (whiteness), *blanco*, *mulato*, and other color terms were not mere essence placeholders in human whitening discussions, but, rather, and often, essences, for they were indistinguishable, in the early modern Spanish mind, from blood. I find further evidence of this pre-theoretical approach to nature and its kinds in Francisco de la Reina's *Libro de albeitería*, especially when he discusses colors and qualities of horses. Instead of approaching color like fierceness, quickness, or any other outward manifestation of a hidden cause or essence (that is, instead of engaging in sign doctrine), the Spanish veterinarian thinks of color as the essence: color's causal agent (gore or *cruor*) is intrinsic to the very notion of blood. My interpretation is supported by the fact that Reina repeatedly uses color terms with the adjectival phrase "de nación," typically used like "de casta" or "de raza" to mean "by birth" or some other phrase indicating one's nature, essence, being, or kind (see folios lv, lvii, lviii). Thus, in early modern Spain and Spanish America, someone might be defined as "español de nación," "mestizo de casta," "de raza castiza," etc. Descriptors such as "blanco de nación" signal that Whiteness is an essence, a nature, a kind or taxon determined by birth or blood, and this cognition of blood is folkbiological.

Returning now to *El Orinoco ilustrado*, the Jesuit missionary warned his readers that Whiteness would be lost in the next generation if the *puchuela* had a child with a mulatto or Black man. Such a reversion to Blackness or Mulattoness constituted another process of degeneration—a deviation from her Whiteness or Spanishness—and henceforth she would belong to the category *salta-atrás*.[48] Gumilla does not explain how, but throughout he assumes that there is an essence that produces a (natural) kind with specific characteristics and behaviors, even if that essence is neither observed nor observable. What has been overlooked by scholars is that the jump-back principle—a staple of colonial whitening equations as far back as the Inca Garcilaso's implicit one—originated in animal husbandry. Sande y Lago noted in his manual on horse medicine that occasionally a colt does not take after his thoroughbred father, and when this happens he must be crossed with mares from his own line (*linaje*). By doing so, he will restore the pure breed (*casta*), and he will correct and replenish it so that it matches the quality of its grandparents. He offers as proof the case of a Greek woman from Elis, a city famous for its breeding of thoroughbred stallions. According to Aristotle, the Greek woman had sex with a Black man but gave birth to a White (rather than a mulatto) daughter. The unhappy hybrid daughter married a Black man and gave birth to a son whose color matched his grandfather's.[49] This return to the *casta* of the grandfather (thoroughbred stallion or pureblooded Black man) in Sande y Lago's *Compendio* conveys the very same *salta-atrás* principle found in colonial Spanish American whitening equations, although in horses it constitutes, ideologically, a desirable regression.

This jump-back taxon appears also in a *casta* painting depicting a *tente-en-el-aire* ("up in the air") girl with her parents,[50] her mother a "jump-back" and her father an *albarrazado* (a term, I suspect, that was derived from the Spanish *alvarraz*, a skin disease in horses, as far back as the thirteenth century).[51] Not by chance the painter chose to place the bad seed, if you will, in the foreground, carrying a woven basket tray with fruit and flowers, while each of the parents has a flower in hand, for the motto that accompanies the scene affirms that the crossing of a *torna-atrás* female with an *albarrazado* male is a "bad graft" (*injerto malo*).[52] The ubiquitous presence of fruits and vegetables in *casta* paintings, at times accompanied by a legend, visualizes the parallel processes of variation in the domains VEGETABLE and HUMAN which transcend *casta* painting.[53]

The mottos that adorn selected series of *casta* paintings were of course proverbial expressions, folk wisdom. Whereas it has been astutely observed that the former were sometimes contradicted by the gentleness of their associated domestic scenes,[54] any such divergence between folk beliefs about human diversity and their visual representation for European buyers cannot deplete the animal–human–plant continuum of commonsense biology. The stylized, *über* exotic settings of *casta* paintings yield new meanings when we remember that the eighteenth-century Spanish Atlantic was a profoundly agricultural society. Fruits, vegetables, horses, and mules surround the participants in and the products of hybridization in humans, which process was analogous, in the pre-theoretical mind, to crossbreeding in plants and brutes.

Both folk and natural philosophers or historians reasoned from the known to the unknown, in contradistinction to modern science, which reasons from the unknown to the known. This is a fundamental difference that has enormous implications for our understanding of early modern ideas about blood and human diversity. On Atran's account, Michel Foucault's epistemology of science, which poured the foundations for so many postmodern discussions of blood, biological essentialism, and scientific racism, ignores a defining characteristic of folk or pre-theoretical understandings of nature:

> [W]hatever scientific epistemology is adopted, there is a methodological presupposition to the effect that science seeks to decompose and explain the known in terms of the unknown. Common sense distinguished itself from science by aiming principally to maintain the familiar composition of the world and, if necessary, to assimilate the unknown to the known ... Science and common sense thus do not presuppose the same ontologies, and their respective semantic frameworks deal, as it were, with "different" worlds.[55]

Should we attempt to align Gumilla's chart for classifying persons with Black blood with the following one for classifying persons with Indian blood, we perforce encounter yet another dimension of that "different world" in which Spanish folk and Thomistic Aristotelian understandings of biological diversity were situated:

I. From European male and Indian female issues the *mestiza* (two quarters from each side).
II. From European male and *mestiza* issues the *cuarterona* (one-quarter Indian female [quadroon]).
III. From European male and *cuarterona* issues the *ochavona* (one-eighth Indian female [octoroon]).
IV. From European male and *ochavona* issues the *puchuela* (completely White female).[56]

Legally, when Indian women had children with Spaniards (*mestizos*), their children's *condition*, or status, was determined by *patrilineal* descent, because Indian women were not slaves or the descendants of slaves. (It was assumed that the mothers of *mestizos* were Indian women, not Spanish women.) That is precisely why *mestizos*, unlike Indians, did not have to pay tribute (that is, perform forced labor) as subjects of the King, much to the chagrin of many Spanish missionaries and jurists.[57] Further, in both British America and Spanish America, the condition (bonded or free) of the child of African blood, even if he or she had entirely cleared African blood, was determined by *matrilineal* descent in accordance with the Justinian codes of imperial Rome. Gender and blood—or its intrinsic and inseparable *cruor* translated out as Blackness or Whiteness or Mulattoness in so many Spanish whitening discussions—were entangled when Spaniards in the Old and New Worlds had to decide who was Black, Indian, or White, and who was free, in temporary servitude, or enslaved.

Finally, there is another, and equally compelling, reason why the ideological apparatus used to activate folkbiological notions of hybridity and degeneration for classifying Blacks and African mixed-bloods could not be replicated in order to confer or remove indigeneity from Indians: the intricacies of conversion. Neophyte status and the applicability of the Indian privileges attendant upon it had long vexed church and crown officials. Why shouldn't a *mestizo* who lived among wild Indians be considered a neophyte like the latter? Hadn't Blacks been categorized in several papal bulls as neophytes? Hadn't at least one bishop of Peru allowed even *puchuelas* the privilege or dispensation to marry blood relatives? These and other questions came out of folkbiological essentialism, specifically, from the idea that the environment shaped the person as much as blood did. Clothing and customs should determine neophyte status, according to many church officials since the sixteenth century who extended Indian privileges to persons who were not full-blooded *indios*, for example, "those *mestizo* men and women who live among Indians, using the latter's dress and customs, as if they were in reality Indian men or women, although they are not fully."[58]

Pope Clement XI had unleashed a flurry of reforms around conversion in China and the Spanish New World. Hence, Father Gumilla's urgent proviso in which he alludes to a papal bull that revised the blood ratios required for persons to be taxonomized as Indians or *mestizos*:

> It should be noted that this charting follows the old standard which was used to determine which of those groups could be included in the term *neophyte* (that is, *newly converted*), so that missionary fathers could according to their privileges dispense with certain degrees of kinship and relationship in order to marry them morally and legally. But following Pope Clement XI's new bull it is evident and declared that *neophytes* is understood to mean only Indians and *mestizos*; as a result, quadroons and octoroons are considered, and must be regarded as, Whites.[59]

In the first half of the eighteenth century, then, a papal bull weighed more heavily than blood in measuring and classifying persons with Indian ancestry in the Spanish Atlantic. Pope Clement XI had decided that *only two crossings with Spanish Catholics were required for degeneration*—that is, for the descendants of full-blooded Indians to depart from their origins, to leave indigeneity, to clear that ancestral blood from their bodies. Henceforth, two generations of Christianity (of mixing Spanish and Indian bloodlines) were enough to make Whites out of the descendants of Indians, in the eyes of the law and society.

Notwithstanding Pope Clement XI's interventions, a preponderance of evidence suggests that precise blood equations for whitening humans *originated* in husbandry and animal husbandry. In "Color etiópico," Feijoo y Montenegro argued that skin color and other distinguishing features in human kinds were not caused by blood or sperm, the two essences commonly adduced in his times: "Neither can it be said that the particular color and configuration of human groups [*naciones*] comes to them hereditarily from fathers and grandparents, through a continuous series of many generations, and proceeding from some unknown principle."[60] Instead, he uses an essence placeholder: there is some unknown quality in every country or region that impacts soil, water, and other elements of nature, and that causes and maintains human variation if it is not interrupted. He cites an example from the vegetable kingdom:

> The seed of wheat planted in less-fertile soil produces a very inferior grain in shape, color, flavor, etc., that they call rye. The seed of cabbage grown in good soil, planted in another that is less suitable, produces in the first generation cabbage not as good as that from which the seed was taken; in the second, it already produces wild cabbage; and in the third and fourth this same plant starts deteriorating so much that these wild cabbages, grandchildren and great-grandchildren of the cabbage, appear to be vegetables of a completely different type compared to their grandparent and great-grandparent. Why couldn't the same thing happen, in proportion, to men?[61]

Note here that *degeneration is complete*—a different kind has been produced in nature by sowing seeds in a non-native soil—*by the third or fourth crossing.*

More than one hundred years earlier, strikingly similar claims about degeneration in plants had appeared in Ramírez de Carrión's *Maravillas de naturaleza*. After asserting that wheat degenerates into oat, barley, or chaff after it is planted in wetlands, he provides a specific equation: "In humid soil, wheat planted from the same germ seed for three years becomes rye in the third year."[62] This is an especially vivid example of folkbiological essentialism. On one hand, Ramírez de Carrión notes that a pedigree wheat will reproduce itself when the preceding generation's or harvest's seed is sown in the same soil; on the other, he warns that sowing that high-quality seed in a soil adverse to its original environment causes the wheat to lose its superior essence—to degenerate, to depart from its origins or nature. It becomes *in the third or fourth generation* a bastardized, or inferior, grain.

Selective dog breeding in the Spanish world generated similar equations. For breeding farm dogs or mutts, rather than hunting dogs, the Catalonian friar Agustín advised that the male not mate with a female that was less than a year old nor with a female that was more than two years old.[63] In sum, the blood of a female dog *without* pedigree survived only two crossings or mating seasons: one-quarter of the mutt's blood is the minimum required to preserve the breed. If one substituted for this undistinguished canine a female who ranked low in the social hierarchy of Spanish America, the equation would be three crossings for that female's blood to be cleared: the first crossing/mating season with a Spaniard yields the *mestizo* or *mulato*; the second, the *cuarterón*; and the third, the *ochavón*, who was no longer Indian or Black, but instead Spanish. Three crossings is the equation for degeneration, whether it is constructed ideologically as a benefit or as a setback, for *by the third generation* (a progeny with only one-eighth of that Indian's or Black woman's blood), *the original essence (Indianness or Blackness) has been replaced with Spanishness*. This certainly does not conform to Gumilla's whitening equation for Blacks, but it does equal the Pope's whitening equation for Indians. Furthermore, it is irreducible to that earlier taxon found in the Inca Garcilaso's account: *cuatralbo*, the person with one-quarter of Indian blood, or (figuratively) four White feet—that is, a Spaniard for all intents and purposes.

Another Spanish proverb about selective dog breeding specified that degeneration was completed in *four crossings*, or by the *fourth generation*: "Raza de can, amor de cortesano y ropa de villano, no dura más que tres años":[64] "A purebred race of dogs, a courtier's love, and a peasant's clothes don't last for more than three years." Beyond three seasons of being used to improve or create other races of dogs (*rehazer las razas*), a pureblood race experiences *degeneratio*: the product of the fourth crossing has totally deviated from the purebred race. Say, for instance, that a farmer decided to improve his native breed of dogs by breeding them with a pedigree dog. The product of the first crossing would be a half-breed (half purebred and half native). Now, if the same farmer were to perform a second crossing, between that half-breed and the native canine race, the result would be a quarter-breed, that is, a pedigree dog with one-quarter of the thoroughbred's blood. A third crossing, between that quarter-breed and the native breed, would yield a dog with one-eighth of purebred lineage, which is the minimum amount of blood required to preserve the prized dog's pedigree. This coordinates neatly with Gumilla's blood equation for whitening Blacks (and Indians, before Clement XI's revisal), in which the octoroon is still a Black, but the fourth crossing produces a Spaniard or White.

Innate or not to humans, folkbiology became harnessed to the ideology of Eurocentrism in the early modern Spanish world, with enduring consequences for people who are—or are believed by others to be—of non-European ancestry, and for which ancestry blood remains the preferred essence placeholder. Given the kinship I have proposed between breeding plants and animals and breeding White people, interrogating when and how that harnessing occurred might be the logical next step for scholars of blood equations in colonial Spanish America, and beyond.

Notes

1. All translations from the Spanish are my own. Scott Atran, "Folk Biology and the Anthropology of Science: Cognitive Universals and Cultural Particulars," *Behavioral and Brain Sciences* 21 (1998), 547–609, 550–1; also see Scott Atran and Douglas Medin, *The Native Mind and the Cultural Construction of Nature* (Cambridge, MA, and London: MIT Press, 2008), 21.
2. It is impossible to review here the hefty critical literature on this debate. To scratch the surface of competing perspectives, see Leda Cosmides, John Tooby, and Robert Kurzban, "Perceptions of Race," *Trends in Cognitive Sciences* 7.4 (April 2003), 73–9; Scott Atran, *The Cognitive Foundations of Natural History: Towards an Anthropology of Science* (Cambridge University Press, 1990); Atran and Medin, *Native Mind*; Lawrence Hirschfeld, "On a Folk Theory of Society: Children, Evolution, and Mental Representations of Social Groups," *Personality and Social Psychology Review* 5.2 (2001), 107–11; Susan Gelman and Lawrence Hirschfeld, "How Biological Is Essentialism?," in *Folkbiology*, ed. Douglas L. Medin and Scott Atran (Cambridge, MA, and London: MIT Press, 1999), 403–46; Francisco Gil-White, "How Thick Is Blood? The Plot Thickens ...: If Ethnic Actors Are Primordialists, What Remains of the Circumstantialist/Primordialist Controversy?," *Ethnic and Racial Studies* 22.5 (September 1999), 789–820; Gil-White, "Are Ethnic Groups Biological 'Species' to the Human Brain? Essentialism in Our Cognition of Some Social Categories," *Current Anthropology* 42.4 (August–October 2001), 515–54; Daniel Kelly, Luc Faucer, and Edouard Machery, "Getting Rid of Racism: Three Proposals in Light of Psychological Evidence," *Journal of Social Philosophy* 41.3 (Fall 2010), 293–322; Julia Shulman and Joshua Glasgow, "Is Race-Thinking Biological or Social, and Does It Matter for Racism? An Exploratory Essay," *Journal of Social Philosophy* 41.3 (Fall 2010), 244–59.
3. Sebastián Covarrubias, *Tesoro de la lengua castellana o española según la impresión de 1611, con las adiciones de Benito Remigio Noydens publicadas en la de 1674*, ed. Martín de Riquer (Barcelona: S.A. Horta, I.E., 1943), 199.
4. Ibid., 925.
5. Ibid., 755.
6. See Ann Laura Stoler, "On Political and Psychological Essentialisms," *ETHOS* 25.1 (1990), 101–6; Mariselle Meléndez, *Deviant and Useful Citizens: The Cultural Production of the Female Body in Eighteenth-Century Peru* (Nashville: Vanderbilt University Press, 2008), 165–9; Ruth Hill, "Caste Theatre and Poetry in 18th-Century Spanish America," *Revista de Estudios Hispánicos* 34.1 (2000), 3–26, 4–6; Atran and Medin, *Native Mind*, 22.
7. Covarrubias, *Tesoro*, 824.
8. Real Academia Española de la Lengua, *Diccionario de la lengua castellana*, 6 vols (Francisco del Hierro, Viuda y Herederos, 1726–39), 3: 500.
9. Gonzalo Correas, *Vocabulario de refranes y frases proverbiales*, prol. Miguel Mir., ed. Víctor Infantes (Madrid: Visor Libros, 1992), 104.
10. Ibid.
11. Altran and Medin, *Native Mind*, 21; also see Douglas Medin and Andrew Ortony. "Psychological Essentialism," in *Similarity and Analogical Reasoning*, ed. Stella Vosniadou and Andrew Ortony (Cambridge University Press, 1989), 179–95.

12. Atran and Medin, *Native Mind*, 21.
13. Atran, *Cognitive Foundations*.
14. Miguel Agustín, *Libro de los secretos de agricultura, casa de campo y pastoril*, facsimile edn (Valladolid, Spain: Editorial Maxtor, 2001), 167.
15. Real Academia Española de la Lengua, *Diccionario de la lengua castellana*, 3: 500.
16. Pedro García Conde, *Verdadera albeitería* (Madrid: Antonio González de Reyes, 1707), 4–5.
17. Miguel Agustín, *Libro de los secretos*, 146–7.
18. Ibid., 146.
19. Ibid., 146, 151, 256.
20. Manuel Ramírez de Carrión, *Maravillas de naturaleza en que se contienen dos mil secretos de cosas naturales, dispuestos por abecedario a modo de Aforismos fáciles y breves de mucha curiosidad y provecho* (Monzilla: Juan Bautista de Morales, 1629), 85, 94, 90.
21. Elio Antonio de Nebrija, *Diccionario Latino–Español (Salamanca 1492)*, introd. Germán Colón and Amadeu Soberanas (Barcelona: Puvill-Editor, 1979), n.p.
22. Covarrubias, *Tesoro*, 199.
23. César Oudin, *Tesoro de las dos lenguas francesa y española. Trésor des deux langues françoise et espagnolle* (Paris: Marc Orry, 1607), n.p.
24. Nicolás Monardes, *La Historia Medicinal de las cosas que se traen de nuestras Indias Occidentales (1565–1574)*, introd. José María López Piñero (Madrid: Ministerio de Sanidad y Consumo, 1989), 228.
25. Real Academia Española de la Lengua, *Diccionario de la lengua castellana*, 2: 156.
26. Ibid., 4: 556.
27. Ramírez de Carrión, *Maravillas de naturaleza*, 77.
28. María Concepción García Sáiz, *Las castas mexicanas: Un género pictórico americano* (Milan: Olivetti-Grafiche Milani, 1990), 104.
29. Fernando Calvo, *Libro de albeitería en el qual se trata del caballo, mulo y jumento y de sus miembros y calidades y de todas sus enfermedades, con las causas, señales y remedios de cada una de ellas*, 6th rev. edn (Madrid: Andrés García de la Iglesia, 1675), 3.
30. Martín Arredondo, *Obras de albeyteríam*, rev. edn (Madrid: Bernardo de Villadiego, 1669), 4–5; García Conde, *Verdadera albeitería*, 3.
31. See Francisco de la Reina, *Libro de albeytería*, rev. and expanded edn (Saragossa: Lorenzo y Diego Robles, 1583), folio 94 verso, folio 101 recto; Benito Jerónimo Feijoo y Montenegro, R.P., "Del descubrimiento de la circulación de la sangre, hecho por un albéitar español," *Cartas eruditas y curiosas* (Madrid: Real Compañía de Impresores y Libreros, 1774), vol. 2: 314–23, in Proyecto Filosofía en Español, www.filosofia.org., Biblioteca Feijooniana 1–6.
32. Francisco García Cabero, *Templador veterinario de la furia vulgar en defensa de la facultad veterinaria o medicina de bestias, y de los albéitares péritos y doctos* (Madrid: Antonio Marín, 1727), 1–18.
33. Francisco García Cabero, *Curación racional de irracionales y conclusiones veterinarias deducidas de diferentes principios philosóphicos con que se prueva la medicina, cirugía y albeitería una misma ciencia o arte* (Madrid: Pedro José Alonso de Padilla, 1728), 155.
34. Ibid., 156–7.
35. Domingo Royo, *Llave de albeitería. Primera y segunda parte* (Saragossa: Francisco Revilla and Joseph Fort, 1734), 475–6.

36. García Conde, *Verdadera albeitería*, Prologue, n.p.
37. Calvo, *Libro de albeitería*, 2; Arredondo, *Obras de albeyteríam*, 4.
38. Fernando Sande y Lago, *Compendio de albeitería sacado de diversos autores*, rev. edn (Madrid: José González, 1729), 7.
39. Ibid., 2.
40. Albert Sicroff, *Los estatutos de limpieza de sangre: Controversias entre los siglos XV y XVII*., trans. Mauro Armiño (Madrid: Taurus, 1985), 131.
41. Correas, *Vocabulario*, 338, 350.
42. Inca Garcilaso de la Vega, *Comentarios reales de los Incas*, ed. Ángel Rosenblat, 2 vols (Buenos Aires: Emecé, 1943), 2: 278–9. Calvo, *Libro de albeitería*, 11; see also Oudin, *Tesoro de las dos lenguas francesa y española*, 1: 790; Francisco Sobrino, *Diccionario nuevo de las lenguas española y francesa*, rev. and expanded edn, 2 vols (Brussels: Casa de Francisco Foppens, 1721), 424; Arredondo, *Obras de albeyteríam*, 6.
43. Correas, *Vocabulario*, 96; Ramírez de Carrión, *Maravillas de naturaleza*, 34; García Conde, *Verdadera albeitería*, 431.
44. Antonio Jiménez, *Colección de refranes, adagios y locuciones proverbiales* (Madrid: Pierart Peralta, 1828), 21.
45. José Gumilla, S. J., *El Orinoco ilustrado*, introd. and notes Constantino Bayle (Madrid: Aguilar, 1946), 86.
46. Ramírez de Carrión, *Maravillas de naturaleza*, 33.
47. Gumilla, *El Orinoco ilustrado*, 86.
48. Ibid., 87.
49. Sande y Lago, *Compendio de albeitería*, 21.
50. García Sáiz, *Las castas mexicanas*, 111.
51. Author Unknown, *El libro de los caballos. Tratado de albeitería del siglo XIII*, ed. and introd. Georg Sachs, Prologue by Rafael Castejón (Madrid: *Revista de Filología Española*, supplement XXIII, 1936), 48.
52. García Sáiz, *Las castas mexicanas*, 111.
53. See Magali Carrera, *Imagining Identity in New Spain: Race, Lineage, and the Colonial Body in Portraiture and Casta Paintings* (Austin: University of Texas Press, 2003), 88.
54. García Sáiz, *Las castas mexicanas*, 102.
55. Atran, *Cognitive Foundations*, 79.
56. Gumilla, *El Orinoco ilustrado*, 85.
57. Ruth Hill, *Hierarchy, Commerce and Fraud in Bourbon Spanish America: A Postal Inspector's Exposé* (Nashville: Vanderbilt University Press, 2005), chs 5, 6.
58. Miguel Olabarrieta Medrano, *Recuerdo de las obligaciones del Ministerio Apostólico enla cura de las almas. Manual moral ordenando primariamente a los señores párrocos, o Curas, de este Nuevo Mundo, en este Reino del Perú y los demás de las Indias* (Lima: Diego de Lira, 1717), 96.
59. Gumilla, *Orinoco ilustrado*, 85. According to a royal decree from 1703, King Philip V had requested from the Holy See the clarification of grave matters including marriage dispensations granted to Indian neophytes by Jesuits in the New World. The bull to which Gumilla alludes was one of three issued by Clement XI in 1701 to arbitrate the term "Indian neophyte," upon which rested the definition of "Indian" in the Spanish New World. See "A los Virreyes, Audiencias, Gobernadores, Arzobispos y Obispos de las Indias, remitiéndoles los trasuntos de los breves de su Santidad tocantes a

los indios," in Antonio Muro Orejón, *Cedulario americano del siglo XVIII. Colección de disposiciones legales indianas desde 1680 a 1800, contenidas en los cedularios del Archivo General de Indias*, 2 vols (Seville: Escuela de los Estudios Hispano-Americanos, 1956), 2: 79–82.

60. Benito Jerónimo Feijoo y Montenegro, "Color etiópico," in *Teatro crítico universal* (Madrid: Real Compañía de Impresores y Libreros, 1778), vol. 7: 66–93, Proyecto Filosofía en Español, www.filosofia.org, Biblioteca Feijooniana 1–14. I quote from section X, article 45, p. 10.
61. Ibid., X, 46, 10.
62. Ramírez de Carrión, *Maravillas de naturaleza*, 135.
63. Agustín, *Libro de los secretos*, 352.
64. Correas, *Vocabulario de refranes*, 433.

3
"Rude Uncivill Blood": The Pastoral Challenge to Hereditary Race in Fletcher and Milton

Jean E. Feerick

In his exposition on modern sexuality, Foucault famously emphasizes the uniqueness of modern cultural forms by seizing on the Renaissance as a point of contrast. This earlier era, he avers, was defined by a distinctly pre-modern episteme in being riveted not by sex but by "the blood relation." In that period, he maintains, "the value of descent lines were predominant" and "blood constituted one of the fundamental values."[1] Foucault was right to draw attention to the signifier of blood for the pre-modern world. At once a material substance—one of the four humors that flowed beneath the skin—blood was also a signifier infused with metaphysical properties, a conduit of quasi-immaterial essences transmitting lineal identity from one generation to the next. In the absence of a theory of genetic transmission, the mechanisms understood to govern the exchange of attributes carrying degrees of gentility were diffuse, thought to be mediated by airy animal spirits that conjoined the material body with a transcendent order. But they nonetheless held a powerful grip on the period and would cast a long shadow over subsequent race systems, which bore the imprint of this pre-modern hereditary order in granting the signifier of blood a position of primacy.

To modern eyes, such a system would seem to evoke and instate a class-based structure, one that naturalizes the social differences we have come to know as contingent and fungible articulations. But it would be more accurate to view Renaissance writers, pace Foucault, as understanding the difference of blood in ontological terms, as imparting rigid distinctions at birth. In that respect this system of identity may be better compared with our modern race system, since it traces deep divisions among people to the body's inner recesses, mystifying and essentializing those markers. And yet, although this ideology enjoyed the status of being dominant during the sixteenth century, it did not go unchallenged, and in this chapter I examine the works of two Renaissance poets—John Fletcher and John Milton—who enfolded a critique of this system of hereditary race into the pastoral plays they wrote on either side of a 25-year period in the seventeenth century.

Fletcher's play, *The Faithful Shepherdess*, was performed in 1609 during the early years of King James I's reign, when it was famously hissed off the public stage, disappointing spectators who expected a representation of rustic revelry.[2] But 20-some-odd years later, the play's sustained pastoral engagement found a more receptive audience at the court of Charles I and Henrietta Maria—a new edition appearing in 1629 and a court performance following soon thereafter on Twelfth Night in 1634. Its revival influenced the young poet Milton, who was invited to write an entertainment to honor the Earl of Bridgewater's installation as Lord President of Wales later that same year and who settled on a pastoral drama for the occasion.[3] Calling his play simply "A Masque Presented at Ludlow Castle," Milton leaned heavily on Fletcher's earlier production, reviving similar characters, plots, and settings for this new dramatic context. Both plays have puzzled critics for being generically atypical for each poet. Fletcher's name would rise to acclaim on the back of that dramatic hybrid—tragicomedy—which he would help to invent for the seventeenth-century English stage, making his early foray into pastoral drama appear as both a representational break and an expression of poetic inexperience.[4] Similarly, Milton's early experiment with a pastoral masque has been a problem needing to be explained, spawning many compelling arguments to the effect that he "reformed" and reshaped this royalist genre from the inside out.[5] I propose that each poet's decision to write a pastoral drama need not be viewed as a deviation or crux, particularly if we view these representations as motivated by pastoral's traditional impulse: the interrogation of ruling ideologies. Indeed, if pastoral is often seen by modern readers as nostalgic and escapist,[6] Renaissance writers viewed it as politically efficacious for being able to conceal a poet's licensed speech.[7] Puttenham, for one, identified pastoral's "eglogue" as being devised "not of purpose to counterfeit or represent the rusticall manner of loves and communication; but under the vaile of homely persons, and in rude speeches to insinuate and glaunce at greater matters."[8] That the pastoral mode had, as early as the Elizabethan period, come to be appropriated by the court as a vehicle of royalist ideology conferred still extra value on it for two poets who have been seen to harbor anti-court and pro-country sentiment.[9] Each was engaged in an act of reclaiming pastoral as an expression of humble life, seeking to return it to its classical origins when it served primarily as a vehicle to represent "herdsmen."

Indeed, it is possible to see both Fletcher and Milton as self-consciously adhering to a pattern established by Virgil and later embraced by Spenser in cutting their poetic teeth on the lowly mode of pastoral. Under that guise, each interrogates and undermines the aristocratic ideology that upheld a view of blood as a transcendent, immaterial, and stable repository of elite identity. In place of this mystified account, Fletcher and Milton portray elite blood—and both plays are extremely attentive to social distinctions—as subject to rapid transformation and requiring extreme effort to temper. They also share a tendency to place the means of such healing in hands far removed from courtly circles, associating such practices with commoners or supernatural forces that are, in turn, aligned with country values.

The implications of this reading are threefold. Fletcher's often neglected, stylized poem-play reclaims its imaginative and ideological links to the tragicomic corpus that was yet to establish a foothold on the English stage at the time of its first performance. Additionally, Milton's reformed masque assumes a still deeper challenge to aristocratic identity than critics have allowed in perceiving the masque as a decorous compliment to the peer for whom it was penned.[10] If scholars have noted how Milton challenged generic conventions with his formal modulations, they have been less willing to consider how generic forms instate and encode other differences of "kind," including those rooted in somatic signifiers. I argue that his challenge to the royal masque also encodes a challenge to a system of blood, but that this challenge need not be so explicit as to render his poem unsuitable for the occasion. Finally, by bringing Fletcher's pastoral production to bear on a reading of Milton's masque, I hope to extend our understanding of the connections between the two poets, moving from the identification of similar characters and motifs to a thicker account of borrowing and adaptation. By providing such a reading, I hope to further dislodge Fletcher's longstanding association with courtly modes and tastes and to draw out his contributions to later revolutionary sensibilities. Indeed I consider it a great irony that the Caroline court identified Fletcher as a spokesman for its Platonic celebration of royal passion. For, with Annabel Patterson, I see his little pastoral play as a satire of precisely those embodied forms, believing the play unworthy of its reputation as a soft, lyrical poem that was too delicate for the rough tastes of the theater.[11] Its representation of a passionate, sexual, and mutable ruling class may have been too subtle for an audience hoping for comic burlesque, but

Milton detected a different motive, simplifying his own plot so as to exaggerate the contours of its emphasis on blood's mutability. I therefore build on the insights of an early reader who claimed "All the very best and sweetest parts of Comus are stolen from [Fletcher's] exquisite Pastoral," in proposing that Milton also adapted Fletcher's critique of elite blood as a crucial premise of pastoral. For it, no less than the royal form of the masque, was in need of reclamation.[12]

Sedimented modes of pastoral

Pastoral's classical origins afforded it considerable openness of application and appropriation diachronically, since, as Paul Alpers has demonstrated, a fundamental duality defines Virgil's *Eclogues*. On one hand the songs give voice to the experience of exile and to disaffection with an abusive ruling class, a condition which the shepherd Meliboeus embodies in Virgil's opening Eclogue. On the other hand they represent the simple lives of herdsmen, humble figures denoting the poet and humankind generally, in the stance that a shepherd like Tityrus of that first Eclogue embodies. He directs our attention to the act of singing itself and to the conditions that enable and constrain his expression when he refers to the "god"—an embedded reference to Virgil's own imperial patron—who has sanctioned for an uncertain time his desire to pipe.[13]

This doubleness of emphasis carried over to the Renaissance amid altered conditions of production. Here, according to W. W. Greg, pastoral took root in and "came to its fairest flower amid the artificiality of a decadent court."[14] Seemingly oblivious to an earlier, Theocritan investment in documenting the contours of the humble and unsophisticated life, pastoral as it took hold at Renaissance courts gave expression to the tension between court and city, frequently voicing a poet's desire to escape the excesses of a ruling class. Sidney's prose pastoral, *The Old Arcadia*, was typical in this regard, portraying aristocrats as shepherds given to express their desires in landscapes removed from court. Indebted in title and form to Sannazaro's much earlier *Arcadia*, Sidney's pastoral romance captures a major modal shift in the emphasis it affords elite characters. That these shepherds denote aristocrats, rather than humble folks, is registered explicitly in Sidney's plot: two princes—Musidorus and Pyrochles—decide to disguise themselves as shepherds in order to win the love of two princesses. There is no question of who and what they represent. Even if and as the "pastoral mask" allowed a poet like Sidney to explore and probe questions of rule, the mode's social function had been fundamentally recalibrated in its transfer from classical Greece and Rome to the Europe of the sixteenth and seventeenth centuries.

The centripetal tug that pastoral witnessed in becoming increasingly aligned with courtly values and personages took even stronger hold in the fledgling genre of pastoral drama, which began to appear in Italy in the early to mid-fifteenth century as part of aristocratic nuptial entertainments.[15] In 1506, in one of the earliest recordings of staged pastoral, Baldassare Castiglione and Cesare Gonzaga dressed themselves as shepherds and recited an eclogue interspersed with songs in a performance of Castiglione's *Tirsi* at the court of Duke Guidobaldo of Urbino. This event concluded with a "panegyric of the court and the circle of the Cortegiano."[16] A pattern was to develop around this courtly mode in which the Arcadian setting of most of these plays served as a kind of mirror for the polite society of the Italian courts. If the court appeared in idealized form in the high mythological subjects that were a regular feature of these pastoral dramas, the plays yet captured a kind of salient social division, since they also featured "low rustic strains" in figures who resembled the "realistic, Paduan-speaking peasants" living beyond the Italian courts.[17]

These Italian pastoral experiments culminated in the great—if quite different—expressions of the form produced first by Tasso, with the *Aminta* performed in 1573 for Lucrezia d'Estee, and then by Guarini, with *Il pastor fido* performed for Charles Emmanuel, Duke of Savoy, and his new bride, Catherine of Austria, in 1585. Both pastorals foreground distinctions of rank, embedding praise for their elite patrons by honoring their noble bloodlines and shadowing them in the deities who preside over the action. Guarini's pastoral, for instance, opens with accolades for the Great Catherine, addressing "the valour of thy noble blood" and predicting that "from thy grand and most illustrious race, / New worlds will be supplied."[18] Tasso, too, had expressed a desire to heighten the generic status of pastoral when he observed in his "Discorsi del Poema Eroica" that he "intends to graft epic elements onto a pastoral tree" by weaving socially elevated characters into his representation, along with the higher style that would accord with their rank.[19] In both Tasso and Guarini, these socially eminent shepherds are juxtaposed with the satyrs who roam the woods and embody a kind of rustic lowness by virtue of their depraved, sexual appetites. If Guarini's more complex narrative muddies this opposition through plot sequences that call for a tempering of the "fury" of noble blood and warn against the possibility of "[degenerating]" from one's "noble race" (31), it yet celebrates the view that the play's noble shepherds carry the imprint of divinity in their blood and that Arcadia will be healed by restoring such connections. We learn that the harsh law that the gods have imposed as a sanction on the Arcadians can be released only when "two of heav'nly race be joined in love" (38). As the play progresses, we discover that the presumably lowborn lover Mirtillo, like his beloved Amarillis, carries this divine bloodline, revealing them to be the "Celestial branches" of a common "celestial root" (39). Later, the cunning Corsica, eager to bait Amarillis in a trap, invokes a similar botanical trope in arguing that "aconite and hemlock ne'er were known / To grow from out a healthy root" (96), implying that plants, like people, assume a predetermined form at birth. The emphasis on a continuity of growth from roots to sprigs works in tandem with the notion of a divinely infused bloodline to underpin a hereditary order, with the result that "the social tensions of pastoral are largely mystified or elided."[20]

Staging English revisions

Fletcher clearly had Guarini on his mind when he set out to rewrite the Italian pastoral play for an English context. Italian editions of *Il pastor fido* were published in England in 1592, with an English translation following in 1601. Just a few years later, in 1606, the pastoral would be performed for King James, making it ripe for imitation on the part of a rising young dramatist who had spent some time in courtly circles before his family's fall from grace. As W. W. Greg has argued, "but for the *Aminta* and *Il Pastor Fido*, *The Faithful Shepherdess* would never have come into being; as a type it ... is a conscious attempt to adapt the Italian pastoral to the requirements of the English stage."[21] To ensure audiences would catch the allusiveness of his project, Fletcher boldly echoed Guarini's title—rendered in English as *The Faithful Shepherd*—while also signaling the originality of his adaptation by shifting the gender of the play's moral axis from a man to a woman. A year or so later, when the play was first published, a dedicatory poem to Sir Robert Townesend provided another sign that Fletcher wished to effect a break with the Italian antecedents in his move to frame his pastoral drama in lowly, rustic terms. Having impugned the public audience's desire for "country hired Shepheards, in gray cloakes" in his letter to the reader, Fletcher here describes his play in precisely those terms, as a "poore Shepheard" in "home-spun gray" (ll. 10–11) which he offers up as meager payment for a debt to his patron. That he knows the fare to be modest, he emphasizes in comparing it to a "sallet" rather than the "great meat" a patron might be accustomed to receive to satisfy his "pallet" (ll. 20-1). As against the feast for royalty that Guarini's play offers, Fletcher provides a modest meal, one that will please "good feasters" (l. 19), if not royal stomachs, implying further that his play will restore measure to the inflated pastoral mode. And yet, by the time his play was reissued roughly 20 years later after its performance at the Caroline court, this ethos was overwritten by a brief dialogue that William Davenant wrote for the occasion and appended to the play's opening. Shifting our gaze from the shepherd in "home-spun gray" to a Priest and Nymph preparing a sacrifice for Pan, the dialogue directs our attention to the "gentler Deity" in the audience, describing King Charles and the Queen in the hyperbolic language typical of the Italian pastorals as "this Islands God; the worlds best King."[22] If Davenant thereby sought to anchor the meaning of Fletcher's homely pastoral in a royal essence, that goal could only be achieved by turning a blind eye on much of the play's representation. For Fletcher indicates at almost every turn that nobility of blood is not the locus of virtue that Guarini imagined it as in construing the faithful shepherd of his play's title to be of "heav'nly race" (38).

Indeed the opening of Fletcher's play blocks precisely the sort of association that Davenant makes in hailing King Charles as an earthly god. Clorin, the figure for whom the play is named, is the first to appear on stage, vowing to remain faithful to her recently deceased beloved and to devote herself to curing those who have "Growne wilde or lunaticke, their eies or eares / Thickned with misty filme of dulling rume" (1.1.37–8)—that is, those in the grip of passions conceived as powerful material agents. As she speaks, a satyr approaches, gathering fruits for his master, Pan, when he sees her and immediately hails her as a "devine" being of "heavenly forme" (1.1.58–9). Echoing the language Guarini uses for his patrons, the rustic man identifies her as a being "Sprong from great immortall race / Of the Gods," observing an "awfull majesty" in her visage that outshines "dull weake mortalitie" (1.1.60–3). Momentarily stunned into silence, Clorin responds only after he departs the scene. Hardly a deity, she reveals herself to be the daughter of two ordinary shepherds, a maiden of flesh and blood like any other mortal. Wondering at the Satyr's unsolicited submission, she concludes it must be the talisman-like effect of her chastity which "bindes fast, / All rude uncivill bloods, all appetites" (1.1.125–6). This emphasis on blood's mutability will be a hallmark of the play, establishing the condition for Clorin's healing powers as she applies her tempering remedies to shepherds suffering from passion's "looser blood" (1.3.66).

If this process of "[binding]" blood appears, in the case of the Satyr, to be effortless, it quickly becomes apparent that the operations on blood's passions and appetites that Clorin performs are laborious and contingent, successful only for those willing to receive them. Indeed, Clorin's opening speech expresses just how much effort goes into maintaining her own chaste disposition. Her determination to free herself "from all ensuing heates and fires / Of love" can be successful only by removing herself from contact with activities that will stir her passions, including "all sports, delights and games," as well as "youthfull coronals" and "merry pipes" (1.1.7–14). By inhabiting a "low Cabin, of cut boughs" (5.4.16) shielded from the sun deep in the woods, Clorin has found a prophylactic against the "vaine illusion[s]" that might compel her "to wander after idle fiers" and lead her to stray "Through mires and standing pooles" (1.1.115–20). In this account of fens, pools, and fires, Fletcher provides a psychomachia, transforming the pastoral landscape into an external version of the "complex internal economy" of the human mind as it rides the liquid waves of passion.[23] Crucially, he also offers an intervention in the ideology that equated noble blood with divinity by revealing Clorin's divine-like aura to be the product of conscious choice. An inherited bloodline has little to do with the virtue of this shepherdess, unlike Guarini's regal pair who alone have the power to heal the social wounds plaguing his Arcadia. Fletcher's virgin, by contrast, is an everyday healer whose acts of moderating blood's tempests—both her own and those of others—involve neither magic nor transcendence but sheer determination.

The tendency to place the control of blood's properties in the hands of ordinary individuals, rather than in noble genealogies, is evident as well in the way Fletcher reorients the botanical motifs present in Guarini, not least the mention of roots, branches, sprigs, and herbs. As we have seen, Guarini favors metaphors of organic growth, insisting on the connection of plant to root. Insofar as Fletcher grants Clorin the ability to know and apply the different powers of herbs to heal others, he foregrounds the dynamic interchange that occurs between herb and human body. As we watch Clorin maneuver through her garden in a scene that echoes (or anticipates) Perdita of *The Winter's Tale*, she praises the distinct powers of different herbs. So she commends "Calamint" for having "vertues" that "do refine / The blood of Man, making it free and faire, / As the first houre it breath'd, or the best aire" (2.2.30–2), while "foule Standergrasse" and "lustfull Turpentine" are "banish[ed]" for the fact that they "intice the vaines, and stirre the heat / To civill muteny" (2.2.35–8). In this anatomy of nature's latent powers and the variously benign and malignant effects they have on the human body, Fletcher captures the fluidity and mutability of blood, which can be enflamed or refined at any given moment by congress with the surrounding world.

In fact, Fletcher achieves comic heights by exaggerating the infixity of blood's qualities in the parade of noble shepherds who wander through the woods surrounding Clorin's cabin, indulging their passions in vignettes indebted to the emblems of distemperance that appear in *Faerie Queene* Book 2. Perigot, who enjoys a position of social preeminence within this community, described as "the top / Of all our lusty Shepheards" (1.2.145–6), is defined by the fiery attributes associated with the aristocracy, propelled first by an excessive love for Amoret and then by an equally excessive anger toward her. Amoret primes us for his vacillation when she warns him to guard against the heat of the day and night which might "moove your blood" (1.2.89). He responds in the high voice of epic, offering her oaths of assurance that he will be constant, urging "Let me deserve the hot polluted name, / Of a wilde woodman" (1.2.130–1) should his desires deviate. Of course it is not long before he twice takes aim with the intent to kill his defenseless beloved in scenes of mistaken identification that draw heavily on the lunar confusions that propel *A Midsummer Night's Dream*. If there is comedy, so there is tragedy. His high temper evokes as well the tragic intonations of Lear on the heath when, believing Amoret to be untrue, he heaps his rheum on the world in a suicidal rant: "Shee's gone, shee's gone, blow hygh thou North west winde, / ... / And shake the world as at the monstrous birth, / Of some new Prodegey, whilst I constant stand" (4.1.2–7). Here the tragic mode rapidly transforms into burlesque. A similar mutability of affect seizes the other gentle shepherds who, one after another, open themselves to the dangers of the night. Echoing the flaming Pyrochles who is propelled by his heat through Spenser's faery land, Alexis receives Cloe's proposal of a tryst with a profusive: "oh how I burne / And rise in youth and fier!" (1.3.190–1). Meanwhile, Cloe, anticipating that this new lover's passion may be unreliable, arranges a backup in the shepherd Daphnis, observing to herself: "He that will use all windes must shift his saile" (1.3.195). A wannabe siren-in-heat, Cloe, too, rides the waves of desire, searching out, but failing to secure, a man up for the task of satisfying her.

The problem, as Fletcher appears to construct it, originates with "ill governance" (5.5.80) on the part of both those ruling the land and those enjoying preeminence therein. Indeed, as critics have observed, it is the figures most removed from the "town" of this pastoral representation who are most effective at moderating their passions, granting them the strongest claim to virtue. Fletcher's Satyr is his most stunning embodiment of this shift in ethos, as he exhibits none of the expressions of "looser blood" (1.3.66) that the elite "owners" of this pastoral realm embody.[24] Breaking with the tendency in Tasso and Guarini to portray the rustic satyr as an emblem of lasciviousness, Fletcher invests his rustic figure with natural gentility. Notably, other than Clorin, he appears to be the only figure who knows any labor. We see him first diligently attending to the needs of Pan and his paramours by gathering fruits and nuts in distant reaches of the woods, and then turning his service to aiding Clorin, by bringing her wounded shepherds in need of ministry. Surprised by his effort, Clorin repays him by observing: "Though thou beest outward rough and tawny hued, / Thy manners are as gentle and as fayre, / As his who bragges himselfe, borne only heyre / To all Humanity" (4.2.63–6). Encoding, perhaps, a reference to the monarch, Fletcher inverts the pastoral hierarchy that he inherited from the Italians, making his rustic figure an emblem of temperance and his elite shepherds "manlike monsters" (5.5.103).

If Fletcher's Satyr redirects his loyalties from Pan to Clorin during the course of the play, his movement expresses a sanctioned moral shift. For the play subtly alludes to a lord of these woods who is defined by his absence and omissions of rule. Indeed, one might read Pan as a more distant version of the villainous Sullen Shepherd who, though an "owner" of land in this pastoral world, yet exhibits extreme negligence in the care of those entrusted to him. His sheep are "nye starved," "alwaies scabby," and "dye before their weaning" (1.2.204–7). So, too, his dog "Lookes like his Maister, leane, and full of scurffe" (1.2.208). If he disregards his own creatures, he also disregards his peers, sabotaging their "holy plighted troths" (1.2.199) and "[lusting] after" all matter of shepherdesses and "smooth" (1.2.200, 204) youths. More than simply being absent, his version of "husbandry" translates into an active assault on the community. A similar state of affairs is embodied by Pan, who not only fails to supervise his realm but refuses to comport himself like a "Master" (1.1.52). We hear that he feasts (1.1.54), frolics (1.1.55–6), drinks the "lusty blood" of grapes (1.1.76), and sleeps "Under a broad beeches shade" (1.1.99). If we follow Patterson's reading of trees in pastoral as emblems of a patron's protection, Fletcher inverts this pastoral trope, too, figuring the supreme patron of Thessaly as offering refuge only to himself.[25]

The Satyr's movement is replicated, if belatedly, by the other guardians of the town—the priest and Old Shepherd. We have already experienced their rule to be ineffectual in the absence of Pan, witnessing how the ceremony they perform to release the shepherds from "hot flames of lust" (1.2.7) has little effect. Immediately following this event, Cloe initiates her hunt for men, and it is that very evening that nearly all of the shepherds set off for the woods, embracing the free rein of passion that it enables. When the Old Shepherd discovers their absence the next morning, he and the Priest wisely set off to find them and are directed to Clorin's cabin where many of them are being healed. As McMullan argues, "the political effect of the transfer of power is highly charged, since figures representative of both spiritual and secular power in the play defer to a virgin for an appropriate solution to the country's ills."[26] Having rehabilitated the lusty Alexis, the wounded Amoret, the frigid Thenot, and the enraged Amoret, Clorin welcomes the converted Amarillis, observing: "we have perfoormd a woorke / Worthy the gods them-selves" (5.5.143–4). Earlier in the play, the God of the River had risen from his watery bed to rescue the wounded Amoret, having been assailed both by her lover Perigot and by the evil Sullen Shepherd and left for dead. This earlier moment seems to suggest that these mortals, like Guarini's before them, need the strength of those of "heav'nly race" (38) to save them. But the moment is quickly neutralized when this heroic gesture on the part of the god slips into a pattern all-too-familiar to us from the male mortals who populate the play: the God of the River's passion gets the better of him and in a lyrical rewriting of Marlowe's "Passionate Shepherd," he invites her to "go with [him], / Leaving Mortall company" (3.1.409–10). Intuiting the ontological difference that makes her "unworthy to be woed, / By thee a God" (3.1.438–9), she declines, and he vanishes from the pastoral. By the end, Clorin is the primary axis of virtue that the play offers, singled out for her devotion to the act of tempering blood-grown-wild, through labors that are "Worthy the gods them-selves" (5.5.144). If she gestures at a god-like restitution of nobility to bloodlines that have declined, the play reminds us that this virgin is anything but a god, and her patients, though elite, little more than earthly creatures subject to blood's mutability. The effect of this lyrical representation of passion is to demystify the signifier at the heart of a system of hereditary race, offering the woods as a release from the myth of a transcendent elite identity.

Milton's dark pastoral glass

The court of Charles I and Henrietta Maria, accustomed as it was to embody the divine quality of royalty in Platonic masques, did not perceive the critique of blood lingering beneath Fletcher's stylized shepherds, and his play was revived for their enjoyment on Twelfth Night in

1634, framed by Davenant's panegyric to royal spectators hailed as deities on earth. Critics suggest that the court seized upon Fletcher's play and its rejection by Jacobean audiences as "an indictment of the earlier age, a proof of the new court's finer morals and esthetics, and a handy answer to Prynne," who had attacked the court's licentiousness as embodied by its penchant for "amorous pastorals" in his *Histriomastix*.[27] But even as the court found *The Faithful Shepherdess* a fitting emblem for reasserting its moral authority, evidence suggests that the young poet Milton, when invited to write a masque to honor the Earl of Bridgewater's installation as Lord President of the Council of Wales, intuited a very different ethos in Fletcher's pastoral. John Carey has already identified Fletcher's pastoral as the single most important source for Milton's *Masque at Ludlow*, tracing many parallels of character and motif between the two dramas.[28] I observe still deeper affinities between the plays on ideological grounds. For Milton's genius is to take the broad indictment of elite blood as variable, immoderate, and inflamed that he found in Fletcher's play—where it appears in a satirical, generically hybrid form—and to funnel it into a more somber and realistic meditation on elite identity, which occurs not in an imaginary Arcadia with stylized shepherds (Fletcher's Thessaly) but in a real time and place inhabited with flesh-and-blood aristocrats who perform their actual social roles.

His masque was presented on 29 September 1634, the Feast of St Michael and All Angels, a day when officials typically were sworn into office.[29] Symbolically, this calendrical pattern served to align those newly instated officials with guardian angels, imagining them as protectors of the people, much as the Earl of Bridgewater's new role invested him with royal power to secure order in a Welsh borderland characterized by social unrest. But Milton's masque flips this equation on its head in that it figures the three children of the Earl who acted in this masque—aged 15, 11, and 9—as themselves requiring protection as they traverse a "drear wood" (l. 37) where forms are never what they appear to be and where good and evil are intermixed in a complex representation of an enmattered world. If Fletcher's Thessaly presents plenty of opportunities for elite shepherds to "Set up their bloods to saile" (1.3.143), it yet offers a universe where evil is easily identifiable and can be avoided or embraced as one might choose. Milton's pastoral world is considerably darker. Here the powers of embodied virtue are less potent than they are in the heavenly realm, and even elite bodies are "unexempt" from the "mortal frailty" (ll. 684–5) that defines all living forms.

If Milton's pastoral landscape builds on even as it complicates Fletcher's vision, it represents a complete reversal of the royal masque that it purportedly exemplifies.³⁰ In February of the year in which Milton's masque was performed, Thomas Carew's ostentatious *Coelum Britannicum* was performed at court, the King's "offering" in response to the Queen's presentation of Fletcher's pastoral on Twelfth Night of that year.³¹ It presented one of this court's most exalted conceptions of the theory of the divine right of kings, imagining the "ruler as a quasi-divine source of national power and unity," whose "immortall bosomes" "burne with emulous fires."³² Significantly, Bridgewater's two young sons acted in this masque, where they assumed the role of torch-bearers who accompany Jove's royal entourage. This casting decision allowed them to embody a version of the King's divine flame, imagining an aristocratic community of blood as affecting an unbroken continuum between earth and heavens.

Cognizant of these two recent court performances, Milton set about to challenge the theory of immanence that Carew's masque celebrated by playing it off against the quite different universe of blood figured in Fletcher's play. Significantly, *Comus* opens with the Attendant Spirit emphasizing the gulf that separates the "serene air" (l. 4) of the heavens from the earth where "rank vapours" "soil" (ll. 16–17) even "aerial spirits" (l. 3) who descend on heavenly "[errands]" (l. 15). If a masque like *Coelum* imagines the heavens as needing reform by earthly deities, Milton's masque allows no such inversion. Here, the earth's inhabitants are described as leading a "feverish" existence, being "confined" and "pestered" (ll. 7–8) by their material world. Ironically, it is Comus who echoes the courtly masque's emphasis on immanence not only by arriving flanked by torches but also in addressing his followers in an exalted description of their embodied form. Drawing on the terms used at court to describe a royal disposition, he asserts "We that are of purer fire / Imitate the starry quire" (ll. 111–12), suggesting that the physical movements of the heavenly spheres can be replicated on earth by demigods like him. Since he enters the stage accompanied by a "rout of monsters, headed like sundry sorts of wild beasts" (s.d., p. 181), the audience immediately perceives this rhetoric as inflated. He and his cohorts instead seem to embody the "rank vapours of this sin-worn mould" (l. 17), a "degenerate and degraded state" (l. 474) enabled by the "loose gestures" (l. 463) of lust to which they dedicate themselves. Comus's misguided tendency to perceive his peers as earthly gods replicates a "Stuart dichotomy between lesser mortals and high-minded masquers," one that Milton actively deconstructs.³³

In fact, Milton exposes Comus's emphasis on embodied differences of rank as mere pretense in his first exchange with the Lady, as he tries to entice her to become his "queen" (l. 264). Here Comus receives her with words that echo Fletcher's well-intentioned Satyr when he stumbles upon the chaste Clorin, professing "Hail foreign wonder / Whom certain these rough shades did never breed" (ll. 264–5). His words flatter the Lady into believing he not only perceives but also reverences her high birth and godly disposition. He speaks of an ontological difference between her and these "rough shades," suggesting that she is unworldly, defined by an ethereal quality. His flattery is juxtaposed with his more modest assumption minutes earlier, upon hearing her song, that her music is exquisite for one who is a "mortal mixture of earth's mould" (l. 243). Although he acknowledges that "something holy lodges in [her] breast" (l. 245), he hardly confuses her with a deity. If the Lady rejects his address as flattery, telling him his words fall on "unattending ears" (l. 271), she will elsewhere reveal, along with her brothers, that she is inclined to share some of Comus's assumptions, particularly his equation of rank with virtue and divinity.

Indeed, when she first hears the revelry between Comus and his crew, she presumes that the "ill-managed merriment" (l. 171) comes from some "loose unlettered hinds" (l. 173) who are partaking of a ceremony in thanks for the harvest. As the word "loose" suggests, she views such revelry as an expression of ill-governed behavior characteristic of the lower orders, concluding: "I should be loth / To meet the rudeness, and swilled insolence / Of such late wassailers" (ll. 176–8). Her brothers subsequently echo her assumption in fearing that an attack on her will come from some "savage fierce, bandit, or mountaineer" (l. 425), that is, from a country rustic. Their tendency to equate low birth with vice was an assumption propounded by many royal masques, which often locate "the evils of social disorder in the lower classes";[34] such views may also have been present in the upbringing of these elite children, having been "nursed in princely lore" (l. 34), presumably fed stories of epic grandeur. If the Lady indicates a readiness to move beyond the idea that only those of high blood embody virtue when she acknowledges that "courtesy" is "sooner found" in "lowly sheds" than in "courts of princes" (ll. 321–4), her brothers are more wed to the Caroline court's Platonic universe where virtue is majestic, visible, and glistering in form. Such a view becomes apparent when the elder brother chides his younger brother for fearing his sister's safety, asserting, "Virtue could see to do what Virtue would / By her own radiant light" (ll. 372–3). Insofar as he imagines virtue as ontologically pure, his words are appropriate for a celestial realm where "air" is "mild" and "calm" (l. 4) and unencumbered by "vapours" (l. 17). Like his conception of "Virtue," he imagines his sister as a Platonic essence, believing her to embody in her sojourn through the woods the posture of allegorical forms of Wisdom (l. 374) and Contemplation (l. 376). Her "unblenched majesty" (l. 429) is such that it affords her, in his view, the powers of a goddess to control the material world and the ability to "converse with heavenly habitants" (l. 458). But such assumptions collide with the Attendant Spirit's account of all earthly life as fundamentally removed from the heavenly sphere. Humankind—that "mortal mixture of earth's mould" (l. 243)—mingles spirit with a material body that is at once "frail, and feverish" (l. 8), open to mists and fogs and other "rank vapours" (l. 17) that confuse perception and impede access to the eternal realm.

That Milton rejects the Elder Brother's Platonic view of the world is evident in the way he makes light of the epic tactics that the brothers lean on to secure the goodness of the world. Hearing an unknown sound, which they assume is some "woodman" or "robber" (ll. 483–4), the brothers wrap themselves in an epic stance, drawing their weapons and threatening the approaching figure with "iron stakes" (l. 490). When they learn that the Attendant Spirit is actually their ally but that a "damned magician" (l. 601) has captured their sister, this posture becomes still more entrenched: "let him be girt / With ... Harpies and hydras, or all the monstrous forms / 'Twixt Africa and Ind, I'll find him out" (ll. 601–5). Admiring their heroic spirit, the Attendant Spirit yet rejects it, telling them "thy sword can do thee little stead" (l. 610) against the magician's charms. They must be prepared instead to fight a more insinuating form of evil, one that does not announce itself in explicitly monstrous form but appears "pranked in reason's garb" (758), a moral composite like all earthly life. Significantly, when the brothers storm Comus's palace, they forget this lesson, drawing their swords for an epic fight but forgetting the second part of the Spirit's instructions—"But seize his wand" (l. 652)—hypnotized, apparently, by the appeal of a "[Bold] assault" (l. 648). By emphasizing a kind of epic entrancement on their part, Milton places them in a position analogous to their sister: all of them are led astray by a conviction that virtue is immanent in the world and enjoys a privileged relation to their elite bodies.

The young Lady, unlike her brothers, is not wed to epic contest, praying that she might pass through the woods "unassailed" (l. 219) or at most be subject to a "trial" tailored to her "proportioned strength" (ll. 328–9). But her exchanges with Comus reveal her mortal flesh is tinctured by frailty that contrasts with the impregnability her brother affords her in associating her chastity with divinity. Notably, though she is guarded in her approach to Comus, she yet cannot help but be open to the material world. Indeed, it is her "listening ear" (l. 202) that first tempts her in Comus's direction, drawing her toward the "loud mirth" (l. 201) she believes originates with "gamesome" (l. 172) farm hands. Subsequently, her eyes deceive her into believing Comus to be a "harmless villager" (l. 166) when she is subject to his "dazzling spells" that aim to "cheat the eye" (ll. 154–5). Later, we see him preying on her sense of touch, when he freezes her "In stony fetters fixed" (l. 818) while he attempts to assail a fourth sense, urging, "Be wise, and taste" (l. 812). Unlike those Platonic forms of virtue—"pure-eyed Faith, white-handed Hope / ... And thou unblemished form of Chastity" (ll. 212–14)—the Lady's powers are constrained by her "corporal rind" (l. 663), which affords her only imperfect access to the purity she seeks.

In the resolution he constructs for the masque, Milton enacts a reordering of the children's moral universe in which divinity attaches to elite bodies and evil to laboring ones, and where the difference between the two is visible and distinct. Part of this resolution is effected by haemony, that mysterious herb that the Spirit retrieves for the brothers to aid their assault. A word derived from the Greek adjective for *blood*, haemony is described as a "small unsightly root" that is yet "of divine effect" (ll. 628–9), and may enfold a reference to Guarini's mention of a "celestial root" that defines his regal lovers. In Milton's hands, the root transforms into a trope for earthly life; if elsewhere the herb yields a "bright golden flower" (l. 632), in Welsh soil its virtuous powers lie hidden beneath a leaf that is "darkish" in color and laden with "prickles" (l. 630). Insofar as we accept the root's connections to the word *blood*, I propose that the herb's meanings resonate with the recalibration of blood's significance central to Milton's pastoral. For haemony is neither an epic weapon nor a magical talisman, but rather a modest power that fortifies the children, guarding them from "all enchantments" and material agents like "mildew blast, or damp" (l. 639). The fact that its origins are humble, lying close to the earth where it is "Unknown" to the "dull swain" who "Treads on it daily" (ll. 633–4), suggests that it offers a theory of embodied virtue at odds with the children's valuation of eminence. Its powers come to the children mediated by a humble source, the "shepherd lad / Of small regard .... yet well skilled / In every virtuous plant" (ll. 618–20). Combining a humble pedigree with a principle of labor, this herb holds the promise of tempering blood in its darker, more deeply enmattered—and less than golden—earthly state.

That the brothers achieve only a partial victory after being fortified with haemony—scattering Comus and his forces but not freeing their sister— may signify that their achievement of its ideal has not been reached and that there is more—even ongoing—work for them to accomplish. The goddess Sabrina, who is called upon to finish the task of releasing the young Lady, replays its meanings in another key, transferring the trope from the herbal to the human realm. Like haemony, whose properties vary with place, Sabrina's ability to be a "virgin pure" (l. 825) has altered with her circumstances. As a mortal, she was the "daughter of Locrine" (l. 826), ensnared by a noble family defined by passionate excess, whose feuding resulted in her death by drowning. And yet, in that other country beneath the river's flood, she discovered a brighter potential for herself, undergoing a "quick immortal change" (l. 840) reminiscent of haemony's "bright golden flower" (l. 632) that blooms in another land. Her transformation allows her to become a healing force, neutralizing "urchin blasts" (l. 844) that infect herds and ministering remedies to "clasping [charms]" and "numbing [spells]" (l. 852), thereby replicating the powers of haemony.

Together these emblems of pastoral—a wild herb mistaken for a weed and a discarded virgin—denote the complex process by which a "clotted" (l. 466) material world transforms into something more perfect. Notably, neither of these figures has embodied a celestial form on earth. Both are defined instead by postures of lowliness, of fallen-ness, pressed to the earth or to the riverbed and seemingly absent any value. But through these figures, who move from darkness and prostration to lightness and preeminence, Milton modifies the principles ordering the world of his elite audience. Critics have tended to see Milton at this early stage in his career as shoring up "aristocratic virtue" and attaching the "moral health of the nation" to a reformed aristocracy.[35] But I propose that this modest pastoral play strikes more deeply at the idea of an hereditary order and the account of immanence on which it depended. Such an order was presumably taken for granted by a peer ascending to the position of Lord President of Wales—a post that made him a surrogate king, a kind of earthly deity. Milton confounds the view of reality upon which such titles rested. For the earthly world that he frames with this dark pastoral is a place where virtue is cloistered, lying in the most unexpected of places, and, conversely, where falsehood is "vizored" in palatial extravagance (l. 697) and fiery eminence. If there is a "golden key / That opes the palace of eternity" (ll. 13–14) for frail mortals, it does not derive "from forefathers"[36] nor from bloodlines tinged with divinity. Far from being something infused at birth, true nobility follows a plodding path defined by "hard assays" (l. 971) and "wandering labours long" (l. 1005). The irony that Milton, echoing Fletcher, captures is that only by stooping—in imitation of the reclaimed form of pastoral that he secures—can one reach such heights.

Notes

1. Michel Foucault, *The History of Sexuality: An Introduction*, trans. Robert Hurley (London: Penguin, 1978), 147.
2. Evidence for this reception appears in Fletcher's letter "To the Reader," appended to the play's first quarto. There Fletcher describes his audience as "missing whitsun ales, creame, wassel and morris-dances." See *The Faithful Shepherdess* in *The Dramatic Works in the Beaumont and Fletcher Canon*, gen. ed. Fredson Bowers, ed. Cyrus Hoy, vol. 3 (Cambridge University Press, 1976). All citations of Fletcher's play are to this edition and will be cited in the text with reference to act, scene, and line numbers.
3. Carey describes the play as "a Platonic pastoral drama"; see *Milton: The Complete Shorter Poems*, ed. John Carey (London and New York: Longman, 1971), 170. All future citations of *Comus* are to this edition and will appear in the text with reference to line numbers.
4. But see Robert Henke's discussion of tragicomedy's links to pastoral in "Pastoral as Tragicomedic in Italian and Shakespearean Drama," in *The Italian World of English Renaissance Drama: Cultural Exchange and Intertextuality*, ed. Michele Marrapodi (Newark and London: University of Delaware Press, 1998), 282–301.
5. For this argument, see David Norbrook, *Poetry and Politics in the English Renaissance* (London: Routledge, 1984), ch. 10.
6. Paul Alpers states that "The most widespread view of pastoral is that it is mere wish fulfillment"; see *The Singer of the Eclogues: A Study of Virgilian Pastoral* (Berkeley: University of California Press, 1979), 4–5. For a discussion of the "self-consciousness" of the pastoral mode, see Alpers, *What is Pastoral?* (University of Chicago Press, 1996), 16.
7. For a discussion of pastoral's social efficacy, see Annabel Patterson, *Censorship and Interpretation: The Conditions of Writing and Reading in Early Modern England* (Madison: University of Wisconsin Press, 1984).
8. George Puttenham, *The Art of English Poesie*, ed. G. D. Willcock and A. Walker (Cambridge University Press, 1936), 38, as quoted in Patterson, *Censorship and Interpretation*, 37.
9. For this view of Fletcher, see Philip J. Finkelpearl, *Court and Country Politics in the Plays of Beaumont and Fletcher* (Princeton University Press, 1990); and Gordon McMullan, *The Politics of Unease in the Plays of John Fletcher* (Amherst: University of Massachusetts Press, 1994). For Milton, see Norbrook, *Poetry and Politics*; and John Creaser, "'The present aid of this occasion': The Setting of *Comus*," in *The Court Masque*, ed. David Lindley (Manchester University Press, 1984), 111–34.
10. For Milton's decorousness, see Barbara Breasted, "*Comus* and the Castlehaven Scandal," *Milton Studies* 3 (1971), 201–24, esp. 202. For the masque as a "conditional" compliment, see John D. Cox, "Poetry and History in Milton's Country Masque," *ELH* 44.4 (1977), 622–40, esp. 623.
11. See Patterson, *Censorship and Interpretation*, 181–2.

12. Alfred Guy Kingan L'Estrange, *Life of Mary Russell Mitford ... Told by Herself in Letters to Her Friends*, 11 October 1827 (New York: Harper and Brothers, 1870), 1: 274.
13. Alpers, *What is Pastoral?*, ch. 1.
14. Walter W. Greg, *Pastoral Poetry and Pastoral Drama: A Literary Inquiry, with Special Reference to the Pre-Restoration Stage in England* (New York: Russell & Russell, 1959), 1–2.
15. Henke, "Pastoral as Tragicomedy," in *Italian World*, ed. Marrapodi, 286. For analysis of Italian pastoral drama, see Louise George Clubb, *Italian Drama in Shakespeare's Time* (New Haven and London: Yale University Press, 1989).
16. Greg, *Pastoral Poetry*, 31.
17. Henke, "Pastoral as Tragicomedy," in *Italian World*, ed. Marrapodi, 286.
18. *The Faithful Shepherd: A Translation of Battista Guarini's Il pastor fido by Dr. Thomas Sheridan*, ed. Robert Hogan and Edward A. Nickerson (Newark, DE: Associated University Presses, 1989), 24. All citations of the play are to this edition and will be cited in the text with reference to page number.
19. As quoted in Henke, "Pastoral as Tragicomedy," in *Italian World*, ed. Marrapodi, 289.
20. Ibid.
21. Greg, *Pastoral Poetry*, 266.
22. See Hoy (ed.), *Faithful Shepherdess*, 499.
23. Less Bliss, "Defending Fletcher's Shepherds," *Studies in English Literature, 1500–1900* 23.1 (1983), 295–310, esp. 302.
24. See Fletcher's "To the Reader," in *Faithful Shepherdess*, ed. Hoy, 497.
25. See her discussion of "the tree of patronage" in *Pastoral and Ideology* (Berkeley and Los Angeles: University of California Press, 1987), 52–6.
26. McMullan, *Politics of Unease*, 67.
27. Patterson, *Censorship and Interpretation*, 181. See also Barbara K. Lewalski, "Milton's *Comus* and the Politics of Masquing," in *The Politics of the Stuart Court Masque*, ed. David Bevington and Peter Holbrook (Cambridge University Press, 2006), 296–320, esp. 302–3.
28. See Carey (ed.), *A Masque*, 170–1.
29. See John Creaser, "'The present aid,'" in *Court Masque*, ed. Lindley, 115.
30. For this argument, see David Norbrook, "The Reformation of the Masque," in *Court Masque*, ed. Lindley, 94–110.
31. See Lewalski, "Milton's *Comus*," in *Politics of the Stuart Court Masque*, ed. Bevington and Holbrook, 304.
32. As quoted in Creaser, "'The present aid,'" in *Court Masque*, ed. Lindley, 118 and 124.
33. Leah S. Marcus, "John Milton's *Comus*," in *A Companion to Milton*, ed. Thomas N. Corns (Oxford: Blackwell, 2001), 232–45, esp. 240.
34. Lewalski, "Milton's *Comus*," in *Politics of the Stuart Court Masque*, ed. Bevington and Holbrook, 313.
35. See Breasted, "*Comus* and the Castlehaven Scandal," 201.
36. This phrase appears in Milton's marginalia to the Commonplace Book; see Creaser, "'The present aid,'" in *Court Masque*, ed. Lindley, 120.

4
African Blood, Colonial Money, and Respectable Mulatto Heiresses Reforming Eighteenth-Century England

Lyndon J. Dominique

Introduction

Elsie B. Michie's *The Vulgar Question of Money* (2011) and Jennifer DeVere Brody's *Impossible Purities* (1998) explore nineteenth-century literature using two very different female tropes. Michie examines novels which "establish a set of values"[1] by contrasting rich women with poor ones. She believes that "the rich woman embodies those behaviors that individuals feared resulted from the increasing importance of money in the nineteenth century. The novel uses her as a foil to propose the psychological and moral stances necessary to counter such material engrossment."[2] For her examination of literature from the period, Brody uses the highly ambiguous trope of the "feminine mulattaroon."[3] She asserts that this hybrid figure, "who can be designated also, if not always alternatively, as a mulatta, an octoroon, a quadroon, a musteee, mestico, griffe, or creole,"[4] "is perpetually being erased or effaced" by nineteenth-century writers, "in an effort to stabilize (reify) the tenuous, permeable boundaries between white and black, high and low, male and female, England and America, pure and impure."[5]

Because the erasure of Brody's "feminine mulattaroon" stimulates readers' desires for clear racial differences and distinctions, and the disgust expressed toward Michie's "rich woman" encourages them to "extol, by negation, the values she does not embody,"[6] we can conclude that nineteenth-century writers use these gendered literary tropes by design, to appeal to readers' prejudices about black blood and female wealth. These tropes are not merely spaces where "the contested struggle to define 'proper' boundaries"[7] about blood and money takes place, however; this 'contested struggle' is actually resolved once the troped figure is narratively negated—an act that is spearheaded or executed by the author in order to establish the "proper" system of prejudices about blood and money that the reader is encouraged to favor.

Examining other connections between these troped figures and their narrative negations, however, also allows for revelations about blood and money that can belie a prejudiced system of belief. For instance, since Michie does not consider Brody's rich female mulattaroons and their peculiar presence within the novel, she misses a unique opportunity to discuss how such women generate specific responses to money when the specter of African blood becomes a central feature in a novel's marriage plot. If "the heiress is associated with endogamy" and "had to be exchanged endogamously"[8] as Michie says, one wonders how Olivia Fairfield (one of Brody's rich mulattaroons) generates disgust about money as well as race within the white English family that *The Woman of Colour*'s (1808) anonymous author requires she marry into. The disgust generated toward this woman might actually perform progressive work, encouraging antislavery sentiment through the rejection of a rich woman of color whose money comes from the slave trade.[9] Additionally, if *The Woman of Colour* is capable of doing this kind of antislavery work, then far from merely being a disgusting hybrid figure whose erasure stabilizes cultural prejudices and boundaries, as Brody suggests, the endogamous marriage of a wealthy female mulattaroon might be a desirable feature in the novel, actively making visible the stabilizing forces that seem intent on Olivia's erasure in a deliberate effort to critique or undermine their power.

These progressive and prejudiced perspectives about blood and money are both informed further by an important distinction that the juxtaposed tropes make clear: the foundations for the "feminine mulattaroon" and the "rich woman" as well as their stabilizing effects on readers derive from very discrete parental sources. The rich woman's money and its ensuing disgust is a financial legacy usually bestowed or bequeathed by a white father in the language of portions, dowries, settlements, jointures, or death entitlements;[10] the mulattaroon's hybridity and the desire to erase or efface its threatening potential are connected to a black mother and the biological legacy of stigmatized blood that appears in "the mulattaroon's 'singed' skin (or comparable dark mark)"[11]—a legacy that must be negated because, as Lynda Boose has argued, it threatens the dominance of the white father and his ability to reproduce himself in his own image.[12]

I have foregrounded the impacts that white and black parents have on Michie's "rich woman" and Brody's "feminine mulattaroon" as well as outlined the ways that terms such as blood, money, disgust, desire, and negation operate in texts that influence readers' beliefs because these terms, tropes, and ideas provide a necessary entree to this essay which examines wealthy women of color and how they function as vehicles for social justice as well as prejudice in English literature. I bring my particular ideas about race, literature, and social justice to bear by examining a female trope derived from Michie's "rich woman" and Brody's "feminine mulattaroon." I refer, of course, to the mulatto heiress of my title—specifically, the daughter of a rich white father and an enslaved black or African mother.

When mulatto women are discussed in contemporary literary criticism it is usually within nineteenth-century American or transatlantic contexts exploring the tragic consequences arising from their mixed black and white blood, as seen in Kimberly Snider Manganelli's *Transatlantic Spectacles of Race: The Tragic Mulatta and the Tragic Muse* (2012), or Brody's *Impossible Purities*. Tia Gafford also identifies this century as one in which "The Reformation of the Mulatto Hero/ Heroine"[13] takes place. Against this critical backdrop, I propose an even earlier and more localized origin for this "reformation," and one that presents another perspective on mixed blood.

I consider the tensions between African blood and colonial money as they are established in England more relevant and insightful for understanding the representation and reformation of the early mulatto heroine. Edward Long's *History of Jamaica* (1774) describes rich mulattoes being sent to England by their wealthy Jamaican fathers in the eighteenth century, and *The Woman of Colour* and Edmund Marshall's *Edmund and Eleonora* (1797) are literary confirmations of this historical fact. The mulatto heiresses Olivia Fairfield and Alicia Seldon display many overt similarities aside from their race, gender, and wealth. They are both Jamaican born, overtly religious, and intellectually accomplished daughters of wealthy white men and enslaved black women with royal African backgrounds. The plots in their novels are also quite similar. At the behests of their fathers' living and dead wills, Olivia and Alicia are forced to travel to England to be educated or to marry. There is a stark difference, however, between these mulatto heiresses and the wealthy women of color seen in George Alexander Stevens's *The Dramatic History of Master Edward, Miss Ann, Mrs. Llwhyddwhuydd, and Others* (1743), Mrs Charles Mathews's *Memoirs of a Scots Heiress* (1791) and Robert Bisset's *Douglas; or, The Highlander* (1800). Where these wealthy women of color are sent to England to partake in authorial acts of narrative negation that stabilize a "proper" system of prejudiced belief for readers, mulatto heiresses are charged with repurposing this system, and encouraging readers to re-view the mixture of African blood and colonial money as a progressive opportunity for Britons to correct their own prejudices. Ultimately, by repurposing Britons' disgust toward African blood as well as their desire to erase it, mulatto heiresses make their most heroic impressions as respectable social justice reformers in eighteenth-century literature.

The eighteenth-century novel and wealthy women of color in England

To understand how wealthy women of color function as figures for social justice as well as prejudice in the eighteenth-century British novel we must, first, review how skin color and complexion operate as separate racial discourses in this period. Under colonial law, the children of slaves legally followed the condition of the mother. This law also pertained to biracial children that enslaved African women bore for white planters. Since the stigmas associated with the African woman's blood are phenotypically as well as legally transferred to her descendants, dark skin color and African identity congeal in black people of all colors to form a physiognomy of inferiority.

Roxann Wheeler's *The Complexion of Race* (2000) complicates our understanding of skin color, however, by highlighting the nuance with which eighteenth-century Britons used the word "complexion." It meant much more to them than mere skin color. Complexion, Wheeler argues, "was connected to familial inheritance and climate but not reducible to them"[14] since it also accounted for one's own unique disposition—one's physiology as well as one's physiognomy. Complexion was often understood independently of skin color, even while being intimately connected to it. For instance, although skin color was fixed and could not change, Wheeler notes that one's "complexion changed over the course of one's life, through age, conditions of living, geographic region, and even according to whether one was a man or a woman."[15]

This fluid, physiological conception of complexion complicates our understanding of the ways in which people born with African blood were seen in the British colonial world. Although stigmatized by their blackness, people of color also had access to this transformative discourse of complexion, which offered opportunities to change their dispositions, and, in turn, redefine how their bodies were seen to function in the world. When rich, white, West Indian planters sent their colored offspring to England, "To learn those arts that high-bred youths are taught,"[16] these ideas about complexion and skin color must have influenced them. Financing an elite education in the metropolis was a way for these men to experiment with complexion—to test whether a change of place and purpose could physiologically alter their children enough for them to claim respectable status by transcending the fixity of their stigmatized skin and the physiognomy of black inferiority. Arranged marriages between mulattoes and whites offered another way for a rich white planter to facilitate his colored offspring's change of complexion and status. Both of these transformations relied on the colonial white man's money as a stimulus for change.

Eighteenth-century novelists are clearly aware of this tension between black skin color as a fixed racial designation determined by an African mother's blood, and the flexibility of black complexions made possible by a white man's money because they utilize it in novels depicting wealthy women of color in England. These women do not appear frequently in literature, and even when they do appear, Felicity Nussbaum asserts that they are "seldom granted personhood or subjectivity."[17] But the few intriguing fictional representations that challenge this trend are broad enough to encompass three distinct female types: the racially ambiguous woman, the unattractive Negro woman, and the beautiful mulatta. These wealthy women of color all have desires to change their complexions and status in England through acquisitions of wealth, husbands, and education—acts that attempt to transform not merely the way their stigmatized bodies are seen in the world, but their very dispositions. I will show, however, that these transformations are actually authorial acts of narrative negation which ultimately present all wealthy women of color as figures of disgust, thereby capturing the novel's role in stabilizing a "proper" system of prejudice and superiority for English readers to live by.

George Alexander Stevens's *The Dramatic History of ... Mrs. Llwhyddwhuydd* is one of the first eighteenth-century novels set in England to feature a female mulattaroon prominently in its plot and title. The novel follows the rising fortunes of the almost unpronounceable Chloe Llwhyddwhuydd as she navigates life, love, marriage, and childbirth. But, before she becomes a wealthy woman in English society, Chloe Justice starts life in much more distant and humble circumstances. "Born in America,"[18] her colonial birthplace is already well represented in literature as a site for criminals, conversion, and race mixing in texts such as *The Widow Ranter* (1690), *Moll Flanders* (1721), and *Colonel Jack* (1722).[19] Indeed, Stevens brings these exact impressions about America to the fore with his revelation that Chloe is "the daughter of a mulatto woman, [and] her papa [is], the famous counselor Justice who was sent abroad on account of some books being missing."[20] These parental stigmas cloud the complexion of this presumed quadroon woman, identifying her as the embodiment of an infamous mix of African blood tainted by an association with American plantation slavery as well as English "Justice" sullied by fraud and criminality. With these inauspicious origins, Chloe starts the novel in the lowly station of "undercook to a West Indian family, who came to England for improvement."[21] In London, she acts deliberately to transcend the stigmas clouding her complexion by orchestrating a conversion from English outsider to insider.

Marriage is one of the strategies she employs to achieve this end. In her bid to marry her landlord, David Llwhyddwhuydd, Chloe uses blatant exaggeration and outright lies to refashion the infamies surrounding her parents' financial and biological pasts. "My papa was a counsellor at law," she tells her friend, Miss Shred,

> "he was prosecuted for his merit, persecuted for the superiority of his talents, and banished from his native country by faction; but I am his own daughter; and can I ever agree to have these arms filled with any body, who is not a gentleman born, or who is not so by profession; or is not a person of consequence."[22]

By presenting herself as the daughter of a persecuted English martyr, this American "undercook" whitewashes the disgust surrounding her father's criminality and audaciously implies that she outranks David since, as a mere landlord, he is not a "person of consequence" or a gentleman "born" or "by profession." Chloe applies this rank-elevating strategy again at the end of the novel when she discusses her mother's "mulatto" bloodline. "*Your* family may be the more ancienter in England," she tells her (now) husband, David, "but *my* family is from America; we were the ancestors of five original nations of the Senekas, from whom came the Chikasas, from whom came the Catawaws, from whom came the Cherokees."[23] The flexibility associated with racial categories in the early eighteenth century makes it conceivable that Chloe's "mulatto" mother could be an Anglo-Indian much like Unca Eliza Winkfield in *The Female American* (1767).[24] But, because she initially whitewashed her father's indiscretion in order to legitimize her own fraudulent claim to genteel status, it stands to reason that Chloe could also be strategically covering up her mother's African bloodline in order to legitimize and improve her own infamous racial connection to

North America. By refashioning herself as the offspring of a persecuted English martyr and esteemed lines of American Indian tribes that span generations, Chloe achieves three notable successes: she whitewashes her associations with African blood and slavery as well as English criminality, she claims rank she has no title to, and she passes herself off as being as genteel as any established white woman in England. In short, she fulfills her desire to change her original complexion as she establishes herself as a respectable, wealthy partner in her husband's thriving public house business.

Readers, however, are encouraged to view with disgust rather than admiration Chloe's passing as a respectable woman of color in England. In their eyes, her scandalous behavior throughout the novel is proof positive of a persistently immoral disposition. Before her marriage to David, Chloe engages in a torrid assignation with Mr Samly, an actor who eventually abandons her. Thereupon, she agrees to marry David, and she makes a "resolution of chastity" to never "grant any other man ... an involuntary embrace,"[25] evidence of an assumed change of disposition. But after their marriage, Chloe immediately lapses, forsakes this resolution, and leaves David to settle into a sexual relationship with the dashing gallant, Corporal Knott. Inconstant to her vows of honor and libidinous with her sexuality, Chloe's scandalous behavior replays the disgusting stigmas associated with her English and mulatto parents, and offers the strongest indication that her original disposition has not changed. Marriage to an Englishman and residence in England might improve her rank, but the same can't be said about her physiology. This presumed quadroon may have the means to pass and prosper as an almost-white wealthy woman in England, but Stevens makes it clear to his readers that she does not have the "proper" respectable English complexion because her mother's and father's blood and status are both tainted. They are both implicated in identifying Chloe as nothing more than a metaphorically black (meaning immoral) almost-white woman in English society.[26] Stevens urges readers to reject Chloe's gentility on the basis of her parents' disgusting dark marks—a narrative negation that encourages them to believe in the superiority of English morality.

The flexibility that Mrs Llwhyddwhuydd has to transform her stigmatized complexion is not available to all wealthy women of color in England. When a wealthy woman of color is coded black or of African descent instead of racially ambiguous, her parental stigmas are more prescriptive and much less transformative, as Robert Bisset's *Douglas* highlights. In this novel, the Negro woman, Mrs Dulman, is introduced as a *"negro-driver's frow* [sic] *fresh from Demarara,"*[27] a Dutch West Indian colony and presumed site of her birth. She arrives in London with "a mint of money from the nigers [sic] in the Vest Indies"[28] and a new white husband whom she displays at the Ranelagh pleasure gardens. Charles, the novel's hero, "cast his eyes on [this] very vulgar couple bedizened out with a most profuse finery."[29] Not merely disgusted with this couple's display of nouveau riche gaudiness, Charles's aunt, Mrs Lighthorse, draws attention to Mrs Dulman's physically unattractive body, describing her as the "lady with the mutton fist."[30] Charles adds to his aunt's impression of Mrs Dulman's muscular brawniness by calling her a "vulgar dowdy, with ... broad shoulders, large face, thick lips, pug nose, and wide nostrils."[31] His repeated use of "vulgar" coupled with his emphasis on the exaggerated dimensions that make up Mrs Dulman's body mark this Negro woman as the complete inverse of all that's feminine, attractive, and white. The "profuse finery" of her dress also adds a class critique to her depiction: Mrs Dulman is just as disgusting a representation of middle-class refinement as she is of white English femininity. Hilary Beckles has discussed the way that depictions of black women in West Indian plantations are deliberately masculinized in order to justify their continued presence and placement there,[32] and Bisset's novel appears to confirm this impression, suggesting that an unattractive, muscular woman with black skin would be better served following the presumed subservient condition of her black parents in Dutch-colonial Demarara rather than that of genteel white women in London, England.

Mrs Dulman, however, is as intent as Chloe Llwhyddwhuydd about improving her complexion in the Atlantic world. She cannot achieve this, as Chloe does, by passing as a respectable, almost-white woman, so she employs another of Chloe's strategies: marrying white men. Mrs Dulman's success in this endeavor can be traced through the names associated with her journey from maidenhood to marriage. Her maiden name, "Dutchsquab," is itself a hybrid combination of two words that identify her physically and geographically. "Squab" in Samuel Johnson's *Dictionary* means "thick and short," suggesting that "Madam Dutchsquab"[33] is an unpleasantly shaped, foreign woman.[34] Yet physical inferiority and foreignness did not prevent her from attracting and marrying the negro-driver, Monsieur Heureux, in the Dutch slave colony where he works. By this marriage, "Madam Dutchsquab" is transformed into *"dame"*[35] Heureux, a French last name meaning "happy," "glad," "lucky," or "fortunate." Certainly, she has had good luck to secure a colonial man of means who transforms the way her stigmatized body is seen in the colony. Her good fortune does not end down in Demarara.

Monsieur Heureux "dying, and leaving her very rich, she married the person with her, Mr. Dulman, Esquire."[36] Despite her physical unattractiveness, then, dame Heureux's financially improved complexion facilitates another upwardly mobile marriage, this time to a white English lawyer or noble as his "Esquire" title implies. Upon "taking a fine house in London ... the *dame* endeavored to learn drawing, music, and all fine accomplishments"[37]—all efforts designed to complete the transformation of her complexion by bestowing on her disposition an air of refinement and respectability.

They are all complete failures. Despite efforts to alter how she is seen and who she is through geographic relocation and accumulations of wealth, female accomplishments, and status through marriage, we are told that "nothing could *whitewash the negro*"[38] in Mrs Dulman. Neither of her marriages improves her disposition. Monsieur Heureux's "Negro driving itself has no great tendency to liberalize the mind and Dutch minds are not the most easily liberalized any more than the Dutch manners are the most easily refined."[39] Her new husband's description as a "poor, mean, pliant creature"[40] offers further confirmation that her current marital connections are equally low and undesirable. And Bisset, himself, negates her current appearance by referring to her with a pejorative mix of *former* marital and rank honorifics ("*negro-driver's frow* (frau)," "*dame* Heureux," "Madam Dutchsquab"), but never calling her by *current* married name (Mrs Dulman). Thus, the wealthy yet unattractive Negro woman in England is exposed and negated as a completely un-transformed and un-transformable figure. Her inferior disposition is fixed by her African parents and reflected in an unattractive body that neither white men nor "a mint of money from the nigers in the Vest Indies" can make respectable. The disgust leveled at Mrs Dulman is the negation by which Bisset stabilizes a belief in the "proper" superiority of white English femininity for his readers.[41]

So far, Chloe and Mrs Dulman have demonstrated that although wealthy women of color have both the desire and the money to transform their complexions, their dispositions are literally or metaphorically as black as, and physically or morally as disgusting as, their parents. The mulatto heiress in Mrs Mathews's *Memoirs of a Scots Heiress* alters these impressions, however. Miranda Vanderparcke is the "niece of a Dutch gentleman who occupied the next house"[42] to the respectable Mrs Semhurst in Harley Street, London. Miranda's father, Mr Vanderparcke, "had settled in Batavia [the capital of the Dutch East Indies, now known as Jakarta], and married a negro woman there."[43] Miranda is introduced to the novel as an orphan whose "riches were immense."[44] As a result of her parents' deaths, she takes up residence with "this uncle ... her faithful guardian, and only relation in England."[45] Miranda's physical description and behavior are the complete inverse of Chloe's and Mrs Dulman's. She is "genteel and elegant in her look and manner,"[46] and "though but eighteen years of age appeared to have reached an uncommon pitch of excellence."[47] She has "great symmetry in her features, and the most pleasing expression in her countenance; she had fine eyes, fine teeth, and long glossy black hair ... her person was exquisitely formed, and her motions highly graceful."[48] And in terms of behavior, "Miranda could do nothing reprehensible while reason was her judge; for she was purity itself, darkly as she was arrayed."[49] By transcending all the physical and moral stigmas that influence her racial counterparts, this beautiful mulatta appears to successfully change her complexion, becoming, in the process, a respectable Africanized figure of desire rather than disgust.

Miranda's geographic and parental origins inform her successful transformation. Gert Oostindie and Bert Paasman have shown that Dutch attitudes toward Africans differed greatly in their East and West Indian colonies:[50] Batavia, where "the Dutch were colonizers without a strong cultural impact, willing to tolerate if not necessarily appreciate the local elites' culture,"[51] produces a very different woman of color than Demarara, where "blatant contempt dominated Dutch and Dutch colonial perspectives towards Africans and their Caribbean offspring."[52] We have already seen the "blatant contempt" leveled at the Dutch West Indian Negro, Mrs Dulman. The toleration of local elites in the Dutch East Indies, however, appears to hold more promise of success for an upwardly mobile Negro woman like Miranda's mother. Such women thrive in this area where "European culture was considered superior to Asian culture, which in turn surpassed African culture"[53] presumably because they are willing to transform themselves and approximate European genteel behavior. As a result, they are able to marry respectable European businessmen like Mr Vanderparcke rather than inferior European negro-drivers like Monsieur Heureux. But, one wonders, does this colonial spirit of toleration extend to wealthy, transformed Batavian women of color who emigrate to European metropoles? In England, is Miranda able to marry someone more respectable than a "poor, mean, pliant" fortune-hunter like Mr Dulman?

The short answer is no. In England, stigmas cloud even Miranda's respectable complexion, the most obvious being that "she was almost a negro"[54]— phrasing that draws explicit attention to her mother's stigmatized blood. Her father's status and colonial money are also stigmatized.

Oostindie and Paasman note that "the Dutch East Indies were known as a place where one could become rich fast (legally or illegally) and where one could send failed and incautious Europeans."[55] Mathews, however, aims her most damaging stigmas squarely at Miranda's disposition. Having been in England for only a year, Miranda knows little English: "Dutch being her native language, she was frequently at a loss to express herself."[56] We are also told, "Miranda's character of mind was as foreign as her hue."[57] This "very extraordinary foreigner,"[58] then, is identified by a series of negations: she cannot speak or look like a traditional British woman and she doesn't even have the capacity to think like one. It is Mrs Semhurst's son, the "very respectable"[59] Mr Cyril, however, who illustrates the degree to which Mathews actually negates this ostensibly desirable "ebon-beauty."[60] Cyril views Miranda with a mix of desire and disgust, claiming that she would be a "complete beauty"[61] if only her skin color (which "wanted some shades of the deep African dye; but ... had passed the degree of copper colour"[62]) could be blanched. *Scots Heiress* resolves his ambivalence not only by killing Miranda off at the marriage altar seconds before she is able to execute her marital vows to Cyril, but also by explicitly referring to this beautiful mulatta as a "negro-bride"[63] in this scene. Reminding readers of Miranda's stigmatized African blood as a bride and then enforcing her death at the altar are authorial acts of negation designed to evoke the Vanderparckes' Batavian interracial marriage and distinguish it from Miranda and Cyril's English equivalent. Europeans in colonial Batavia may accept marriage to a "negro-bride," but in England such behavior is a disgusting and unacceptable breach of respectability that will not be tolerated. The authorial erasure of the beautiful mulatta is a narrative negation that stabilizes readers' "proper" belief in the superiority of pure, working-class whiteness embodied by Mary Hamilton, whom Cyril subsequently pursues.

Just like her wealthy counterparts in other novels, Miranda is ultimately recognized as a disgusting black woman who lacks a "proper" English complexion *because* of her parents' foreign blood and possibly tainted colonial money and *despite* her own desirable attributes. From this repetitive pattern of observations, a clear premise emerges: this literary trope involving wealthy women of color is designed to break down rather than reinforce the distinction between a respectable black complexion and a stigmatized black skin color. Though marriages to white men, displays of wealth, female accomplishments, and relocations to England go some way to achieving Chloe, Miranda, and Mrs Dulman's desires to claim respectable status, eighteenth–century novelists are invested in negating the idea that black women of different colors can change their fundamental dispositions as the period's discourse about complexion would suggest. In order to convincingly stabilize a reader's belief in the "proper" superiority of English femininity, morality, and whiteness, novelists must prove that even the most successful and respectable women of color cannot escape the disgusting stigma of black inferiority attached to their African mothers' blood. In *Mrs. Llwhyddwhuydd*, *Douglas* and *Scots Heiress* physiognomy trumps physiology, and the novel's role in using women of African descent to reinforce prejudices about skin color "as the most important component of racial identity in Britain"[64] becomes clear.

With a father's colonial money and a mother's African blood, mulatto heiresses reform England

While some eighteenth-century novels are intent on proving that a white man's money and status cannot eradicate stigmas associated with an African woman's blood, and that black inferiority is imprinted on dark skin, others are equally as intent on challenging these ideas. *The Woman of Colour* and *Edmund and Eleonora* are both aware of, and invested in correcting, prejudiced attempts to establish black inferiority as fixed and "proper." The mulatto heiresses Olivia Fairfield and Alicia Seldon are employed to do this work by standing at the forefront of authorial attempts to use wealthy women of color in ways that achieve a different transformative end. Instead of representing black women's desires to transform their own complexions, Olivia and Alicia are employed to change the way respectable Britons should see themselves.

This work begins with the different ways Olivia and Alicia are seen in England. Where Alicia is repeatedly called a "young beautiful Mulatto girl" throughout the first volume of her text,[65] Olivia, in her tale, understands that she is "little less than a disgusting object to an Englishman"[66] who is usually "mortified at his folly when he has caught a view of [her] mulatto countenance."[67] The same impression holds true for English women. When she is introduced to her prospective sister-in-law, Mrs Merton, Olivia remarks: "I held out my hand, and that lady was very near taking it in hers; but I fancy its colour disgusted her."[68] Despite having the same complexion, then, mulatto heiresses aren't seen in the same way: some, like Olivia, are objects of disgust, while others, like Alicia, are figures of unabashed desire. I have already mentioned that *Scots Heiress* presents Miranda Vanderparcke as another embodied mix of desire and disgust. But where she is erased to appeal to English readers' prejudices, Alicia and Olivia[69] repurpose desire and disgust by redirecting these terms toward debates about African blood and colonial money, doing so in ways that seek to correct rather than stabilize the "proper" system of prejudiced beliefs that English readers encounter and are encouraged to favor in other texts featuring wealthy women of color.

Mulatto heiresses generate both disgust and desire about their fathers' money and its involvement in the slave trade. In her tale, Olivia becomes emotional when she recounts her father's role as a slave master: "he contented himself ... with seeing that slaves on his estate were well kept and fed, and treated with humanity," she explains, "but their minds were suffered to remain in the dormant state in which he found them!"[70] In her eyes, her father's benevolence cannot belie his failing with respect to his slaves' intellectual stagnation. "My father ... could not give the tone of morals to an island," she repines, "he could not adopt a line of conduct which would draw on him the odium of all his countrymen."[71] By recounting her father's dogged adherence to an oppressive regime that denies slaves the right to improve, Olivia expresses dissatisfaction with the moral impotence of his benevolence as well as disgust toward the Jamaican code of social conduct to which he submits. Alicia's father financially impacts the lives of slaves in a far more decisive way. Not only has "every negro upon his plantation ... received baptism"[72] from a paid clergyman, but he has also "liberally given freedom to every slave upon the plantation."[73] By financially freeing himself from the institution of slavery, paying his Negro workers, and bankrolling their religious salvation, Mr Seldon stands as the correct exemplar of strong and morally principled benevolent paternalism that Mr Fairfield is not. Together, these texts politicize desire and disgust by promoting the belief that abolition is a desirable and respectable use of an English paternalist's money, while the amelioration of slavery is more dissatisfying and less respectable since it financially maintains a morally disgusting status quo.

Admiration for, and dissatisfaction with, the colonial paternalist's use of money also relates directly to Olivia's and Alicia's lives and their freedom of choice in marriage. Mr Fairfield's West Indian estate generates almost 60,000 pounds upon his death.[74] This money is used to arrange Olivia's marriage to her English first cousin, Augustus Merton, a man who is physically described as the "image"[75] of Olivia's father. As an "unportioned girl,"[76] Olivia has no individual share in this money. Augustus will control it if he marries her, and if he does not, his elder brother, George, will get it. Olivia accepts, but is clearly dissatisfied with, this financial arrangement as her dream of financial independence illustrates: "I sometimes think, that had my dear parent left me a decent competence, I could have placed myself in some tranquil nook of my native island."[77] By contrast, Mr Seldon's wealth is specifically used for Alicia's education, personal expenses, marriage, and beyond. Initially, he "had placed a considerable sum of money in the hands of the captain *for her use,* and to pay the expenses of her passage and her education in an English boarding school."[78] This "captain," charged with conveying Alicia from Jamaica to England, actually kidnaps his charge and attempts to rape her. But Alicia escapes and finds a safe haven with a family of benefactors who take charge of her education. To them, Mr Seldon writes "that, as he wishes no proper expense may be spared in the education of Alicia, he has inclosed *for her use* a draft upon a capital banking-house of London for five hundred pounds."[79] Additionally, for her future comfort, Mr Seldon promises Alicia 3000 pounds "whenever she should marry with her mother's and my approbation, and as much more at my death."[80]

The difference here is stark: where the paternalist in *Edmund and Eleonora* allocates funds specifically for his daughter's use and personal improvement, the paternalist in *The Woman of Colour* makes his money more for the use of his white relations. These are politicized as well as instructive positions that each novel takes. In an ostensible effort to protect her fortune after her father's death, Olivia is exchanged endogamously between white men of the same family and features. However, this endogamous exchange proves dissatisfying because it involves a free mulatto woman essentially bequeathed by a white colonial Englishman to other white Englishmen in a system of ownership that continues from the cradle to the grave, essentially mimicking the disgusting transactions peculiar to colonial slavery. In this way, *The Woman of Colour* re-politicizes disgust, encouraging readers to see beyond the stigmas associated with Olivia's mulatto complexion, and to recognize, instead, how disgustingly akin to slavery the institution of marriage looks when money rather than affection is the primary object of exchange. *Edmund and Eleonora* politicizes desire when its mulatto heiress chooses to marry for love rather than for money. It is important that "these good people, the Seldons, who wished only to see their children happy, acquiesced without the least hesitation"[81] to Alicia's choice of Mr Adderley, her older, dancing and music master, for a husband. By choosing to marry exogamously, Alicia goes against the heiress's traditional endogamous role as well as the expectations legally inscribed in the Hardwicke Act of 1754 that a father's will should largely govern a woman of wealth's marital interest.[82] When viewed together, the money that white fathers spend on their mulatto daughters offers a politicized correction to contemporary debates about women's rights and roles in society. Freedom of education and choice in marriage are other desirable and respectable acts that a benevolent paternalist should finance, whereas forcing alliances for money's sake are his most dissatisfying, even disgusting, legacy.

The instructive ways in which Olivia and Alicia generate disgust and desire about their African mothers' blood, however, are even more socially progressive and culturally transformative than ideas about abolition, female education, and marital choice that responses toward their fathers' money encourage respectable readers to believe in. *The Woman of Colour* and *Edmund and Eleonora* are both actively invested in making mulatto complexions visible in English society because they want to perform something that other novels featuring women of color fail to do: de-stigmatize African blood. Olivia becomes involved in this work during her first English ball at Clifton in the coastal, slave-trading city of Bristol, where Englishmen saunter up to gawk at *"Gusty's* [Augustus's] black princess" as if she were an *"untamed savage."*[83] A similar scene of public humiliation and dehumanization occurs in Jamaica when Olivia's mother, Marcia, who "sprung from a race of native kings and heroes" in Africa, is "exhibited on the shores" like a "wild and uncivilized African"[84] for the viewing pleasure and purchase of Mr Fairfield. These two disparate scenes of royal African women put through inhuman displays of public humiliation before two white men who look alike indicate that the shores of Bristol and Jamaica as well as the experiences of mother and daughter are identical. Although Marcia has long since died, her disgusting experience of public humiliation on the slave block is deliberately reincarnated for English readers through her daughter's public appearance in Bristol. *Edmund and Eleonora* uses the trope of reproduction to make a desirable rather than a humiliated African woman visible in England. Alicia's mother is actually the second Mrs Seldon. While his first wife was still alive, Mr Seldon had purchased Alicia's mother in a slave auction when she was ten years old. The first, and presumed white, Mrs Seldon took it upon herself to educate the African girl. Her efforts are so successful that she transforms this "young princess"[85] into an accomplished figure of desire attractive enough to secure Mr Seldon in marriage after his first wife dies. It's no coincidence that this accomplished figure of African desire shares the Christian name "Alicia" with the daughter who resides in England. A white woman in Jamaica produced the first desirable "Alicia Seldon";

but the second is explicitly reproduced in England as an exact replica of her already desirable African mother.

More than mere biological reproductions of their parents, then, Olivia and Alicia are acknowledged as English reincarnations or reproductions of respectable African mothers who, when formerly exhibited as slaves in Jamaica, were colonial figures of humiliation or desire. *Oroonoko* is an important precedent for these female representations since Aphra Behn's royal African hero also experiences humiliation and desire when he is first exhibited as a slave "at the mouth of the river of Surinam."[86] As the most desirable commodity, "Oroonoko was first seized on" by slave traders; yet despite his worth, he also suffers the humiliation of being sold with an undistinguished batch of common slaves—"not one of quality [sold] with him." The heroic manner in which this desirable yet humiliated royal African responds to his fallen state has two narrative effects: forcing "blushes on [the] guilty cheeks" of the white slave trader and presumed outrage in the white reader. In Surinam, enslaved royal African blood is exhibited to recognize the African hero's ability to transform a respectable slave's humiliation and attraction into English emotional reaction.

The exhibition of royal African female blood in England is designed to advance Oroonoko's colonial heroism in a domestic setting. Olivia and Alicia are reproduced or reincarnated forms of respectable African heroines exhibited in ways that prove to be both emotionally reactive and culturally transformative for the English. When Mrs Merton's son, George, runs into the breakfast room screaming that Olivia's Negro maid, Dido, is "dirty"[87] because of her black complexion, Marcia's deceased African blackness shares in this public scene of racial humiliation. George's unfiltered choice of word openly expresses disgust about blackness that the adults in his family share but deem improper to articulate. Olivia, however, uses her "olive"[88] complexion to influence George's views about Dido's color, and, by extension, African women like her mother. Through experimentation, she proves to him that black complexions are not the result of dirt but the fixity of race, as witnessed in this exchange about differing skin colors: "'won't yours and hers (Dido's) rub off' said [George]. 'Try,' said [Olivia] giving him the corner of [her] handkerchief; and to work the little fellow went with all his might."[89] Instead of African blood being used as an opportunity to reinforce a "proper" system of prejudiced beliefs about skin color, Olivia facilitates George's contact with, and reading of, her stigmatized skin in a heroic attempt to silence his emotional outburst and correct his belief system before it becomes as "proper," silent and irredeemably prejudiced as his elders'. Through Olivia's act of domestic heroism, George transforms an initial disgust that humiliates living and dead blacks into a cleansed belief in racial diversity that acknowledges them as being as phenotypically fixed and pure as himself.[90]

The reproduced "Alicia Seldon" also corrects the stigmatized way African blood is commonly acknowledged in England. Alicia is an attractive figure of female accomplishment, demonstrating "the greatest genius for music, drawing, and dancing that Mr. Adderley had ever met with."[91] Neither the English arrivals of her mother, nor her African uncle, the monarchical "Maraboo of Senegal,"[92] negatively impact her female "genius." In fact, these royal African relations actually enhance Alicia's desirableness. Marshall explains that Alicia's benefactors had "the highest opinion of the excellence of Mrs. Seldon's heart,"[93] and her uncle (who favors Oroonoko[94]) had "formed" a "wise and benevolent plan ... for the conversion and civilization of his subjects"[95] in Senegal. This information establishes the reproduced Alicia Seldon's royal African blood and bloodline as both civilized and civilizing, and involves her in a poignant act of domestic heroism. She corrects the prejudicial reaction to skin color promoted in other novels featuring wealthy women of color by proving that a respectable African complexion can trump a stigmatized black skin color, and a female disposition influenced by African blood can be as successful and desirable as any other.

Conclusion

If *Mrs. Llwhyddwhuydd*, *Douglas*, and *Scots Heiress* prove that washing a blackamoor respectable is a fruitless undertaking in England, *The Woman of Colour* and *Edmund and Eleonora* prove that washing a mulatto heiress black (meaning African) is an endeavor worth performing there. Within these novels, African blood is a self-reflexive, rather than self-loathing, political tool specifically designed to shore up rather than dismantle notions of African respectability. Marshall and the anonymous author attempt to confront prejudice and transform readers' beliefs by repurposing disgust and desire as socially instructive and politically progressive reactions to a black mother's African blood and a white father's colonial money—reactions that correct rather than stabilize prejudiced English beliefs about the respectability of marriage, slavery, blackness, female education, and the role of paternalism. But correcting English prejudices about African blood appears to be their larger collective aim. Olivia's tale is designed to "teach one skeptical European to look with compassionate eye toward the despised native of Africa"[96]—a clear indication that the anonymous author uses this mulatto heiress to force a British reader's confrontation with a figure of disgust as a means of correcting national misconceptions about Africans. *Edmund and Eleonora* goes even further in this respect. Having legitimized the English presence of the Africans, Alicia, Mrs Seldon, and the Maraboo, Marshall's novel ends with a startling vision of interracial felicity. Alicia and Mr Adderley have "several children of either sex," and a thriving ferme ornée in England where "everything flourished in their hands";[97] Mr and Mrs Seldon move back to Jamaica where their sons become "excellent young men ... married well ... and were happy";[98] and Alicia's uncle, the Maraboo, makes Mrs Tomlyn (Alicia's white French instructor) his "African Princess"[99] who "blessed him with two sons, and two daughters"[100] in Africa. In these productive and reproductive examples of interracial felicity, presented well before the abolitions of the slave trade and slavery, Marshall corrects Britons' beliefs in biological and social degeneration, encouraging them not to fear a nation diminished and weakened by the appearance of African blood, but rather to foresee the desirable attractions of an entire British empire emotionally connected and biologically strengthened by its full, free, and equal inclusion.

Notes

1. Elsie B. Michie, *The Vulgarity of Money: Heiresses, Materialism, and the Novel of Manners from Jane Austen to Henry James* (Baltimore: Johns Hopkins University Press, 2011), 17.
2. Ibid., 16.
3. Jennifer DeVere Brody, *Impossible Purities: Blackness, Femininity, and Victorian Culture* (Durham, NC: Duke University Press, 1998), 18.
4. Ibid., 15–16.
5. Ibid., 18.
6. Michie, *Vulgarity*, 2.
7. Brody, *Impossible Purities*, 18.
8. Michie, *Vulgarity*, 10.
9. Michie's silence is striking because her book discusses Jane Austen's oeuvre without mentioning the provocative depiction of Miss Lambe, the "half Mulatto" heiress in Austen's *Sanditon*. See Sara Salih's "The Silence of Miss Lambe: *Sanditon* and Contextual Fictions of 'Race' in the Abolition Era," *Eighteenth Century Fiction* 18 (2006), 329–53.
10. See "Marriage, Property, and the Common Law," in Lawrence Stone's *Uncertain Union and Broken Lives* (Oxford University Press, 1995), 18–19 for a brief discussion about the financial negotiations between families leading up to and after a daughter's marriage.
11. Brody, *Impossible Purities*, 17.
12. See Lynda Boose, "The Getting of a Lawful Race," in *Women, "Race," and Writing in the Early Modern Period* (New York: Routledge, 1994), 43–53 for more early modern examples of biologically threatening black women.

13. "Split at the Root: The Reformation of the Mulatto Hero/Heroine in Frances E. W. Harper's *Iola Leroy*," *AmeriQuests* 6 (2008), http://ejournals.library.vanderbilt.edu/ojs/index.php/ameriquests/article/view/154/173.
14. Roxann Wheeler, *The Complexion of Race: Categories of Difference in Eighteenth-Century British Culture* (Philadelphia: University of Pennsylvania Press, 2000), 22.
15. Ibid.
16. Edward Long, *The History of Jamaica* (London: Lowndes, 1774), II.xiii, 329.
17. "Women and Race: 'a difference of complexion,'" in *Women and Literature in Britain 1700–1800*, ed. Vivien Jones (Cambridge University Press, 2000), 69–90, 77.
18. George Alexander Stevens, *The Dramatic History of Master Edward, Miss Ann, Mrs. Llwhyddwhuydd, and Others* (London: Waller, 1743), 13.
19. See Dennis Todd's *Defoe's America* (Cambridge University Press, 2010), ix, for a very brief explanation of how America features in the debates about race, conversion, slaves, and slavery in the decades between Behn's death and the publications of Defoe's first novels.
20. Stevens, *Dramatic History*, 13.
21. Ibid.
22. Ibid., 29.
23. Ibid., 187–8.
24. See the Introduction to Wheeler's *The Complexion of Race* for a discussion of the elasticity associated with racial designations. For more on Unca Eliza Winkfield see ibid., 167–73, especially where she asserts that *The Female American* "promotes the notion that however unsettling dark color may be, it is ultimately insignificant" (168). This idea contrasts with my understanding of complexion in *Mrs. Llwhyddwhuydd*.
25. Stevens, *Dramatic History*, 31.
26. Bellamora from Aphra Behn's *The Adventure of the Black Lady* (1697) is another interesting example of a metaphorically black (immoral) white woman in London.
27. Robert Bisset, *Douglas; or, the Highlander* (London: Anti Jacobin Press, 1800), I.312.
28. Ibid., II.133.
29. Ibid., I.311.
30. Ibid.
31. Ibid.
32. Hilary Beckles, *Centering Woman: Gender Discourses in Caribbean Slave Society* (Kingston: Ian Rand, 1999), xx.
33. Bisset, *Douglas*, II.133.
34. In *The Woman of Colour*, Dido, Olivia Fairfield's Negro maid, lists "Squabby" as one of the derogatory epithets used against her in England. See Anonymous, *The Woman of Colour, A Tale*, ed. Lyndon J. Dominique (Peterborough: Broadview, 2007), 100.
35. Bisset, *Douglas*, I.312.
36. Ibid.
37. Ibid.
38. Ibid.
39. Ibid.

40. Ibid., I.312–13.
41. Bisset's condemnation of Mrs Dulman has many overtones with Thomas Jefferson's famous condemnation of black people in *Notes on the State of Virginia* (1785).
42. Mrs Charles Mathews, *Memoirs of a Scots Heiress* (London: Hookham, 1791), II.184.
43. Ibid.
44. Ibid., II.184–5.
45. Ibid., II.185.
46. Ibid., II.183.
47. Ibid., II.185.
48. Ibid., II.183–4.
49. Ibid., II.194.
50. Gert Oostindie and Bert Paasman, "Dutch Attitudes towards Colonial Empires, Indigenous Cultures, and Slaves," *Eighteenth-Century Studies* 31 (1998), 349–55.
51. Ibid., 354.
52. Ibid., 353.
53. Ibid., 351.
54. Mathews, *Memoirs*, II.184.
55. Oostindie and Paasman, "Dutch Attitudes," 351–2.
56. Mathews, *Memoirs*, II.185.
57. Ibid., II.190.
58. Ibid., II.183.
59. Ibid., II.184.
60. Ibid., II.185.
61. Ibid., II.219.
62. Ibid., II.184.
63. Ibid., II.235.
64. Wheeler, *Complexion*, 9.
65. Edmund Marshall, *Edmund and Eleonora; or Memoirs of the Houses of Summerfield and Gretton* (London: Stockdale, 1797), I.98, 99, 122, 131, 145.
66. Anonymous, *Woman of Colour*, 90.
67. Ibid., 83.
68. Ibid., 71.
69. See Sara Salih's *Representing Mixed Race in Jamaica and England* (New York: Routledge, 2011), 70–82 for a wonderful comparative reading of Miranda and Olivia. Undoubtedly, Alicia adds another layer of complexity about mulatto women in eighteenth-century England and English literature.
70. Anonymous, *Woman of Colour*, 55.
71. Ibid.
72. Marshall, *Edmund and Eleonora*, I.109.
73. Ibid.
74. According to the Economic History Association's purchasing power index calculator (eh.net), £60,000 in 1808 is equivalent to £3,650,000 in 2010 income/wealth.
75. Anonymous, *Woman of Colour*, 59.
76. Ibid., 56.
77. Ibid., 55–6.

78. Marshall, *Edmund and Eleonora*, I.96 (my emphasis).
79. Ibid., I.128 (my emphasis).
80. Ibid., I.180–1.
81. Ibid., I.180.
82. See my *Imoinda's Shade* (Columbus: Ohio State University Press, 2012), 33–4 for a brief discussion of the paternalist bias inherent in Hardwicke's 1754 Marriage Act.
83. Anonymous, *Woman of Colour*, 85.
84. Ibid., 54–5.
85. Marshall, *Edmund and Eleonora*, I.125.
86. Aphra Behn, *Oroonoko: or, The Royal Slave. A True History*, ed. Joanna Lipking (New York: Norton, 1997), all quotes 34.
87. Anonymous, *Woman of Colour*, 78.
88. Ibid., 53.
89. Ibid., 79.
90. Also see Dominique, *Imoinda's Shade*, 254–5 for an explanation of the way Olivia's transformation of George Merton can be read as the anonymous author's attempt to place a mulatto woman in a politically instructive position over the English monarch, George III.
91. Marshall, *Edmund and Eleonora*, I.102.
92. Ibid., I.123.
93. Ibid., I.122.
94. Ibid., I.149.
95. Ibid., I.244.
96. Anonymous, *Woman of Colour*, 189.
97. Marshall, *Edmund and Eleonora*, II.368.
98. Ibid.
99. Ibid., I.181.
100. Ibid., II.367.

Part II
Moral Constitution

5
"His blood be on us and on our children": Medieval Theology and the Demise of Jewish Somatic Inferiority in Early Modern England

M. Lindsay Kaplan

Theology inflects and promotes the importance of blood across a range of discourses in the culture of medieval Europe.[1] Among them arises an association of blood with human difference in religious and medical texts that distinguishes male Jewish bodies less in terms of blood lineage, as is the case in the later Iberian context, than in terms of a hereditary bleeding disease. Christian exegetical writings of the thirteenth century represent contemporary Jews as punished with a periodic bleeding resulting from their ancestors' alleged role in the crucifixion. The biblical prooftexts of Matthew 27:25, "His blood be on us and on our children," and the mark of Cain (Genesis 4:15) confirm the idea of a hereditary cursed disability.[2] This theological concept migrates into "scientific" and medical discourses of the thirteenth and fourteenth centuries, where it is explained as hemorrhoidal or menstrual bleeding. The expanded interest in this idea coincides with contemporary efforts to translate the spiritual doctrine of Jewish inferiority into the social and legal spheres of medieval Europe. The Church's frustration at the failure to implement and enforce such a hierarchy contributes to the construction of a cursed, bleeding Jewish body that attempts to render the subjection of the Jews in "real," material terms.

The medieval idea of Jewish bleeding continues to circulate in the early modern period; several scholars have cited this evidence to argue that it powerfully influences sixteenth- and seventeenth-century English representations of Jews. However, a closer examination of discussions of this phenomenon in early modern English primary sources reveals that they largely ignore or discredit the notion. The theological and political circumstances that produced the concept of Jewish corporeal difference no longer operate in the early modern English context. Its decreasing significance can be explained by the perception that Jews no longer pose a threat to Christians, a function of the disempowered status of the small Jewish community dwelling as aliens in early modern England. When Jewish inferiority no longer needs to be fixed in a somatic locus, Jewish bleeding loses its cultural usefulness and salience.

* * *

The notion that God inflicted a physical mark or debility upon the Jews as punishment for the crime of killing Jesus develops, initially, in theological texts. One form that this sign frequently takes is that of a bleeding illness that occurs at regular periods. An early example appears in an exemplum, or sermon story, of Caesarius of Heisterbach (d. 1240); he describes how the Jews "in a certain city of England" experience an annual bleeding: "in the night of the sixth day, which precedes Easter ... the Jews are said to labor under a sickness called a bloody flux, with which they are so much occupied, that they can scarcely pay attention to anything else at that time."[3] While neither the cause nor the site of the bleeding is specified, the timing of this infirmity coincides with the evening of Good Friday, the day on which the crucifixion is commemorated, thus linking the illness of the Jews with the death of Jesus. Other authors are more explicit in describing the nature of the bleeding and explaining it as a punishment that resulted from the crucifixion of Jesus. In addition to Matthew 27, a version of Psalm 78:66 is also cited to provide testimony on the nature of the bleeding: "And [God] smote his enemies in the hinder partes, and put them to a perpetuall shame."[4] In his thirteenth-century *History of Jerusalem*, Jacques de Vitry (d. 1240) employs both verses to articulate the idea of Jewish bleeding:

> All of those Jews whose fathers called out: "His blood [be] upon us and upon our children", they are dispersed generally through the whole world ... they are everywhere slaves, everywhere subjects ... Indeed, unwarlike and weak they have become, like women. Whence, it is said, they suffer a flow of blood each month. God smote them on their bottoms and put them in eternal shame [Psalm 78:66]. Indeed after they killed Abel ... they became rovers and fugitives on the earth (Gen 4:12) just as the curse of Cain, that is, having a trembling head, a terrified heart, fearing day and night, not believing their life.[5]

De Vitry represents the consequences of the Jews' participation in the crucifixion as exile and slavery. Incorporating Cain's murder of Abel as a figure for Jews crucifying Jesus, Jacques extends the idea of a hereditary servitude to include the physical consequences of weakness, trembling, and, drawing on the Psalm verses, anal monthly bleeding for the descendants of those who cried out in Matthew's text. Their weakness is gendered as feminine and their bleeding follows a menstrual cycle. The condition embodies their servitude by rendering them inferior, like women, and in shaming them.

This construction of the Jewish body as cursed with unnatural bleeding appears in texts by other Christian theologians. The Dominican cardinal Hugh of St Cher (d. 1263) extends the bleeding affliction to all Jews, not only the descendants of those who accepted responsibility at the time of the crucifixion. *"He smote his enemies in their posteriors ... And some say that the Jews bear this shame, that in vengeance for the passion of the Lord they suffer a flux of blood."*[6] While the frequency of the bleeding is not specified, its cause, punishment for the crucifixion of Jesus, is cited. The effect of shame caused by this punishment implies a humiliating inferiority. Thomas of Cantimpré (d. 1263) restricts this bleeding to Good Friday, and directly references the crucifixion and the verse from Matthew:

> the impious Jews cried out: "His blood is on us and on our children" (Matt. 27:25) ... because of the curse of the ancestors a vein of villainy runs in the children up to the present, by means of a defect of their blood. And by this inconvenient flow, the impious progeny are incurably afflicted, until such time as the sinner, repenting, acknowledges the blood of Christ and is healed.[7]

This account posits a congenital defect in the blood, brought on by the Jews' participation in the crucifixion, as the source of the bleeding disease. While Thomas suggests that the disease is incurable, it nevertheless seems to be healed by conversion through acknowledging the efficacy of Jesus's sacrifice. However, the condition constructs the Jews' subordination through the text's references to Jewish blood as defective and criminal.

These examples clearly rely on theological texts and concepts to invent a Jewish physical difference that is simultaneously an illness, a punishment, and a distinguishing sign of God's disfavor. While not always explicitly linked to perpetual servitude, this debilitating condition is consistently connected to the crucifixion, the event that sealed the Jews' subjection to Christians in Augustine's authoritative view. In addition, the authors suggest the shaming nature of this illness, implying Jewish abjection. This nexus of ideas alleging a distinct hereditary physical infirmity functions to subordinate Jews to Christians. Its subsequent adoption by authors of texts on natural philosophy and medicine demonstrates the influence of theological concepts on the "scientific" construction of the Jewish body.[8]

It is instructive to consider briefly the tradition of scientific discourses on anal, particularly hemorrhoidal, bleeding, in order to understand how it was utilized in combination with religious formulations. Classical medical opinion associates hemorrhoids with the melancholy humor, beginning with Hippocrates's *Aphorisms*: "XI. Hemorrhoids supervening on melancholic or kidney affections are a good sign."[9] A surplus of melancholy, which can operate like a disease, is relieved by the onset of hemorrhoids, thus the designation of this condition as a "good sign," not a disease. Avicenna, the influential early eleventh-century CE Muslim physician, also correlates black bile and hemorrhoids, but views the latter as a disease caused by the excess of melancholy. This disorder is understood as the result of eating food considered coarse, dense, or difficult to digest; he therefore recommends a change in diet to effect a cure.[10] As for the illness itself, John Arderne sums up earlier opinions in his account of hemorrhoids in his treatise *Fistula in Ano* (1376). For Arderne, hemorrhoids have one, definitive cause:

> Emeroideȝ ar caused of malencolious blode, which is þe fece [feces] of clene blode aboundand in our body; which blode, forsoþ, for his yuel qualitè and odious to nature, discretyue vertu enforceþ for to cast out to þe helpyng of al þe body ... If, forsoþ, þe blode brist out it is called þe emoroydeȝ; but if þat it flowe temperately it doþ many helpyngs and preserueþ þe body fro many sekeneȝ.[11]

Arderne suggests that the melancholy humor is excrement carried in the blood that, owing to its evil and odious nature, must be discharged from the body. The process of expelling this blood corrupted with black bile creates hemorrhoids. Arderne sees hemorrhoids as a disease produced by an excess of this humor, but also one that when alleviated by moderate anal bleeding prevents the development of other diseases.

None of the discourses on bodily function in classical and early medieval scientific texts single out Jews as subject to a distinct type of bleeding; this idea only appears subsequent to the theological development of this Jewish condition. In his thirteenth-century lectures on Aristotle's *De Animalibus*, Albert the Great provides an analysis of hemorrhoids that includes an account of Jewish bleeding absent from prior medical and "scientific" discussions:

> hemorrhoids are caused by a superfluity of thick blood ... Therefore, this occurs by nature particularly among those living on gross and salty nutriment, like the Jews. And because this blood is thick and has an earthy nature, the moon does not have dominion over its flow in the way it does over the menses.[12]

Here, the idea of excessive Jewish bleeding is naturalized and explained in terms of diet, following the opinion of Avicenna, though Albert makes no explicit mention of melancholy here. While the origin for this idea of anomalous Jewish body function is theological, the author is concerned to explain its "natural" cause, and thus omits the religious discourse from his discussion. In providing a natural explanation, he also implicitly dismisses the association of Jewish bleeding with menstruation offered in Jacques de Vitry's account. Although Albert singles out Jews to explain this phenomenon, he does not suggest that their body function differs significantly from that of Christians.

Concurrent with and subsequent to Albert's lectures, medieval universities held quodlibets in which questions were debated on a range of theological and "scientific" topics; the latter were considered in the faculty of arts or medicine.[13] An early fourteenth-century debate by the faculty of arts at the University of Paris took up the question of whether Jews suffered from a particular and distinct form of bleeding:

> [First response] no because Christians and some Jews are of the same complexion ... [Second response]: the opposite appears in truth, because for the most part, [the latter] suffer ... a flux of blood of the haemorrhoids, ... [which] is caused by gross indigested blood which nature purges ... [T]he melancholic shuns dwelling and assembling with others ... Jews naturally withdraw themselves from society ... therefore they are melancholics. Item, they are pallid, therefore they are of melancholic complexion. Item, they are naturally timid, and these three are the contingent properties of melancholics, as Hippocrates says. But he who is melancholic has a lot of melancholic blood, and manifestly must have a flux of blood, but Jews are of this sort ... because they use roast foods ... and these are difficult to digest.[14]

The text hones in on the question of Jewish difference that Albert leaves unexplored. The author first proposes that Jews and Christians possess the same complexion, but rejects this position on the basis that Jews suffer a flow of blood, implying that they do not share a complexion with Christians. The flow of blood is identified as hemorrhoidal and associated with the melancholy humor, identified as a Jewish complexion. The qualities of melancholics here attributed to Jews – shunning society, fear, and pallor – are commonly found in classical medical and natural philosophy texts, which, however, omit any reference to Jews. In this quodlibet they are identified as "natural" conditions of Jews. Dietary factors identified in Avicenna's and Albert's works are also brought in to explain the excess of melancholy in Jews. The debate concludes with the statement that some Christians also suffer a flux of blood, but explains vaguely that "they have some aids whereby they repel" it.[15] This discussion attempts to identify and explain the natural causes of a distinct Jewish bleeding; however, the received and empirical evidence considered ultimately fails to distinguish Jews from Christians in somatic terms. This scientific discourse alone does not formulate the necessary concepts to construct the distinct and inferior Jewish body.

However, subsequent texts considering Jewish bleeding in "scientific" terms adduce its theological explanation in order to suggest a somatic difference between Christians and Jews. Here we see an explicit combining of exegetical and medical ideas on the subject, suggesting that the supernatural is at work in the natural realm. A later commentary on the late thirteenth- or early fourteenth-century *De Secretis Mulierum* introduces, in the context of a discussion on menstruation, a theological explanation for Jewish bleeding:[16]

> Melancholic males generate a good deal of black bile which is directed to ... veins located around the last intestine which are called hemorrhoids. After these veins are filled they are purged of the bile by this flow, which if it is moderate, is beneficial. This is found in Jews more than in others, for their natures are more melancholic, although it is said that they have this flow because of a miracle of God, and there is no doubt that this is true.[17]

The physical description of the development of hemorrhoids follows Albert in part but omits discussion of diet and focuses instead on the susceptibility of men possessing a melancholy complexion, following Avicenna's view. Jews initially seem to differ only in quantity, not quality, from Christian melancholics. However, their condition is given a supernatural etiology, a miracle of God. Jews experience this condition more than others, but the distinct cause differentiates their bodies from

Christian ones. While the author does not provide a reason for this miracle, its citation augments the "scientific" discussion and enables a construction of Jewish bodily difference.

Other texts considering Jewish bleeding in "scientific" terms expand upon its theological origin in articulating the reason for this condition. Bernard of Gordon, a member of the medical faculty of the University of Montpellier, cited both causes in his early fourteenth-century *Lilium Medicinae*:

> Note ... that Jews for the most part suffer a flux of haemorrhoids for three reasons. Generally they are in [a state of] idleness, and for this reason superfluities of melancholy are gathered. Secondly, they are generally in [a state of] fear and anxiety, and for this reason melancholic blood is multiplied, according to ... Hippocrates ... Thirdly, this [occurs] because of divine punishment [according to {Psalms 77.66:} And he struck in the bottom of the back; he consigned them to eternal shame.][18]

Social and scientific reasons are first offered for this Jewish condition, again drawing from empirical and textual sources. However, the third reason offered provides the explanatory cause: Jewish idleness and fear are a result of divine punishment; this is supported with the familiar prooftext from Psalm 77:66. These statements present what is in fact a suppressed tautology: religious ideas conjure a chimerical Jewish difference that is articulated in terms of cultural and "biological" phenomena, which in turn is explained in theological terms. This construction functions both as a projection onto and a lens for reading Jewish bodies as rendered inferior through the punishment and its attendant shame.

The *Problemata varia anatomica*, also known as *Omnes homines*,[19] an anonymous text probably contemporary with Gordon's *Lililium* and the Paris quodlibet, offers the most transparent account of the theological origin of this Jewish malady:

> Why do Jews indiscriminately suffer this flux? One should reply first of all theologically, because at the time of Christ's passion they cried out, "Let his blood {be upon us and upon our children." For that reason it says in Psalms: "He struck them in their posteriors"} [The text considers other contributing factors, such as diet and lack of exercise.] But because the Jews are not in work or motion nor in converse with men, and also because they are in great fear because we avenge the passion of Christ our redeemer—all these things produce

coldness and impede digestion. For this reason much melancholic blood is generated in them, which is expelled or purged in them at the menstrual time.[20]

Initially, the text raises the question why Jews, like Christians, "indiscriminately" experience this bleeding. The specter of Jewish parity, however, is immediately dismissed with the introduction of the primary cause establishing Jewish difference: the theological explanation. Here we see a synthesis of most of the texts considered above: divine punishment for the death of Jesus in the form of anal bleeding, the generation of melancholy blood—explained as a result of fear of Christian revenge—and the association with menses. Again, the Jewish body is characterized by its abjection: subject to humiliating bleeding as specified in the Psalms text, likened to the menstruating female body, and cowed in fear of Christian violence. While "natural" causes for Jewish bleeding are offered, theology nevertheless provides the divine origin of this condition that constructs and secures the inferior Jewish body.

The subjection of Jews, as implied by the logic of supersession in Christian theology, animates the idea of a cursed, bleeding Jewish body. New Testament texts articulate figures of Jewish spiritual inferiority and hereditary guilt that provide the rationale for later medieval attempts to subordinate Jews in social and political terms, as well as corporeal ones. Paul draws on the text of Genesis 21 to present the conflict between Hagar and Sarah as an allegory for the Jews' supersession by the Christians, articulated in terms of servitude.[21] Hagar, signifying "Jerusalem which now is; ... is in bondage with her children," the Jews (Galatians 4:25). Sarah, who figures the "Jerusalem which is above, is free"; her offspring are the Christians (4:26, 31).[22] Here Paul represents the Jews as slaves, born of a slave woman, in contrast to free Christians. The Gospel of John connects the idea of Jewish servitude to sin in Jesus's accusation of his Jewish followers: "whosoever committeth sin, is the servant of sinne" (John 8:34).[23] The Gospel of Matthew introduces the important concept of congenital guilt; when Pilate refuses to preside over Jesus's crucifixion, the Jews accept the role: "Then the people as a whole answered, 'His blood be on us and on our children!'" (27:25).[24] Here blood operates as a metaphor; the Jewish people as a whole take responsibility for Jesus's death, implicating themselves and their children as well, although the text does not link this culpability directly to subjection.

Augustine synthesizes these ideas to develop a notion of hereditary subjection that affects contemporary Jews.[25] The *Reply to Faustus* draws on the Genesis narratives of Cain and Ham, combining them with ideas about Jewish servitude and guilt nascent in the texts of Paul, John, and Matthew to establish his premise. He relies on John's text to elaborate his discussion of Cain as denoting the Jews:

> If Cain had obeyed God ... he would have ruled over his sin, instead of acting as the servant of sin in killing his innocent brother. So also the Jews ... in subjection to sin reigning in their mortal body, ... have been inflamed with hatred against [Abel/Jesus] ... Christ, the head of the younger people, is killed by the elder people of the Jews.[26]

Here we see the New Testament topoi of elder and younger brothers[27] and servitude to sin incorporated into Augustine's thinking. The Jews become servants or slaves to sin through their alleged participation in the crucifixion of Jesus. However, this action has consequences not only for the Jews implicated at the time of the crime, but for contemporary Jews as well. Augustine reads Cain's punishment as signifying the subsequent exile of the Jews.

> in every land where the Jews are scattered, they mourn for the loss of their kingdom, and are in terrified subjection to the immensely superior number of Christians ... the continued preservation of the Jews will be a proof ... of the subjection merited by those who, in the pride of their kingdom, put the Lord to death.[28]

The hereditary guilt for the crucifixion results in the Jews' continued subjugation, not only spiritually, but politically, as they are dispersed among other nations and ruled over by the Christians; these ideas clearly influenced de Vitry's account of Jewish bleeding. The *Reply to Faustus* similarly employs Noah's son Ham as a figure for hereditary Jewish servitude:

> [Ham] is the Jewish people ... [who] saw the nakedness of their father, because they consented to Christ's death ... And thus they are the servants of their brethren. For what else is this nation now ...?[29]

Again, Augustine does not only condemn the transgressing Jews who lived at the time of the crucifixion, but claims that contemporary Jews remain slaves to Christians.

In his development of Augustine's discussion of Ham, Isidore of Seville brings in the verse from Matthew as a prooftext in support of the hereditary nature of Jewish servitude. He notes that Noah cursed Ham's offspring to servitude:

> Ham sinning, his offspring were damned; it signifies insofar as the reprobates here indeed offend, but in the coming generation, that is, in the future, they receive the sentence of condemnation. And in the same way the Jewish people, who crucified the Lord, even now transmit the penalty of their damnation to their children. For they said: "His blood be upon us and upon our children" (Matthew 27:25).[30]

Isidore uses the quotation from Matthew to explain how the curse passed from the Jews who participated in the crucifixion to their offspring living "even now." The punishment of slavery is inflicted on both the perpetrators and their children, contemporary Jews. The figure of Cain and the prooftext from Matthew variously appear as evidence in subsequent medieval articulations of Jewish inferiority, including the texts delineating Jewish bleeding.

The notion of Jew as spiritual slave becomes institutionalized in ecclesiastical law as well as in state regulations throughout medieval Europe. As Shlomo Simonsohn notes, "The theological concept of Jewish subjugation, inferiority and servitude, while not legal per se, carried weight inside and outside the Church; ... canon law actually includes some papal pronouncements describing the Jews as slaves of the Christians."[31] It first appears in the form of law in the Seventeenth Council of Toledo (694) which resolves that the Jews "should be ... subjected to perpetual slavery, granted to those people whom [the king] shall order them to serve, and they should remain dispersed everywhere."[32] The theological justification for Jewish servitude is not given here, but the order that they be subjected and dispersed seems to echo Augustine's use of the Cain narrative to explain God's punishment of the Jews. The phrase "perpetual servitude" indicates the hereditary quality of this condition, also intimated in the figural interpretations of Cain and Ham. Pope Innocent III employs this concept of perpetual servitude in a letter of 1205, which includes his request of the French king to suppress the Jews:

> the Jews ... by their own guilt, are consigned to perpetual servitude because they crucified the Lord ... [In response to reports of Jewish acts against the Christian faith] ... We, therefore, asked ... [the king] to restrain the excesses of the Jews that they shall not dare raise their neck, bowed under the yoke of perpetual slavery, against the reverence of the Christian Faith.[33]

The divine nature of the Jews' sentence of servitude and their perceived insults to the Christian faith authorize the Pope to order secular leaders to exercise dominion over Jews and punish their transgressions.

However, authorities both inside and outside the Church cite Jewish subservience with only intermittent success, despite its divine origin, in their attempts to realize a social hierarchy which establishes Christian superiority. In a subsequent letter (1208), Innocent laments the failure of secular rulers to ensure Jewish servitude, drawing explicitly from figures and texts set out in patristic writings:

> The Lord made Cain a wanderer ... but set a mark upon him ... Thus the Jews ... as wanderers ought they to remain upon the earth ... [and] be forced into the servitude of which they made themselves deserving when they raised sacrilegious hands against [Jesus] ... thus calling down His blood upon themselves and upon their children.[34]

Innocent figures Cain and his mark as Jews who have become wanderers and slaves as a result of the crucifixion.[35] He iterates the hereditary nature of this punishment, as Isidore did, in the indirect quotation of Matthew 27:25. However, Jews "ought" to be wanderers and must be "forced" into their servile status, an indication that they are not properly inhabiting their inferiority. The Pope synthesizes these elements—Cain's mark, hereditary servitude, the blood curse—in order to emphasize Jewish subjugation which has been neglected by temporal rulers.[36] By the beginning of the thirteenth century, the idea of Jewish subordination to Christians is a widely, albeit anxiously, held belief. The frequent papal reminders to secular rulers and repeated issuing of conciliar laws bear testimony to their failure to prevent Jews from exercising authority over Christians.[37]

One strategy to position Jews as distinct and inferior to Christians was the requirement that the former wear a distinguishing badge, as promulgated in Canon 68 of the Fourth Lateran Council. Although the ostensible rationale for this ruling was to prevent mixing between Christians and infidels, its implementation marked only the latter, and in a derogatory manner.[38] Evidence that the Jews felt the stigmatizing nature of this distinction can be found in records of their efforts to negotiate or pay for release from the order.[39] On the other side appears an almost hysterical effort to enforce this law; in England alone, the first country to implement the canon, it was enacted at least six times between 1218 and 1287.[40] Based on this evidence, Geraldine Heng argues that:

> Canon 68 thus in effect instates racial regime, and racial governance, in the Latin West through the force of law ... The coalescence of

England's identity as a national body ... pivoted on the politico-legal emergence of visible and undifferentiated Jewish minority into race, under forms of racial governance supported by political Christianity, and sustained through the mobilization of affective communities enlisted through stories of race.[41]

Heng understands the construction of a "visible" Jewish racial identity as serving to create a unified Christian English identity. While she also notes the various "biomarkers" by which medieval "Jews were systematically defined and set apart," she does not focus on the shaping role the theological concept of inferiority plays in this process.[42]

However, any consideration of the medieval emergence of Jewish embodied difference needs to take into account the influence of religious ideas. While religious identity has traditionally been understood as culturally rather than corporally constructed, Ania Loomba has recently argued that

> the history of racial formations testifies not to a neat separation between these categories [faith and the body] but to their deep interconnection; without such interconnection we cannot understand the very development of what is now referred to as "scientific" racism. Indeed, "biology" itself has a history, and one that is profoundly shaped by the history of racial ideologies. The separation of "biology" from "culture" is the outcome of this very history. A mechanical and historically uninflected understanding of "biology" obscures the complex contours and genealogies of racial ideologies and reinforces the divisions between "culture" and "nature," which were not established in an earlier period and should not be understood as absolute in our own ... "biology" and "culture" should be understood as terms whose meaning fluctuates contextually and in relation to one another.[43]

While medieval "scientific" thinking distinguished itself from theology, Christianity pervaded the culture of the period; the clergy staffed and attended university courses in natural philosophy and medicine. The production of the bleeding Jewish body out of exegetical texts serves as evidence for Loomba's argument that cultural assumptions influence the shape of scientific knowledge. The theological trope of Jewish servitude powerfully influences medieval attempts to imagine a Jewish physical difference that would materialize the inferior status already denoted in spiritual and legal discourses.

The concept of a Jewish body bearing a divine mark of subjugation emerges in the thirteenth and fourteenth centuries, just subsequent to the Church's increased emphasis on Jewish inferiority and its attempts to subordinate Jews to Christians in social and legal contexts. Jewish servitude provides the logic for the imposition of a badge intended to distinguish and disparage Jews; the unreliability of this detachable sign might have inspired the formulation of a corporeal one. The notion of Jewish bleeding offers an ideal solution: the disease itself functioned, in its embarrassing menstrual/hemorrhoidal form, to demean Jews. Although virtually invisible, its divine origin secured its "reality." The authoritative discourses of theology and natural philosophy were marshaled to provide indisputable evidence for the abjected Jewish body at a time when both ecclesiastical and secular authorities struggled with the challenges of implementing Jewish subordination.

The continued acceptance of this idea in early modern England would seem evident, given its circulation in the form of numerous editions of *Problemata varia anatomica*, published under the title *Problemata Aristotelis* or *The Problemes of Aristotle*. At least 12 editions were printed in London between 1583 and 1684, the first in Latin, with subsequent versions in English.[44] In addition to this republication of a medieval text, Thomas Calvert's 1648 *The blessed Jew of Marocco* also draws on the work of earlier authors that refers to the curse of bleeding visited upon the Jews. James Shapiro notes that versions of this idea circulated in sixteenth- and seventeenth-century European texts.[45] David Katz bases his reading of Shakespeare's *Merchant of Venice* on the supposition that Calvert's text reflects a widespread acceptance of this notion in early modern England.[46] Janet Adelman similarly assumes the continued belief in the veracity of the Jewish bleeding illness, relying on evidence from the popular *Problems of Aristotle*, in her reading of *Merchant*.[47] Both she and Katz argue that the play deploys this idea to "recuperate blood difference" between Christians and Jews.[48] Most recently, Irven Resnick has argued, again on the basis of Calvert's text, that "The Reformation brought about no real change to the anti-Jewish stereotype of the foul-smelling, menstruating Jewish male in need of Christian blood."[49] However, a closer examination of early modern English primary texts demonstrates the extent to which they discredit or ignore the idea of a humiliating blood condition divinely imposed upon the Jews.

The medieval discourse of Jewish bleeding largely sought to subordinate Jews by marking them with divinely imposed shameful hemorrhoids and by analogizing this condition to the bleeding of inferior female bodies. The original and the early modern Latin versions of the *Problemata* both cite the verses from Matthew and Psalms that have been used as prooftexts for this degrading Jewish bleeding, and similarly conclude that the blood flow "in ipsis tunc tempore menstrui expurgator [then at the very time of menstruation is purged]."[50] The text considers Jewish body function as a special condition of the larger question of the causes of bleeding hemorrhoids. However, the general discussion of the etiology of the disease in both the medieval and early modern texts includes the notion that hemorrhoidal blood also flows like menses in (Christian) male bodies: "& exit ille sanguis semel in mense, vt menstrua mulieris [and that blood issues out once a month like a woman's menstruation]."[51] Interestingly, the English translation of the section on Jewish hemorrhoids omits the phrase "in ipsis tunc tempore menstrui expurgator." The notion of divine punishment remains, but the idea of Jewish physical abjection is lessened since only male Christian bodies are explicitly likened in their functioning to inferior women's bodies.[52]

Furthermore, there is no evidence that the idea of a divinely imposed Jewish bleeding expressed in the *Problemes* was widely received. Although its multiple editions attest to the text's overall popularity, its discussion of Jewish hemorrhoids does not reappear in English medical texts composed in the early modern period. While Adelman relies on the *Problemes* in her reading of *The Merchant of Venice*, she does not conclusively demonstrate that Shakespeare references the text to prove Shylock's blood difference from that of the play's Christians.[53] Calvert's text does include a discussion of Jewish bleeding, but he does not cite *The Problemes* in the list of his sources in the margins of the text. Calvert names "Cantjpratanus" as his authority for his discussion of "this shameful punishment" of the Jews, although his own description specifies a monthly flow of blood, a point that does not appear in Thomas of Cantimpré's text.[54] "This punishment so shamefull they say is, that Jews, men, as well as females, are punished *cursu menstruo sanguinis*, with a very frequent Bloud-fluxe."[55] In Calvert's account, all Jews are punished with menstrual bleeding; he makes no mention of biblical prooftexts or hemorrhoids.

Although Calvert iterates the idea of a cursed Jewish bleeding, a closer examination of his text demonstrates that he questions its veracity.[56] He prefaces his discussion of Jewish bleeding with the caveat that it "is an excellent relation, if it can be proved to bear its weight with truth" and qualifies his description of the punishment with "they say."[57] Additionally, Calvert follows this account of Jewish bleeding with other accusations made against the Jews; after mentioning the alleged *foetor judaicus* (Jewish stench), he comments, "Every one exercised his witt to lay some folly or vanitie to the Jews charge."[58] He assumes the fictitious nature of these claims and proceeds to debunk a number of them. He concludes: "I leave it to the learned to judge and determine by writers or Travellers, whether this be true or no, either that they have a monthly Flux of Blood, or a continuall mal-odoriferous breath."[59] Empirical evidence, not theology, will decide this question; however, given its placement in the context of other discredited "follies" laid to the Jews' charge, the notion of Jewish bleeding seems equally dubious.

Published shortly before Calvert's text, Thomas Browne's *Pseudodoxia Epidemica* similarly dismisses the existence of the "received opinion ... that Jews stinck naturally":[60]

> Another cause ... much received by Christians, that this ill savour is a curse derived upon them by Christ, and stands as a badge or brand of a generation that crucified their Salvator; but this is a conceit without all warrant, and an easie way to take off dispute in what point of obscurity soever: a method of many Writers, which much depreciates the esteeme and value of miracles, that is, therewith to salve not onely reall verities, but also non-existences.[61]

Browne's text never raises the possibility of a cursed Jewish bleeding, but in dismissing the notion that the Jews were punished with a type of mark at the time of crucifixion, he removes the basis for such a supposition. He locates this idea as a Christian belief, reducing it from an assumed universal truth to a specific theological understanding. This secular assessment of the claim informs Browne's criticism of authors who cite miracles as evidence in order to avoid advancing an argument based on demonstrable fact. In pointing out the ease with which such claims can be made, since they do not require further support, he suggests that explanations of miraculous agency can just as easily be used to prove "non-existences" as "reall verities." Although he voices concern about the depreciation of their "esteeme and value," his observation that miracles cannot offer conclusive proof indicates how much the authority of this type of explanation has diminished by the early modern period.

The virtual discrediting of the notion of cursed Jewish bleeding in England is certainly overdetermined; the disruption of the transmission of Catholic exegetical traditions occasioned by the Reformation, the rise of travel accounts describing encounters with Jews, and the increased reliance on empiricism in the scientific method all alter earlier attitudes toward Judaism and Jews. However, for the purposes of this argument, the absence of a threatening Jewish presence in early modern England offers the most compelling explanation for the desuetude into which the idea of Jewish bleeding falls. As Shapiro notes, though a small community of Jews dwelled unofficially in sixteenth- and seventeenth-century England, "in case after case, the English show little surprise or concern at the discovery of Jews living in their midst."[62] As illegal aliens, many of whom professed Christianity, Jews already occupy a subordinate position in early modern English society. In these circumstances, there was no need for a construction that initially arose to secure the hierarchy of Christians over Jews. The conditions of its creation no longer obtaining, the idea of the inferior, bleeding Jewish body subsides into obsolescence in early modern England.[63]

An historical analysis of the concept of Jewish bleeding demonstrates the importance of theology for the construction and deployment of this racializing discourse. When the medieval Church failed to ensure the subordination of Jews to Christians through social and legal means, theologians developed "scientific" discourses that marked Jewish bodies as distinct and inferior. However, medical formulations explaining Jewish bleeding were similarly unable to distinguish it from Christian bodily function; the theological rationale of a divine curse provided a specific etiology that secured hierarchical difference. When the theological conditions subtending this fiction altered in the course of the general expulsion of Jews from most of Western Europe and the advent of the Reformation, the discourse of Jewish bleeding no longer served a purpose. Initially developed to make manifest the immutable truth of Jewish spiritual, physical, and thus social subordination, this construct, in its subsequent abandonment, instead discloses the opportune deployment of a contingent construct.

Notes

1. For example, see Caroline Walker Bynum, *Wonderful Blood: Theology and Practice in Late Medieval Northern Germany and Beyond* (Philadelphia: University of Pennsylvania Press, 2007) and David Biale, *Blood and Belief: The Circulation of a Symbol between Jews and Christians* (Berkeley: University of California Press, 2007).
2. *Geneva Bible* (1583).
3. Caesarius of Heisterbach, *The Dialogue on Miracles*, trans. H. von E. Scott and C. C. Swinton Bland, 2 vols (London: Routledge, 1929), 2.23:102.
4. *Geneva Bible* (1583).
5. Quoted in Peter Biller, "A 'Scientific' View of Jews from Paris around 1300," *Micrologus* 9 (2001), 158, trans. mine.

6. Quoted in Willis Johnson, "The Myth of Jewish Male Menses," *Journal of Medieval History* 24 (1998), 281.
7. Quoted in ibid., 288.
8. Most of the contemporary scholars who write on Jewish bleeding assume that theology influences this new medical knowledge: see Irven Resnick, "Medieval Roots of the Myth of Jewish Male Menses," *Harvard Theological Review* 252 (2000), 241–63. Peter Biller initially concurred with this assumption in his essay "Views of Jews from Paris around 1300: Christian or 'Scientific'?," in *Christianity and Judaism: Studies in Church History* 29, ed. Diana Wood (Cambridge, MA: Blackwell, 1992), 196. While he argues in a later essay that this belief is "scientific" and free from theological influence, Biller suggests that Jacques de Vitry's account of Jewish bleeding, which cites Matthew 27.25, Psalm 77.66, and refers to Cain's murder of Abel, probably influenced Albert's non-theological account. "'Scientific' View," 143, 153, 158. Johnson states that a "literalization of an exegetical motif" occurs when the idea of Jewish bleeding developed in Christian discourse engenders the "fact" of Jewish bodily difference in medical texts. "Myth," 286.
9. Hippocrates, *Hippocrates*, trans. W. H. S. Jones, 6 vols (Cambridge, MA: Harvard University Press, 1959–67), 4.6.10–14:183.
10. Biller "'Scientific' View," 139.
11. John Arderne, *Treatises of Fistula in Ano, Haemorrhoids and Clysters*, ed. D'Arcy Power (London: Kegan Paul, Trench, Trübner, 1910), 57.
12. Albert the Great, *Questions Concerning Aristotle's On Animals*, trans. Irven Resnick and Kenneth F. Kitchell, Jr (Washington, DC: Catholic University of America Press, 2008), 310.
13. Biller, "Views," 188–9.
14. Quoted in ibid., 192–3.
15. Ibid., 193.
16. While the main text was probably composed by a student of Albert's in the late thirteenth or early fourteenth century, the date of the commentary is probably later and its country of origin unknown. Helen Rodnite Lemay, *Women's Secrets: A Translation of Pseudo-Albertus Magnus's De Secretis Mulierum with Commentaries* (Albany: State University of New York Press, 1992), 1–2.
17. Ibid., 73–4.
18. Quoted in Biller, "Views," 198; the final section in brackets is my rendering of the Latin which Biller leaves untranslated.
19. Irven Resnick describes this text as an "influential fourteenth-century ... collection of pseudo-Aristotelian questions" in his *Marks of Distinction: Christian Perceptions of Jews in the High Middle Ages* (Washington, DC: Catholic University of America Press, 2012), 111.
20. Quoted in Biller, "Views," 199. In his English translation, Biller omits some of the prooftexts provided in the original Latin; I have inserted them in brackets {}.
21. While contending that Paul's texts do not advocate for supersession, John Gager acknowledges that Paul has been consistently (mis)read as privileging Gentiles over Jews. *Reinventing Paul* (Oxford University Press, 2000).
22. *Geneva Bible* (1583).
23. Ibid.
24. Ibid.

25. For a discussion of the Christian concept of Jewish servitude see Shlomo Simonsohn, *The Apostolic See and the Jews*, 8 vols (Toronto: Pontifical Institute of Mediaeval Studies, 1991), 7:94–102.
26. *Reply to Faustus* 12.9, www.newadvent.org/fathers/140612.htm.
27. Ishmael is the elder brother to Isaac; see also Romans 9:12, in which Paul likens the Jews to the elder brother, Esau.
28. *Reply to Faustus* 12.12, www.newadvent.org/fathers/140612.htm.
29. Ibid., 12.23. In *The City of God*, Augustine also explains the general origin of slavery as a result of sin with recourse to the story of Ham, 19.15, www.newadvent.org/fathers/120119.htm. For Augustine's doctrine of Jewish witness, see Jeremy Cohen, *Living Letters of the Law: Ideas of the Jew in Medieval Christianity* (Berkeley: University of California Press, 1999), 19–65.
30. *Mysticorum Expositiones Sacramentorum Seu Quaestiones In Vetus Testamentum, Patrologia Latina Online*, vol. 83.8.13, my translation.
31. Simonsohn, *Apostolic See*, 7:96.
32. Ibid., 7:97; Amnon Linder (ed.), *The Jews in the Legal Sources of the Early Middle Ages* (Detroit: Wayne State University Press, 1997), 537.
33. Solomon Grayzel, *The Church and the Jews in the XIIIth century*, 2 vols (New York: Hermon Press, 1966), 1:115, 117.
34. Ibid., 1:127.
35. Augustine also considers Cain's mark in relation to the Jews in his commentary on the Psalms: "Not without reason is there that Cain, on whom, when he had slain his brother, God set a mark in order that no one should slay him (Genesis 4:15). This is the mark which the Jews have." *Expositions on the Psalms*, 59.18–19, Part I, www.newadvent.org/fathers/1801059.htm.
36. The 1222 Council of Oxford echoes the text of Galatians in implementing the notion of Jewish inferiority into ecclesiastical law: "Since it is absurd that the children of a free woman shall be slaves to the children of a bondswoman, ... we decree that in the future Jews shall not possess Christian slaves." Quoted in Grayzel, *Church and the Jews*, 1:315.
37. I would argue that the expulsion of the Jews from most of Western Europe by the end of the medieval period is partly a result of the inability to keep Jews in their proper inferior place in Christian society.
38. For a discussion of the badge and the text of Canon 68 see Grayzel, *Church and the Jews*, 1:60–70, 309.
39. Ibid., 1:64–5.
40. Cecil Roth, *A History of the Jews in England* (Oxford: Clarendon Press, 1964), 95–6. Grayzel provides the texts of 29 laws requiring Jews to wear the badge. See also the discussion of the badge in Geraldine Heng, *Empire of Magic: Medieval Romance and the Politics of Cultural Fantasy* (New York: Columbia University Press, 2003), 88.
41. Geraldine Heng, "The Invention of Race in the European Middle Ages II: Locations of Medieval Race," *Literature Compass* 8 (2011), 280.
42. Ibid., 278–9.
43. Ania Loomba, "Race and the Possibilities of Comparative Critique," *New Literary History* 40 (2009), 503.
44. There may be more; I found these on the Early English Books On-Line database. There is an additional English translation of the text published in Edinburgh in 1595; it appears identical to the 1597 London version. The

early modern Latin edition (1583) is not identical to the text of a medieval manuscript available in a modern edition, but the two versions are very close, and concur on all the relevant points. See Pseudo-Aristotle, *Problemata varia anatomica after University of Bologna MS 1165 (2327)*, ed. L. R. Lind, University of Kansas Publications Humanistic Studies 38 (Lawrence: University of Kansas Publications, 1968).

45. James Shapiro, *Shakespeare and the Jews* (New York: Columbia University Press, 1996), 38.
46. David S. Katz, "Shylock's Gender: Jewish Male Menstruation in Early Modern England," *Review of English Studies*, n.s. 50 (1999), 441.
47. Janet Adelman, *Blood Relations: Christian and Jew in the Merchant of Venice* (University of Chicago Press, 2008), 193, n. 94.
48. Ibid., 127.
49. Resnick, *Marks*, 203.
50. *Problemata Aristotelis* (1563), 69. The wording is slightly different in the fourteenth-century version: "in ipsis tempore menstruali expellitur seu expurgator." Quoted in Resnick, *Marks*, 188, n. 59.
51. *Problemata Aristotelis*, 68. While some medieval texts focusing on health in the general population noted the difference between menstruation and hemorrhoids, others analogized hemorrhoidal and menstrual bleeding, including the *Problemata varia*, 38. Resnick cites the twelfth-century works of Roger de Baron, Theodorich Borgognoni, and Gilbertus Anglicus, but ignores its appearance in the *Problemata varia*. See *Marks*, 182, 204, n. 116. In this context it was naturalized and not associated with divine punishment or humiliation.
52. The English translation of the *Problemata* does make one change that intensifies the association of this condition with the Jews; while the Latin text asks "quare Iudaei indifferenter patiuntur talem fluxum?" [why do the Jews indiscriminately suffer such a flow], the English text queries "Why are the Jews subject unto this disease very much?" *Problemata*, 68; *Problemes*, D6r.
53. In fact, she acknowledges that in *The Problemes*' "final mix of dietary, social, and psychological causes for the Jews' subjection to this disease, Jewish blood turns out to look very much like Christian blood—in fact to look specifically like Antonio's blood." Adelman, *Blood Relations*, 126. Adelman contends that Shylock paraphrases Matthew 27 in crying out "My deeds upon my head!" during the trial scene (4.1.201) and argues that this brings the theological justification of Jewish bleeding, and the blood libel to which it is sometimes attached, into the play to prove the Jew's blood difference. Adelman, *Blood Relations*, 127; William Shakespeare, *The Merchant of Venice: Texts and Contexts*, ed. M. Lindsay Kaplan (Boston: Bedford/St. Martin's, 2002). However, Shylock makes no mention of Christian blood anywhere in the play; his bond is for a pound of flesh, not blood.
54. Neither does it appear in Heinrich Kormann's *Opera curiosa*, which he cites in the margin. See Heinrich Kormann, *Opera curiosa, in tractatus 6 distributa* (1694), 128–9. Korman also relies on "Cantipratanus" for his information.
55. Thomas Calvert, *The blessed Jew of Marocco ...; to which are annexed a diatriba of the Jews sins ...* (1648), 20.
56. Shapiro notes Calvert's skepticism about this condition: *Shakespeare and the Jews*, 37–8.

57. Calvert, *Blessed Jew*, 19–20.
58. Ibid., 29.
59. Ibid., 31.
60. Thomas Browne, *Pseudodoxia epidemica, or, Enquiries into very many received tenents and commonly presumed truths* (1646), 201.
61. Ibid., 205.
62. Shapiro, *Shakespeare and the Jews*, 76.
63. The continued circulation of the idea of Jewish bleeding in early modern Germany and Spain supports my claim, since sizable numbers of Jews or recent converts from Judaism still lived in both countries. See Resnick, *Marks*, 203–6. My argument demonstrates why he is wrong about England.

6
Sor Juana's Appetite: Body, Mind, and Vitality in "First Dream"

Anna More

Throughout the seventeenth century, even as mechanical philosophy and the new science developed in Europe, Iberian regions remained tied to a Thomist Aristotelian heritage. With the broadening of the history of science in both disciplinary and geographical terms, the relationship between these two philosophical cultures has been reconsidered in several ways, beyond their traditional opposition. Not only did Aristotelianism provide a language that continued to undergird Cartesian and post-Cartesian philosophy, but late Aristotelianism itself developed beyond its fundamentally Thomist outline.[1] The Jesuits were particularly important in making links, however tenuous, between the new science and scholasticism and their presence throughout the Iberian empires meant that a wide range of positions on issues of empirical and theoretical science circulated and were even publicly debated.[2] In the permeable boundaries between confessional and regional scientific debates, whether these were located in the court or in a more amorphous Republic of Letters, it is possible to find places where Thomist metaphysics and post-Cartesian philosophy continued to share fundamental problems, such as the relationship between mind and body or the place of emotions in action and cognition.[3]

It nonetheless may be surprising that one of the most original approaches to these issues in Iberian regions may be found in the writings of the seventeenth-century poet and cloistered nun Sor Juana Inés de la Cruz (1648–95). While Sor Juana lived most of her life in a cloistered Hieronymite convent in Mexico City, her extensive writings were published in Spain even in her lifetime and she was widely celebrated on both sides of the Atlantic.[4] Even among the wide array of Sor Juana's works, however, the poem published under the title "First Dream" ("Primero sueño") (c. 1691) is unique in its ambition. A 975-verse *silva*,

"First Dream" was clearly modeled on Luis de Góngora's famous *silva*, "Solitudes" (*Soledades*) (1613). Yet unlike "Solitudes," whose elemental narrative of a shipwrecked pilgrim provides little more than the barest scaffolding for Góngora's poetic experiments, "First Dream" is a complete and concise narrative of the nocturnal journey of a soul in search of absolute knowledge. Along the way, the poem recounts in physiological detail the sleep of the body and the operations of the intellect, even as it integrates mythological figures as moral examples and plays with the Gongoresque poetic devices of meter, metaphor, and hyperbaton. Although clearly inspired by philosophical and physiological literature, the ultimate meaning of "First Dream" has proven enigmatic, not the least because the exact contours of Sor Juana's intellectual context continue to elude scholarship. Like many works from Iberian regions, "First Dream" displays a mixture of Neoplatonic and Aristotelian influences, although the exact relationship among these has been the subject of intense scholarly debate. Late twentieth-century criticism focused closely on the Neoplatonic aspects of the poem, finding a probable source in the widely disseminated works of Athanasius Kircher.[5] Despite the poem's consistent use of Thomist language, moreover, scholars have often downplayed the importance of Aristotelian metaphysics for the poem's meaning.[6]

Recently, however, several studies have outlined the elaborate influences of Thomism on Sor Juana's works, arguing that her engagement with Aquinas was deeper and more coherent than had previously been thought.[7] Yet these readings often assume that Sor Juana read a strict Thomist corpus, without considering the possibility that she followed one of the late Aristotelian variants or the textbooks common in the seventeenth century.[8] One of the most thorough and convincing accounts of Thomism in "First Dream," moreover, entirely ignores the pressures that the poetic form exerted on the philosophical and moral subject of knowledge.[9] By considering Sor Juana's use of Aristotelian language in another of her well-known writings, "La Respuesta de la Poetisa a la muy ilustre Sor Filotea de la Cruz" ("The Response of the Poetess to the Very Illustrious Sor Filotea de la Cruz"), it becomes clear that Thomist philosophy provides the framework for the problem that she addresses in "First Dream": the soul's desire for knowledge. In this epistolary defense of her learning and writing, mostly likely written shortly after she penned "First Dream,"[10] Sor Juana responds to a request from the Bishop of Puebla that she turn her talents to sacred rather than profane subjects. To defend herself, Sor Juana emphatically distinguishes what she represents as a natural desire for knowledge (*inclinación*) from her will (*voluntad*). Although this distinction clearly responds to the particular circumstances of the "Response," it nonetheless also exposes a problematic tension in Thomism between a teleological determinism and free will. Aquinas's metaphysics distinguishes between three appetites, or inclinations, that drive objects toward teleological ends: the natural, the sensory or animal, and the rational.[11] Sor Juana's distinction between "inclination" and "will" employs the language of two extremes in this scale: the natural appetite, as a fundamental teleological inclination possessed by all objects, and the will, as rational appetite found only in humans.

While Sor Juana uses these extremes to structure her ingenious argument that what appears to be a disobedient will is rather a natural impulse or "inclination," she also introduces a striking third term often tied to a more sensory, or animal appetite. Used as a slippery synonym for both "will" and "inclination," "pleasure" (*gusto*) combines the language of natural determinism and voluntarism. One of the most striking applications of this term occurs when she introduces "First Dream" as the only work she has written for her "pleasure" (*gusto*). In what follows, I will argue that this association between "First Dream" and pleasure allows Sor Juana to find in poetry a language for an ateleological desire or drive. Key to this ontopoetics of "First Dream" is the confluence of sleep as a reduction of physiology to its basic vital functions and the night as a space of poetry. Through its combination of Galenic physiology, Thomist categories of faculty psychology and Ovidian and Gongoresque poetics, "First Dream" inserts an internal drive into cognition, yet one that avoids the teleology implicit in both natural and rational appetites. Rather than a discovery of the limits of the human intellect, as it has most often been read, "First Dream" is therefore an invention of a language of desire.

Appetite in the "Response"

Most likely written shortly after "First Dream," the "Response" employs a language of defense honed during at least the previous decade.[12] The letter's most immediate provocation, however, was a dispute over the publication of Sor Juana's critique of a sermon by the renowned Portuguese Jesuit Antonio Vieira. The exact circumstances that led to the dispute are still unclear. What is known is that at some point before 1690 Sor Juana had been asked to transcribe her elaborate critique of Vieira's Maundy Thursday sermon (*Sermão do Mandato*), preached 40 years before in Lisbon. In 1690, without Sor Juana's consent, the critique was printed in Puebla under the title "Athenagoric Letter" (*Carta atenagórica*).[13] This publication was accompanied by a short prologue signed by another nun, Sor Filotea de la Cruz. While praising the nun's prodigious talents, the prologue requested that Sor Juana turn them toward sacred subjects: "If other religious women sacrifice their will [*voluntad*] out of obedience, you capture understanding, which is the most arduous and agreeable holocaust that can be offered on the pyres of Religion. Following this observation, I am not asking that you change your nature [*genio*], renouncing books, but rather that you improve your selection, once in a while reading that of Jesus Christ."[14]

The true identity of Sor Filotea de la Cruz, while known to Sor Juana, would have to wait until 1700 to be publicly exposed in the posthumous publication of her unprinted works.[15] The author of the prologue was none other than the Bishop Fernández de Santa Cruz of Puebla, a supposed supporter of Sor Juana's. While the politics that led the Bishop to reprimand Sor Juana publicly may never be fully understood, her response makes clear the gravity with which she treated the charge.[16] Sor Juana's argument, deftly expounded on various levels, turns on an inversion of the very terms by which the Bishop had sought to turn her away from profane letters. As the Bishop had noted, Sor Juana's "nature" (*genio*) implied a gift of an order other than that bestowed on ordinary nuns. While others could only "sacrifice their will out of obedience," Sor Juana's talents had the potential either to undermine or to improve obedience. Citing the Apostle Paul's dictum *mulieres in ecclesiis taceant* (may the women be silent in the church) (I Corinthians 14:34), the Bishop had exhorted her: "God does not wish women to exercise letters that engender pride; but the Apostle does not condemn them when they do not undermine women's obedience."[17]

Sor Juana's response plays upon this implied breach between "will" (*voluntad*) and "nature" (*genio*) that the Bishop himself had introduced. Rather than accepting the conditions that the Bishop presents, however, Sor Juana insistently divides her intellectual drive, variously termed "inclination" (*inclinación*), "drive" (*impulso*), or "desire" (*deseo*) from her will. If she had read, written, and published, she argues, it had never been by virtue of her own will but rather following the natural impulse that God had placed in her. Sor Juana's argument therefore contrasts two types of authority and obedience: the first, obedience to a church authority and the second obedience to a divine cause. Both, however, attribute her writing to obedience to others rather than to a disobedient will. By contrasting her "inclination" as obedience to the will of God, to obedience to the will of others, including members of the Church, Sor Juana bifurcates the Bishop's argument. What would otherwise have been a concession to the idea of feminine obedience, a "sacrifice of the will" in the Bishop's words, avoids entirely the issue of voluntarism by substituting expressions of natural determinism for those of free will.

As the series of equivalences that pepper her letter suggest, Sor Juana deploys these terms rhetorically rather than philosophically, most likely in response to the Bishop's own distinction. Even so, the Thomist Aristotelian resonances are unavoidable. "Inclination" was a central component of Aquinas's teleological natural order, philosophically translated through the concept of "appetite." In Aquinas's metaphysics, all objects, whether animate or inanimate, possess an appetitive power that "inclines" them toward determined ends, or, as he writes: "an operation of an appetitive power is completed when the agent is inclined toward its object (ST 1a 75–89)."[18] As Robert Pasnau has written, "inclinations" were tied up in the identity of objects: "An inclination, as we might say, is a tendency toward a certain sort of action, or a disposition to do a certain thing (or be affected in a certain way) in certain circumstances."[19] This disposition toward an end, in turn, was determined by a natural order itself that was in the last instance teleologically orchestrated by divine will. As Aquinas writes: "All natural things are inclined toward their ends through a certain natural inclination from the first mover, which is God, and consequently that toward which a thing is naturally inclined must be that which is willed or intended by God (QDV 22.1c)."[20]

While everything in nature thus possesses an appetite, however, the Aristotelian tripartite soul demanded further differentiation between inanimate and animate objects and between animate beings with or without cognition.[21] Thus, Aquinas distinguished between three types of appetite: natural, animal or sensory, and rational. All beings possessed the basic "natural appetite" as an inclination toward a determined end, but only animals possessed "sensory" or "animal" appetite and within animals only humans possessed a "rational appetite" or a "will." Natural appetite is limited to the most basic functions of animate and inanimate objects: "Thus any capacity desires, by natural appetite, that which is appropriate for it (ST 78 I ad 3)."[22] By contrast, sensory appetite, while passive because it depended upon external stimuli, was linked to the higher cognitive functions of animals and could be affected by information.[23] The rational appetite, or will, was the most flexible as it worked through a complex mechanism of judgment and decision. As Robert Pasnau describes this difference: "Our sensory appetites are little more than mechanical reactions to sensory stimuli. A rational appetite, in contrast, selects its goals from among various alternatives, measuring the extent to which a given alternative satisfies the agent's conception of what is good."[24]

Perhaps the most fundamental distinction between the will, as rational appetite, and the sensory appetite was a diverse relationship to matter. As purely immaterial, like the intellective soul to which it was linked, the will was not directly affected by material transformations while the sensory appetite, or passions, was fully material. This distinct functioning of the sensory and the rational appetites corresponds to one of the most difficult aspects of Aquinas's metaphysics. While for Platonists, the intellect as soul was independent of the body, for Aristotelians, during life the soul was bound tightly to the body. Christianity challenged this concept of the substantive soul, reaffirming once and again the immortality and thus non-corporality of the soul. The result, as Dennis Des Chenes writes of the sixteenth- and seventeenth-century Aristotelianism, was that "the strategy of Aristotelianism was to bind the human soul as tightly to the body as possible, even while committing itself, in accordance with the expressed wish of the Church, not only to granting the possibility of immortality but to showing this to be rationally demonstrable."[25]

Even so, in wake of the 1513 Lateran Council pronouncement on the immortality of the soul and subsequent scandal caused by Pietro Pomponazzi's declaration that the immortal soul could not be rationally proven, the question continued to plague late Aristotelianism.[26] One of the central questions posed by the ambiguous confluence of the immaterial intellect and material and sensory stimuli was the question of whether the will could have something like passions. Even given his rigid distinction between the immaterial intellect and its dependence upon the sensory, Aquinas himself admitted that God and other bodiless beings must express something analogous to the sensory passions of happiness.[27] But while Aquinas equivocated on the hierarchy of the rational and sensory appetites, he clearly circumscribed reason's domain over bodily functions. In fact, as Aquinas writes, "the will moves all the powers of the soul to their respective acts, except the natural powers of the vegetative part, which are not subject to our will" (ST 82.4 reply).[28] Aquinas's rigid segregation of the material and immaterial was particularly challenged by Duns Scotus's position that volitions themselves were immaterial.[29] Late Aristotelianism had to take all of these positions into account and on this and several other issues diverged from strict Thomism. One of the most radical positions was that of

Francisco Suárez, who posited a vital power as the active principle of the intellective soul, thus making it parallel to but independent from external causes.[30]

Whether or not Sor Juana had read Suárez and other late Aristotelians has been an open question.[31] In the case of the "Response," however, the strong distinction between "inclination" and "will" most likely responded to the circulation of these terms in a late seventeenth-century context still imbued with general Thomist and Aristotelian notions. The will continued to be central to discussions of religious practice, as the Bishop's description of feminine religiosity as the "sacrifice of will" indicates. "Inclination" also maintained a strong inflection of natural teleology. The eighteenth-century *Diccionario de Autoridades*, for instance, includes "nature" (*genio*) in one of its definitions: "another [acceptable definition] is a propensity or nature [*genio*] dedicated to a thing."[32] The importance of both of these terms and their synonyms in the "Response" suggests that Sor Juana wished to invoke their deep resonances with fundamental metaphysical concepts. From its first lines, the "Response" introduces the tension between free will and obedience: "Neither my will, nor my weak health, nor my justifiable apprehension have caused my response to be delayed so many days."[33] The opening paragraph, one of the most famous in the "Response," underscores the implication of this initial statement, that Sor Juana's response is an act of obedience to her religious superior and thus an indication of a will complicit with church mandates. From this point on, Sor Juana takes every opportunity to insist on her obedience, understood as a will dedicated to the correct end. As she claims, this obedience has marked all of her practice, even the very act of writing that has occasioned the Bishop's reprimand: "And truth be said, I have never written anything without having been violently forced to do so, and only for others' pleasure; not only [has this been] against my will, but even with a positive repugnance because I have never judged myself to possess the literary skill and source of wit that is the obligation of one who writes."[34] If she has written, then, it is not a sign of disobedience, but rather of a will dedicated to others, in this passage glossed as the intent "to give others pleasure" (*dar gusto a otros*).

By divorcing her actions from her will, Sor Juana sets up a crucial pivot in her argument when she defends her drive for knowledge as an "inclination." This pivot occurs directly following the statement renouncing her will to knowledge:

What is true and I will not deny (first because it is widely known and second because, even though it does not benefit me, God has graced me with a very great love of truth): that from the moment the first light of reason struck me, my inclination to letters was so forceful and strong that neither external reprimands—of which I have received many—nor my own reflections—which have not been few—have been enough to divert me from that natural impulse that God placed in me.[35]

Employing the term "inclination," Sor Juana diverts her renunciation of her will to a divine cause, the "natural impulse that God placed in me." While writing has thus been against her will ("violently forced" [*violentada y forzada*]), then, it has been in accordance with her "inclination." Sor Juana's "inclination to letters" overcomes both "external reprimands—of which I have received many" and her "own reflections—which have not been few."

The term "inclination," then, exists beyond human will altogether, whether Sor Juana's or others'. Indeed, its divine provenance is practically established by the very impossibility of human intervention in its processes. This, indeed, will be the story that Sor Juana weaves, beginning with an autobiography that she glosses as the "narrative of my inclination" (*narración de mi inclinación*).[36] In this, she describes through a series of anecdotes the signs that her "desire for knowledge" is beyond her will. In a narrative that runs from a precocious desire to learn through her inability to maintain the precepts of her superiors in the convent when they prohibit her from studying profane subjects, Sor Juana's anecdotes establish her natural inclination toward learning as a resistence to all human mandates. In essence, what Sor Juana will variously call her "inclination" or her "natural impulse" may be defined as a Thomist "natural appetite." In her "Response," indeed, Sor Juana employs "appetite" as a hinge between divine impulse and private desire, in a more bodily than rational sense: "I thought that I fled from myself, but (oh misery!) I carried myself with myself and brought along my greatest enemy in this inclination that I don't know whether to judge a gift or a punishment from the heavens, since when suppressed and thwarted by all the exercises of religious life it exploded like gunpowder and I became the proof that *privatio est causa appetitus*."[37]

Sor Juana's language in the "Response," in which "inclination" and "appetite" are used as slippery synonyms, reflects an informal use of these terms rather than a strict adherence to Thomist categories. The distance between the demands of philosophical coherence in Thomism and the rhetorical context of Sor Juana's letter may be measured in her peculiar use of the term "pleasure" (*gusto*) as the mediation between individual "will" and divine "inclination." In Aquinas's metaphysics, emotions such as pleasure would be the province of the sensory appetite, which passively received ("apprehended") objects to which it reacted by moving toward the object, as one of pleasure, or away from it, as one feared.[38] Sor Juana's "Response" had already linked the immaterial will to the material effect of "pleasure" in her declaration that she writes not out of her own will but to "give others pleasure." In this first case, the apparent purpose of the suppression of the term "will" in favor of "pleasure" is to divorce human requests from the divine teleology implicit in the term will but absent from the idea of pleasure. At the end of her "Response" Sor Juana will employ the same term to the opposite effect when she describes "First Dream" as the one exception to the rule that she has written to obey the will of others: "What is more, I have never written anything out of my own will but only after being begged and cajoled by others' requests; thus I do not remember having written anything for pleasure except for the little trifle they call the Dream."[39] While she begins the statement by denying her will, she ends with the positive statement that she has written "First Dream," ironically described as "that little trifle" (literally, "little paper" [*papelillo*]), for pleasure. If writing for others' pleasure had divorced these requests from divine teleology, in this case "pleasure" wrests her natural appetite away from the determinism of divine mandate. With this, Sor Juana marks "First Dream" as a space of exceptional freedom outside of the strictures of Thomist teleology.

Desire in "First Dream"

While in the "Response," Sor Juana establishes a strong dichotomy between teleological inclination and the freedom of the will, by designating "First Dream" as the product of pleasure she suggests that the poem is a product of the sensory rather than the natural or rational appetites. Unified as "inclinations," the three appetites in fact differ greatly in relation to freedom. Natural appetite is fully teleological and, indeed, might be equated with teleology itself. The will, or rational appetite, however, preserves the freedom of choice or decision under the rubric of an inclination that must ultimately decide in favor of the good. Sensory or animal appetite, while fed by cognition, is passive and limited in its reactions. Animals might react to objects, but these reactions are not willed. Thus, while the natural appetite is an internal determination in which the inclined object holds no agency, the will is a free choice that is restricted nonetheless by its dependence on intellect and its need to decide in favor of a correct end. Sensory appetite, by contrast, avoids teleology of the good or the correct, but is limited in scope to primary reactions of pleasure or aversion. Furthermore, the inclination itself is not in the power of the animal that is inclined, but is determined to it from without: "For an animal, at the sight of something desirable, cannot not want that thing, because those animals do not have control over their inclination" (QDV 23.1c).[40] Thus, as Pasnau has written, in Thomist faculty psychology the sensory appetite is "a murky gray area between freedom and necessity, responsibility and determinism."[41]

Sor Juana's use of the term "pleasure" to describe the impetus behind writing "First Dream" signals the collapse and contamination of the extremes of inclination and will. In essence, the concept of writing for "my own pleasure" appears to be both determined, as it partakes in what Sor Juana has called her inclination toward knowledge and writing, and free, in the sense that it avoids the terms of obedience that restrict the will. It is this determined and necessary, but ateleological and free impetus or drive, that Sor Juana attempts to discover in "First Dream." The poetic structure, devices, and tropes are essential for this definition of what in Thomism would be an oxymoron: the sensory appetite of the intellect. "First Dream" develops this radical exploration through the poetic elaboration of Thomist and Galenic categories of physiology and cognition in the narrative of the soul's desire for knowledge as the body sleeps. The poem may be divided into six sections: the fall of night; the sleep of the body; the description of the soul's ambition for knowledge; the soul's failed attempt at intuitive knowledge; the soul's attempt at deductive knowledge; the break of dawn and awakening of the body.[42] Sor Juana's account appears to have been directly inspired by Luis de Granada's *Introduction of the Symbol of Faith* (*Introducción del Símbolo de la Fe*), a late sixteenth-century treatise that drew heavily on the Galenic account of physiology.[43] While the description of the physiological and intellectual processes that lead to the sleep of the body and the soul's search for knowledge in the poem correspond to Galenic and Thomist faculty psychology, however, the fact that the soul's journey takes place at night, when sleep has reduced the body to a minimal function of the vegetative soul, disrupts sensory effects on cognition. While the intellective soul remains tied to the sleeping body, it is shorn of its external senses and thus cognition takes place according to its interior senses. The phantasms that mediate cognition are informed by external stimuli during the day, but at night rely on stored images, two times removed from their material object albeit still with a material residue.[44]

While Sor Juana may have used the trope of night and sleep to explore the humoral constitution of the mind, her purpose in "First Dream" is not to trace the roots of her exceptional nature, as she will later have to defend in the "Response," but to investigate cognition as a universal process. In the place of the external stimuli that would provoke cognition during the day, at night the soul's impetus is understood poetically, through personification and metaphor. These are necessary supplements to the immaterial processes that would normally be filtered through the will. Indeed, together with external stimuli, it is reasoned decision that seems suspended during the night. In the first section of the poem, as has often been remarked, the description of the night that falls upon earth is followed by a series of Ovidian references to transgression, often feminine. Nightfall and sleep, by contrast, are repeatedly referred to as inevitable and that except no one, from beast to king. Following the Aristotelian and Galenic account of the physiology of sleep in Luis de Granada's *Introduction of the Symbol of Faith*, Soriano has argued that the logic of the link between nightfall and the body's sleep rests on the fact that both unleash "vapors" that block external senses. According to this tradition, sleep occurs because the blood carries vital spirits or "vegetal warmth" formed by the digestive process from the heart to the brain.[45] The effect upon the body is to depress its functions to those of its basic organs, thus purifying the soul of its engagement with the external world: "The soul, then, unburdened of / external rule that holds sway when, occupied / by material concerns, for good or ill / she spends the day, remote now / but not completely apart, / confers benefits of vegetative warmth / to languid limbs and tranquil bones oppressed / by that passing death, for the body serene / is like a corpse with a soul, / dead to life, alive to death" (vv. 192–203).[46]

Here, a crucial distinction must be made between Sor Juana's account and the poetic tradition in which sleep purifies and directs the humoral constitution of the mind.[47] While Sor Juana may certainly have ascribed to a humoral account of her nature (*genio*), in the tradition of Juan de Huarte's influential *Examen de ingenios para las ciencias* (1574), her purpose in "First Dream" is not to explore her cognition as a humoral pathology.[48] Rather, "First Dream" reduces all physiological processes to their barest functions to focus entirely on the internal processes of cognition. In these processes the vision, as image, is the meeting point between the intellect and matter. During sleep, when the external senses are suppressed, the intellect works from images that have been stored in memory.[49] Thus she describes the brain receiving "the damp but most clear vapors / of the four tempered humors / that not only did not cloud the semblances / the intellect gave to imagination / which, for safer keeping and in purer form, / presented them to diligent memory / that etched them, tenacious, and guards them with care, / but permitted fantasy / to form diverse images" (vv. 255–65). In this passage, blood is reduced to a physiological function as conduit for the vital spirits formed by the "tempered humors" (*atemperados humores*), whose neutralization permits "clear vapors" (*claros vapores*) that heed rather than hinder cognition.[50]

Sor Juana's reduction of cognition to an internal process parallels one of the most heated debates in late Aristotelianism: the interaction between the sensory image and the intellect. Sor Juana diverges from a rigid Thomist materialism, describing the process of the dormant imaginative faculty through the metaphor of the Alexandrine lighthouse that shines its beacon on the sea such that on its surface are clearly visible "the size, number, and fortune" of the boats that have risked sail. The imaginative faculty is thus an "invisible brush" that paints figures that are "without light, bright mental colors / the figures not only of all sublunar / creatures but also of those / clear hues that are intellectual stars" (vv. 282–7). As has often been commented, the "clear hues that are intellectual stars" refer to the concepts to which the soul, unburdened by external senses, has clearer and more direct access than during the day.[51] Indeed, immediately following this description of the mental images, Sor Juana describes the soul's approximation to pure intellect: "She, meanwhile, transformed into / immaterial being and beautiful / essence, contemplated that / spark shared with highest Being and cherished it / deep inside, in His image; / and deeming herself almost separated / from the coarse corporeal chain that keeps her / ever bound, hinders the flight of intellect / that measures the immensity of the sphere" (vv. 292–301).

While the impediment of "the coarse corporeal chain" has often been cited as proof of the influence of Neoplatonism on "First Dream," the Thomist intellective soul was similarly immaterial and, as Soriano has pointed out, the soul is described as *almost* shorn of its corporal chain rather than completely distinct.[52] Immediately following this approximation of the immaterial limit in the Aristotelian intellect, Sor Juana represents cognition as mediated not by images received from external senses, but by "intentional species." As Soriano has pointed out, the section that has often been seen as a strange pause in the narration, in which Sor Juana describes the soul's ambition for knowledge through the metaphor of the Egyptian pyramids, is in fact a central component of the poem's narrative.[53] After recounting the beginning of the soul's "intellectual flight" as an eagle "now raking / the air with its talons, struggling to break that / impunity with two wings, / weaving steps out of atoms" Sor Juana introduces the pyramids as a metaphor for the intellect's ambition. The soul who contemplates the pyramids cannot even see the "subtle tip that appears to touch the first sphere / until, wearied by wonder, / it did not descend but plunged, / found itself at the foot of the ample base, / scarcely recovered from vertigo, a harsh / retribution for winged, visual daring" (vv. 361–8).

While for Soriano this section establishes the moral limits of human intellect, in the Thomist tradition, there is reason to revisit this conclusion. Attributing the idea to Homer, Sor Juana calls the pyramids "external signs of internal dimensions / that are the intentional species of the soul" (vv. 401–3).[54] The term "intentional species" is a clue that Sor Juana may have been aware of late Aristotelian debates on the mediation of material stimuli and immaterial intellect. While "sensible" species were the images generated by sensory input, these still contained material effects. "Intentional" species were secondary, images of images, that the intellect accessed in place of the sensory images. A subject that generated much debate in late Aristotelianism, intentional species were a limit point of the substantive soul, "said never to be without conditions of materiality, even while being forms 'without matter' (that is, forms that are not conjoined to matter in the usual way)."[55] Francisco Suárez's account of cognition was the most radical attempt to resolve the contradictions in the Thomist account of the immaterial process of cognition. With Suárez, "intentional species" became the medium for a cognitive process in which the agent intellect worked in parallel but was not dependent upon the sensory image "so that, whenever the imagination produced a phantasm, the intellect produced an intelligible species and vice versa."[56] As Eckhard Kessler points out, while still maintaining the essential material element of the Thomist substantive soul, Suárez came close to the Neoplatonic position of an autonomous intellect.[57]

In "First Dream" the pyramids are "intentional species" that permit the soul to reflect upon cognition itself, defined as an "aspiration" toward the Aristotelian "prime mover" or "first cause": "as the ambitious flame burns upward in a / pyramidal tip toward heaven, so does the / human mind, miming that form, / ever aspire to the First Cause." While this identity defines the intellective soul's search for knowledge as an attraction to the First Cause, and thus a teleological end, the pyramids are the form of the soul's cognition of this identity, rather than the act of knowledge itself. In essence, through the pyramid, the soul is contemplating an image that will allow it to understand its own ambition.[58] The narrative of this ambition elaborates upon knowledge not as a divine end but as a repeated and incessant desire. As the soul attempts to climb the "soaring mental pyramid," compared to which the pyramids and even the Tower of Babel were diminutive, its ambition to climb to the highest point "of its own mind" is termed an "ambitious desire." Once there, "the piercing gaze / of her beautiful eyes of intellect, no / spyglass before them" had present before it "all of creation." This excess causes the "understanding" to fall back "cowardly." "Not so," Sor Juana continues, "did she, repentent, revoke / her intent from its bold goal" (vv. 454–5). Indeed, after describing in exquisite detail the attempt of the soul to understand intuitively, but "looking at everything, saw nothing," Sor Juana recounts a slow but insistent recovery through the metaphor of a ship that has run ashore but will be repaired, in this case by reflection and method: "prudent reflection usurped repairs" (vv. 573–4).

While the "intentional species" therefore confirm the immateriality of the soul, which has approached complete separation from the body and thus angelic knowledge, the drive of the soul is personified, material, and corporeal. The paradox of immaterial intellect and sensory appetite is resolved by allegorizing the desire for knowledge in narrative and metaphor. The narration of the soul's "ambition" and "desire" forms the backbone of the poem, carrying the soul forward as it attempts a discursive approach to knowledge. If the problem in the intuitive approach had been the immediacy of all creation, the discursive approach will work through mediation, or "levels," just as, Sor Juana explains, the hand shields the sun as a "pious mediation" to protect the eyes from direct sunlight. This scaled method is also understood as an ascent ("ascending scales") with the idea of finally reaching, "immaterially," the "honorable summit." After investigating the base, in vegetative life, and moving through brutes, the soul finally arrives at man, "compendium that resembles the angels, / the plants, the brutes; whose exalted lowliness / partakes of all of nature" (vv. 692–3).

But just as the excess of intuitive knowledge had caused the soul's first fall, now the vertigo of the material detail will stop it in its tracks: "These, then, were the steps I wished at times to roam, / but at others I forswore / that wish, judging it excessive boldness / in one who failed to comprehend the smallest, / the most simple part of natural effects / nearest at hand, to attempt / to apprehend everything" (vv. 704–9). As Christopher Johnson has written, "First Dream" works through a constant trope of hyperbole, expressed in images of excess, immensity, and daring.[59] These all point to a place beyond, however, a limit that the soul never surpasses. As Soriano points out, the poem remains loyal to the inability of the corporeal soul to pass over into the realm of angelic knowledge, in the Thomist system reserved for the soul completely divorced from the body. It is this limit point, indeed, that will define desire itself in the turn to poetic metaphor. Thus the soul becomes mesmerized in its attempt to understand two objects: the stream that runs in eddies through a complex course, including a passage through Hades, and the flower, itself a metaphor for desire, delicately described as inexplicable mixture of contrasting white and purple and through the erotic metaphor of a "sweet wound of the Cyprian goddess" (v. 743).

The substantive soul is the object of a drive that, when teleological, is constitutionally unable to surpass matter. As much as its drive is described as moving forward, then, it is also confused and circular. At these times, metaphor is a digression that serves as a bridge, allowing respite while the soul catches its breath before continuing. The inability to understand minimal details is offset, for instance, by the irrational push of the soul, compared in one last metaphor to Phaeton. This "daring example" who drove his car into the air, against Zeus's orders, animates the soul: "his gallant if luckless high impulse firing / my spirit to the place where valor finds more / open paths of daring than fear encounters / examples of chastisement." The metaphor of Phaeton for the soul's ambition has been the center of debate over the meaning of the poem, with Octavio Paz finding in it the model for Sor Juana's own bold desire.[60] Arguing that Paz has misread the metaphor, Soriano points to the poem's negative assessment of Phaeton: "Rather it is a model: a pernicious / example that engenders the wings for a / repeated flight of the ambitious spirit / that flatters terror itself / to compliment bravery, / and spells glories among letters of havoc" (vv. 803–9).[61]

It is hard to determine, however, with what voice the poem criticizes ambition through the figure of Phaeton. Although the poem states that Phaeton serves as a "pernicious" model for the soul's "repeated flight," this statement is just as likely descriptive as moralistic. In fact, this moralizing does not represent a closure. The soul's movement forward is rather interrupted by the return of the physiological: "But while choice foundered, confused, among the reefs, / touching sandbars of courses impossible / to follow each time it tried" (vv. 827–9) the stomach runs out of the food that had maintained the state in which "the vapors / ascending, damp and soporific, beset / the throne of reason" (vv. 846–7). If the Alexandrine lighthouse had provided the first metaphor for the imaginative faculty, it is now the magic lantern that represents the images fleeing, "en fácil humo, en viento convertidas" as the external senses return to the body. While the reference to the magic lantern, most certainly inspired by Athanasius Kircher's *Ars magna lucis* (1646), has fueled the Neoplatonic interpretation of the poem, it is important to note its intermediary function, not only as a reference to the vanishing intentional species that had presented themselves before the soul, but to the poetic images, in metaphors such as the pyramids, the flower, and the stream.[62]

Day breaks in one last glorious metaphor, as the battle between the sun and the night personified as women warriors. While the physiological process of the body is Aristotelian, the drive of the immaterial intellect and diurnal cycle that drives the poem's narrative are both figured poetically. Through this negative space filled in by poetics, Sor Juana seeks a language to describe what in Thomist metaphysics had no place: the drive of the intellective soul. At the suppression of reason and the will during sleep there was no possibility of answering for the soul's drive to knowledge in a Thomistic language of appetite. The attempt to purify the intellect of its material bonds to the body, as seen in the technical account of the intentional species that form cognition, gives way to a poetic narrative of the soul's search for knowledge. While the poem's famous ending has proven exceedingly enigmatic, in its most basic configuration it returns to the body its external senses and ordered reason. Yet it also clearly replaces fragmented Thomist faculty psychology with a unified self, as the sun "with judicious light / and distributive order, began to give / all visible things their colors / and entirely restoring their operation / to the exterior senses / the World illuminated with a surer light and I awake" (vv. 969–75). The limitation of human knowledge, recounted in accordance with Thomist categories, is not the true narrative of "First Dream." Rather, the poem is a tale of a self, defined not as matter formed by the soul, but as matter impelled by desire.

Notes

Preliminary versions of this chapter were presented at Notre Dame University and the University of California, Berkeley. I would like to thank the public at those venues, particularly Ivonne del Valle, Francine Masiello, Emilie Bergmann, and Estelle Tarica. I would also like to thank Orlando Bentancor, Kimberly Anne Coles, and Ralph Bauer for their generous and insightful comments on drafts. All translations, except where otherwise noted, are mine.

1. For fundamental studies, see Dennis Des Chenes, *Life's Form: Late Aristotelian Conceptions of the Soul* (Ithaca, NY: Cornell University Press, 2000) and *Spirits and Clocks: Machine and Organism in Descartes* (Ithaca, NY: Cornell University Press, 2001); C. B. Schmitt, "The Rise of the Philosophical Textbook," in *The Cambridge History of Renaissance Philosophy*,

ed. Quentin Skinner, C. B. Schmitt, Eckhard Kessler, Jill Kraye (Cambridge University Press, 1988), 792–804; Katharine Park, "The Organic Soul," in ibid., 464–84; and Eckhard Kessler, "Psychology: The Intellective Soul," in ibid., 485–534.
2. On Jesuit science, see Mordechai Feingold (ed.), *Jesuit Science and the Republic of Letters* (Cambridge, MA: MIT Press, 2003); Steven J. Harris, "Mapping Jesuit Science: The Role of Travel in the Geography of Knowledge," in *The Jesuits: Cultures, Sciences, and the Arts, 1540–1773*, ed. John O'Malley (University of Toronto Press, 1999); and Rivka Feldhay, "The Cultural Field of Jesuit Science," in ibid. For Jesuit science in late seventeenth-century New Spain, see Anna More, "Cosmopolitanism and Scientific Reason in New Spain: Sigüenza y Góngora and the Dispute over the 1680 Comet," in *Science in the Spanish and Portuguese Empires, 1500–1800*, ed. Daniela Bleichmar et al. (Stanford University Press, 2009).
3. Aside from the works cited in note 1, above, see Susan James, *Passion and Action: The Emotions in Seventeenth-Century Philosophy* (Oxford University Press, 1997) and Martin Pickavé and Lisa Shapiro (eds), *Emotion and Cognition in Medieval and Early Modern Philosophy* (Oxford University Press, 2012).
4. The widely disseminated work by Octavio Paz, *Sor Juana Inés de la Cruz o las trampas de la fe*, 3rd edn (Mexico: Fondo de Cultura Económica, 1983) has been recently questioned on several fronts, but continues to provide a good general overview.
5. For Sor Juana's Neoplatonism, the many works by José Pascual Buxó are indispensable. See, for instance, José Pascual Buxó, *Sor Juana Inés de la Cruz: Amor y conocimiento* (Mexico: UNAM, 1996), 181–203. See also, Paz, *Sor Juana Inés de la Cruz*, 475–81.
6. As one of the most meticulous scholars of Sor Juana's influences states, "the poetess 'speaks in a Thomist style,' because this was the usual language in her context. It is clear that this in itself does not prove anything." Marie-Cécile Benassy-Berling, *Humanismo y religión en Sor Juana Inés de la Cruz*, trans. Laura López de Belair (Mexico: UNAM, 1983), 136.
7. See Alejandro Soriano Vallès, *El Primero sueño de Sor Juana Inés de la Cruz: Bases tomistas* (Mexico: UNAM, 2000) and Alberto Pérez-Amador Adam, *De finezas y libertad: acerca de la Carta Atenagórica de Sor Juana Inés de la Cruz y las ideas de Domingo de Báñez* (Mexico: Fondo de Cultura Ecónomica, 2011).
8. For late Aristotelianism, see note 1, above. For textbooks, see Schmitt, "The Rise of the Philosophical Textbook."
9. Soriano, *El Primero sueño*.
10. As the first volume of Sor Juana's poetry was published in 1689 without "First Dream," it is probable that she wrote the poem after that date but before she penned the "Response."
11. Robert Pasnau, *Thomas Aquinas on Human Nature: A Philosophical Study of Summa Theologiae 1a 75–89* (Cambridge University Press, 2002), 237–9.
12. The tension between Sor Juana and members of the church hierarchy began as early as 1682, the date of a letter to her spiritual advisor, the Jesuit Antonio Núñez de Miranda, in which she responds to his attempts to rein in her writing. See Antonio Alatorre, "La Carta de Sor Juana al P. Núñez (1682)," *Nueva Revista de Filología Hispánica* 35.2 (1987).

144 Sor Juana's Appetite

13. For the latest and most informed reconstruction of these events, see Pérez-Amador, *De finezas y libertad*, 21.
14. Sor Juana Inés de la Cruz, *Obra selecta*, 2 vols (Caracas: Biblioteca Ayacucho, 1994), 1:448.
15. Pérez-Amador, *De finezas y libertad*, 20.
16. For a summary of the history of this debate see ibid., 33–76, esp. 74–6.
17. Cruz, *Obra selecta*, 1:448.
18. Thomas Aquinas, *Summa Theologica*, 81.1c. Cited in Pasnau, *Thomas Aquinas on Human Nature*, 201.
19. Ibid., 206.
20. Cited in ibid., 208.
21. For a concise explanation of the tripartite soul and faculty psychology, see Park, "The Organic Soul," 465–7.
22. Cited in Pasnau, *Thomas Aquinas on Human Nature*, 209.
23. Ibid., 210.
24. Ibid., 241.
25. Des Chenes, *Life's Form*, 69.
26. Kessler, "Psychology: The Intellective Soul," 507.
27. See Peter King, "Dispassionate Passions," in *Emotion and Cognition*, ed. Pickavé and Shapiro, 9–31.
28. St Thomas Aquinas, *Summa Theologica*, trans. Fathers of the English Dominican Province, 4 vols, vol. 1 (New York: Benziger Bros, 1947), http://dhspriory.org/thomas/summa/FP/FP082.html#FPQ82A4THEP1.
29. Ian Drummond, "John Duns Scotus on the Passions of the Will," in *Emotion and Cognition*, ed. Pickavé and Shapiro, 53–74.
30. Simo Knuuttila, "Sixteenth-Century Discussions of the Passions of the Will," in ibid., 116–32.
31. Sor Juana did not cite extensively and so any speculation must rest on the inflections of her argument. See Pérez-Amador, *De finezas y libertad*, for one example of a sustained argument on the influence of the "second scholastic," in this case Domingo de Báñez, on Sor Juana.
32. *Diccionario de la Lengua Castellana [Diccionario de Autoridades]*, vol. 4 (Madrid, 1734), 240, http://buscon.rae.es/ntlle/SrvltGUIMenuNtlle?cmd=Lema&sec=1.0.0.0.0.
33. Sor Juana Inés de la Cruz, *Obra selecta*, 2 vols (Caracas: Biblioteca Ayacucho, 1994), 1:450.
34. Ibid., 1:453.
35. Ibid., 1:454.
36. Ibid.
37. Ibid., 1:456.
38. Pasnau, *Thomas Aquinas on Human Nature*, 240.
39. Cruz, *Obra selecta*, 1:476.
40. Cited in ibid., 214.
41. Pasnau, *Thomas Aquinas on Human Nature*, 213.
42. It should be noted that although these sections are consistently found in criticism, at times they have been subdivided.
43. Soriano, *El Primero sueño*, 35–6.
44. Ibid., 36–7, 54–60. See also Gary Hatfield, "The Cognitive Faculties," in *The Cambridge History of Seventeenth-Century Philosophy*, ed. Daniel Garber and Michael Ayers, 2 vols (Cambridge University Press, 2008), 2:955–61.

45. Soriano, *El Primero sueño*, 54–60. See Fray Luis de Granada, *Introducción del Símbolo de la Fe* (Madrid: Cátedra, 1989), 430–6.
46. Sor Juana Inés de la Cruz, *Obras completas*, 4 vols (Mexico: Fondo de Cultura Económica, 1951), 1:340. All further citations from "First Dream" come from this edition. Verses will be indicated in the text. The translation of "First Dream" is from Sor Juana lnés de la Cruz, *Selected Works*, trans. Edith Grossman (New York: Norton, 2014), 77–110.
47. See Garrett Sullivan Jr, *Sleep, Romance, and Human Embodiment: Vitality from Spenser to Milton* (Cambridge University Press, 2012), 17–18.
48. For an excellent summary of Galenic influences in the early modern period, see Angus Gowland, *The Worlds of Renaissance Melancholy: Robert Burton in Context* (Cambridge University Press, 2006), 40–9.
49. See the citation from Averröes in Simo Knuuttila and Juha Sihvola (eds), *Sourcebook for the History of the Philosophy of Mind: Philosophical Psychology from Plato to Kant* (Heidelberg: Springer, 2014), 189.
50. Aquinas, *Summa Theologica*, 84.8 ad 2.
51. Aquinas, *Quaestiones disputatae de veritate* (12.3 ad 1), cited in Knuuttila and Sihvola (eds), *Sourcebook*, 191.
52. Soriano, *El Primero sueño*, 133–4.
53. While perfectly mapping the rest of the poem onto Thomist categories, Soriano surprisingly misses the reference to "intentional species." Soriano, *El Primero sueño*, 169–72.
54. The source of Sor Juana's attribution of this idea to Homer has not been found. See Soriano, *El Primero sueño*, 168.
55. Hatfield, "The Cognitive Faculties," 2:958.
56. Kessler, "Psychology: The Intellective Soul," 515.
57. Ibid., 516.
58. Soriano, *El Primero sueño*, 178.
59. Christopher Johnson, *Hyperboles: The Rhetoric of Excess in Baroque Literature and Thought* (Cambridge, MA: Harvard University Press, 2010), 222–77.
60. Paz, *Sor Juana Inés de la Cruz*, 496, 503–5.
61. Soriano, *El Primero sueño*, 348–9.
62. On Kircher's magic lantern as a metaphor, see Koen Vermeir, "The Magic of the Magic Lantern (1660–1700): On Analogical Demonstration and the Visualization of the Invisible Author," *The British Journal for the History of Science* 38.2 (2005).

7
Blood and Character in Early African American Literature

Hannah Spahn

James McCune Smith's unique series of character sketches, "Heads of the Colored People" (published in *Frederick Douglass' Paper* between 1852 and 1854), opens with the character of a maimed "black news-vender" who bears a remarkable resemblance to Thomas Jefferson:

> Our colored news vender *kneels* about four foot ten; black transparent skin, broad and swelling chest, whose symmetry proclaims Virginia birth, fine long hooked nose, evidently from the first families, wide loose mouth, sharpish face, clean cut hazel eyes, buried beneath luxuriantly folded lids, and prominent perceptive faculties. I did not ask him to pull off [his] cloth cap [covering] long greasy ears, lest his brow should prove him the incontestable descendant of Thomas Jefferson and Black Sal.[1]

Following this description of the news-vendor's facial traits, which, even to the point of hiding what may have been "long greasy ears," rather accurately corresponds to many portraits of the historical Jefferson, Smith's literary persona Communipaw presents a narrative of the third president's interracial family whose fuller recognition in historical scholarship would be a long time coming. According to Smith, Jefferson "contradicted his philosophy of negro hate" by begetting a number of "crocus-colored" slave children, whom he allowed successively to "run away." As a result, Communipaw declares, Jefferson's "mixed blood" descendants "are to be found as widely scattered as his own writings throughout the world."[2]

Jefferson's blood, as presented here, is in some measure comparable to, and competing with, the ink of his writings, just as his head and chest, documented by so many paintings and engravings, have left their imprint on later generations of an interracial America. Far from being unique, Smith insinuates, Jefferson's bodily characteristics might be copied by his own polygraph, printed on multicolored paper, and distributed "throughout the world." Describing the physical similarities between Jefferson and the subject of his character sketch, Smith is manipulating the conventions of a genre that had a long tradition of reflecting on its medium—*character* in its modern reference combining the senses of personality and of letter, in the original Greek referring to a stamp or imprint, for instance, on a coin—and thus on questions of typicality and "generalizability" as such.[3] Transposing the self-reflexive aspects of *character* into the mid-nineteenth century was especially useful in a literary endeavor that Carla Peterson has analyzed as Smith's ironic questioning of identity and "race."[4]

In this first character sketch of "Heads of the Colored People," Smith's aim is twofold. On the one hand, he ironically projects the logic of blood purity and of his contemporaneous phrenology back on Jefferson's "brow." And on the other hand, he gives dignity and individuality to a working-class African American, opening the doors to an alternative "Republic of Letters" that is "free from *caste*."[5] Smith's Republic of Letters (or, in a sense, of Characters) focuses on characters who professionally embody a quasi-literary medium: apart from the news-vendor, who seems "almost part of" the papers he sells,[6] the "Heads ... Done with a Whitewash Brush" include a boot- (or word-) polisher, a whitewasher, and an editor. While Smith thus stresses the literary "stamp" imprinted on American culture by African American individuals,[7] he illustrates, conversely, that Jefferson's writings were far less influential than were his physical attributes: in the final analysis, it was Jefferson's blood, not his ink, that had made Smith's character sketch possible.

With his series of "heads," Smith was not just writing against contemporaneous notions of racial types, he was also contesting and amending arguments from Enlightenment moral philosophy.[8] In his presentation of a "mixed blood" version of Jefferson "razed to the knees" by his social position as an escaped slave, who elicits "human sympathy" and "warm sympathy" not only from Communipaw himself but also from potentially racist bystanders, the Glasgow-educated McCune Smith points to a complex eighteenth-century heritage of imagining the interconnections among sympathy, character, and blood. This essay begins by sketching these relationships in the Enlightenment—how precisely could discourses of sympathy be aligned with discourses of blood?—proceeding to such African American writers as Lemuel Haynes, David

Walker, William Wells Brown, and, again, James McCune Smith, who effectively refer to Jefferson's "imprint," not simply to "refute" it, as Walker claimed, but rather to employ it as a shorthand for larger problems in American culture that could be useful in fleshing out their alternative visions.

Moral causes

While the historical Thomas Jefferson in one of his rare self-characterizations boasted of a "sanguine" temperament and a hopeful, optimistic nature,[9] his political imagination is more accurately described as oscillating between the sanguine, as he understood it, and the sanguinary. His Scottish-inflected attitude toward individual and national character can be summarized, roughly, as one that started out on the premises of an ostensibly "bloodless" discourse, but that allowed blood to creep in, as it were, through the back door. The occult nature of this blood flow made Jefferson's reasoning not so much self-contradictory as challenging for his critics to diagnose and counteract, from his reflections on blood and character in his draft of the Declaration of Independence onwards. Although it is most often assumed that the Declaration's universalist language bypassed these questions altogether, it can be argued that blood and character were indeed central to its argument, if they are explained by factors akin to what David Hume called "moral causes," as opposed to "physical" ones.[10] Since the tensions within Hume's approach foreshadowed, to a certain extent, problems that were to shape Jefferson's conception of blood and character in the second half of the century, they are worth being sketched here. Most succinctly in his 1748 essay "Of National Characters," Hume had criticized the relatively strong role assigned to the soil and climate in the formation of national character by thinkers from Aristotle to Montesquieu and their followers. For Hume, national characters were shaped by "moral causes" that included "all circumstances, which are fitted to work on the mind as motives or reasons, and which render a peculiar set of manners habitual to us," such as the form of government and, on a more general level, sympathy.[11]

On the face of it, this emphasis on "moral causes" in a sympathetic process of character formation relied on a conception of blood that presupposed a fundamental distinction between the breeding of the "races" of horses, for instance, and human procreation. As Hume famously put it, "[t]he races of animals never degenerate when carefully tended; and horses, in particular, always show their blood in their shape, spirit, and swiftness: But a coxcomb may beget a philosopher; as a man of virtue may leave a worthless progeny."[12] With his conception of "moral causes" in the formation of character, Hume opposed early modern traditions of attributing physical and moral character to blood in transfers from domestic animals to human beings.[13]

On the collective level, however, despite this universalist move to transcend the significance of "blood" relations, "moral causes" could nevertheless contribute to particularist and quasi-exceptionalist arguments about national uniqueness. For example, Hume distinguished between the French, whose authoritarian form of government could be seen to produce a dominant, monolithic national character, and the "great liberty and independency" of the English who, as he saw it, "of any people in the universe, have the least of a national character; unless this very singularity may stand for such."[14] As the connection between English "liberty and independency" and a flexible and unobtrusive, almost "characterless," national character suggests, "moral causes" shaped the structure of character as well as its actual content or characteristics. In both eighteenth-century Britain and revolutionary America, structural "characterlessness," as Susan Manning has analyzed it, could be part of "an ideologically privileged position," from which the existence of peculiarities of character could be depicted as problematic in itself, "as a manifestation of incomplete civility."[15] Conversely, Hume presented the very malleability of English national character as both a product and a sign of civilization and political progress, effectively singling out the English among all other nations.

The universalism of Hume's "moral causes" had further limitations. In his reflections on the peculiarities of national character, he repeatedly used the metaphor of "tincture" and "dye,"[16] evoking a complex set of associations oscillating between the color used for printing characters and the different "complexions" of mind as well as skin. "Of National Characters" has become notorious for a footnote Hume added in 1753 (and revised decisively at the end of his life),[17] in which he voiced a "suspicion" of racial inferiority:

> I am apt to suspect the negroes, and in general all the other species of men (for there are four of five different kinds) to be naturally inferior to the whites. There never was a civilized nation of any other complexion than white, nor even any individual eminent either in action or speculation ... Such a uniform and constant difference could not happen, in so many countries and ages, if nature had not made an original distinction betwixt these breeds of men.[18]

As eighteenth-century critics such as James Beattie or François Xavier Swediaur pointed out, this passage revealed a narrow-minded Eurocentrism while supporting a mistaken view of Hume as a defender of slavery.[19] It remains controversial whether Hume's claim of the existence of different "breeds of men" presented a logical departure from his larger argument in this essay—he had distinguished sharply, after all, between the "breeding" of horses and of men of virtue, respectively—or whether parts of this claim were, in fact, endemic to it. Taking into account Hume's correspondence with Montesquieu in 1749, Silvia Sebastiani has recently stressed the latter possibility: "Contra Montesquieu, Hume liberated humankind from the prison of climate but evoked an undercurrent of natural diversity, which logically preceded any sociological explanation, siding with the supporters of polygenesis."[20] This polygenetic stance, suggested by the original formulation of the note, appears closely connected to the part of the essay most often criticized by his contemporaries, his anticlerical satire of the "priestly character."[21] Seen from this perspective, the essentialist "undercurrent" of Hume's essay can also be read as a paradigmatic case when a new notion of "moral causes" resulted in the displacement of religious and dynastic discourses of blood, indirectly engendering new modes of assigning significance to blood in conceptions of national and "racial" character.

The voice of consanguinity

To return to Jefferson and thus to James McCune Smith's more immediate target of criticism, the American Declaration of Independence transferred to the political stage this partial displacement of religious and dynastic discourses of blood to national and racial contexts. Its performance of a radical break of familial and dynastic ties is enacted by a collective character shaped by "moral causes" and defining its own "liberty and independency" against the foil of "[a] prince whose character is ... marked by every act which may define a tyrant" and whose list of crimes, according to the double sense of *character* as personality and literary sign,[22] wants "no fact of distinguished die."[23] While George III is thus characterized by a concrete series of supposedly consistent actions, in an ironical reversal the American speaking subject assumes the malleable and potentially progressive "characterlessness"[24] of Hume's paradoxically exceptionalist account of English national character. Accordingly, the only attribute in the American self-characterization—one that is mentioned twice—is their "manliness." In a Scottish rather than Continental understanding of the term,[25] this manliness implied, not self-control, but precisely a shifting emotional status, a potentially unstable sensitivity to opposed perspectives. At least according to Jefferson's draft, Americans still felt tormented by their "former love" and "agonizing affection" for their English "brethren." And it is in the context of a "manly" longing for familial harmony that the collective American subject begins to refer to "our own blood," "our common blood," "common kindred," and "the voice of ... consanguinity."[26] At this point, after the list of the king's iniquities has been completed and the voice of consanguinity has fallen on "deaf" ears among their English brethren, Americans' emphasis on their "English" blood (opposed to the blood of "Scotch & foreign mercenaries" in Jefferson's draft) can only indicate a nostalgic longing for an impossible state in the past. The oscillation between expressing and restraining this nostalgia is nothing less than a performance of the "manly," of the shifting national character described above—a character that, in a Humean reading, is distinguished by its "liberty and independency." In other words, the particularist evocation of blood ties by "the voice of consanguinity" (retained in the Congress version) could not be separated from the seemingly universal foundations of American nationhood.

Jefferson's account of the birth of the nation is thus more of a "bloodstained gate"[27] into nationhood (including slavery) than may at first glance be visible. To be sure, Continental and Loyalist approaches tended to establish different, more direct connections between blood and character. J. Hector St John de Crèvecoeur, to quote a famous example, defined the "race now called Americans" in a more straightforward manner as a "strange mixture of blood" and a "promiscuous breed" of Northern and Western Europeans.[28] Since such immediate identifications of blood and character ran the risk of being contaminated by dynastic questions, the revolutionary side had a stake in relying on alternative, sympathy-based accounts of character that either abstracted from the body altogether or privileged aspects other than blood, the nervous system in particular.[29] Nevertheless, the ambivalent evocation of blood in the Declaration of Independence—documenting the "voice of consanguinity" in the very act of severing all dynastic and familial ties to the mother country—conveys a strong sense of the nation-making potential of blood.

In this context, it is important to consider *whose* blood is shed in the Declaration. Jefferson's draft version describes a transatlantic world of war and violence in which the destruction of the Anglo-American family is immediately preceded by the destruction of what for Jefferson was a single "African" family (and potential nation)[30] by the king's "cruel war against human nature itself" in the Atlantic slave trade. Although this passage represents the emotional climax of the enumeration of the king's crimes and a relatively high point in Jefferson's opposition to slavery, its writer could not bring himself to mention physical bloodshed in the king's violation of "the most sacred rights of life and liberty in the persons of a distant people who never offended him." In contrast to this rather abstract language, he reserved the concrete "expense" of blood[31] for internal disputes among members of the Anglo-American family.

Jefferson's fixation on "English" blood, which failed to do metaphorical justice even to the genealogies of all Congress members (including his own), thus points to a more far-reaching omission. In his subsequent writings, he was inclined to illustrate revolutions and the founding of republics by gory imagery, as when he spoke of a "tree of liberty" that had to be "refreshed from time to time with the blood of patriots & tyrants."[32] Nations and national characters supposedly emerged out of, and were consolidated by, the voluntary bloodshed of citizens: to quote a retrospective assessment of the revolutionary wars from a period when this idea still retained some of its appeal, nations "drank the blood of free personalities, as it were, to attain personality as well."[33] Slaves were by definition excluded from this logic of national character, and so were, for Jefferson, all American blacks until they founded their own nation elsewhere. Since this moment seemed to recede further and further into the future, he at best associated them with another bodily fluid: as he put it on a rare occasion when he presented greater personal sympathy with the slaves themselves, they were merely waiting for the "measure of their tears" to fill up.[34]

This tendency to exempt blacks from rhetorical blood baths was aggravated in Jefferson's nineteenth-century writings, when after what he regarded as a glorious consolidation of American nationhood (his election to the presidency) he tried to construct an exceptionalist narrative of the United States as a realm of peaceful progress.[35] This fateful shift in his argument implied channeling the revolutionary blood flow to the other side of the Atlantic, for example in his prediction of Virgilian "rivers of blood" that had yet to "run out" before Europeans could achieve republicanism.[36] Yet, it also meant a departure from the ancient idea of slavery as a latent state of war—"but the State of War continued," as John Locke put it[37]—whose injustice Jefferson had recognized and made a major argument against slavery in his eighteenth-century writings. According to this line of thought, slaves were captives whose lives had been spared; hence they were slaves precisely because they had *not* shed their blood. Should, however, the latent war of slavery rekindle and slaves begin to engage in bloodshed, Jefferson's insight that "[t]he Almighty has no attribute which can take side with us in such a contest" made him fear the "extirpation" of the unjust slaveholders.[38] Faced with this uninspiring prospect, the nineteenth-century Jefferson tried his best to construct an alternative to this argument in an exceptionalist rhetoric that downplayed all possibilities of war on US soil. Instead of captives, he sought to depict American slaves as members of an antiquity-inspired idea of a peaceful "family":[39] that is, not of a community of "blood" relations, but of a paternalist household. In this position, there was no good reason for their blood to leave their collective body—or rather, it became a national problem when it did.

Staining the blood

This background might suggest an additional reason why Jefferson in his well-known supposition of blacks' inability to blush stressed the utter invisibility of his slaves' blood through the skin, hidden as he supposed—or wanted it to remain—behind an "immoveable veil of black."[40] Blood leaked through only two, albeit crucial passages of Jefferson's oft-discussed "suspicion only" of racial inferiority in Query XIV of *Notes on the State of Virginia* (1787). In this text, whose intellectual limitations even by eighteenth-century standards have been described as those of "a home-grown interpretation" of an "increasingly obsolescent" natural history, Jefferson puts relatively little emphasis on questions of anatomy.[41] Initially, he mentions blood only in passing, and inconclusively, as a possible source of skin color: "whether it ['the black of the negro'] proceeds from the colour of the blood, the colour of the bile, or some other secretion, the difference is fixed in nature, and is as real as if its seat and cause were better known to us." In this remarkable phrase, the naïve suggestion that "black" blood itself might have different coloring characteristics, as if a bleeding slave had never been sighted in Virginia, suggests a Jefferson who quite literally tried to avert his eyes from his slaves' blood. That he professed indifference to the much-debated question of the origin of skin color is an indication of his priorities. Conveying precise anatomical information was far less important than stressing the "reality" of an aesthetic problem whose moral implications, for him, legitimated a regression behind Hume's distinction of horses and coxcombs ("The circumstance of superior beauty, is thought worthy attention in the propagation of our horses, dogs, and other domestic animals; why not in that of man?").[42] This connection of blood and "racial" purity anticipates the conclusion of Jefferson's racist musings:

> Among the Romans emancipation required but one effort. The slave, when made free, might mix with, without staining the blood of his master. But with us a second is necessary, unknown to history. When freed, he is to be removed beyond the reach of mixture.[43]

If it mattered whose blood was spilled during the nation's collective birth, it followed that a "stain" on this blood could alter national character. As might be suggested by the metaphor of the stain, Jefferson was continuing and democratizing dynastic discourses of blood purity in this passage as well as in his stress on "English" blood in the Declaration of Independence. On the European stage, he employed the metaphor to describe the immoral behavior of Marie Antoinette, whose production of an illegitimate heir, as he saw it, would "forever stain the pages of modern history."[44] Yet while the queen's stain had a literary-historical quality, the one potentially left by American slaves directly interfered with their masters' blood. Thus, instead of characters in writing, their stain concerned that which was becoming a quasi-natural foundation of (national) character. At the same time, this stain on Virginia's "natural aristocracy"[45] was not as personal as was Marie Antoinette's on the history of the House of Bourbon. Since Jefferson must have known that feminine pronouns would have reflected a more widespread practice in Virginia and elsewhere, his use of masculine pronouns ("*his* master," "*he* is to be removed") may have resulted from his attempt to make a general statement on "the" slave's blood. Such a statement would have gone along with the increasing tendency in the years and decades following the publication of *Notes* to attribute a decisive role to "racial" percentages of blood, for instance, in the definition of franchise requirements.[46] After Jefferson had finished voicing his elaboration of Hume's polygenetic "suspicion," it could be said that what had begun as the aesthetic "reality" of blood and skin color was in the process of congealing into a legal one.

Jefferson's personal solution to the problem of "mixture," apart from his never-to-materialize "expatriation" plans, is readily suggested by a letter he wrote in 1815 in response to the question "what constituted a mulatto by our law?"[47] Explicating the 1792 Virginia law describing those persons as mulattoes who had "one-fourth part or more of negro blood," he admitted that "a variety of fractional mixtures" could cause "a mathematical problem of the same class with those on the mixtures of different liquors or different metals." Corresponding to his comparison to metallurgy (evoking the newly coined term *amalgamation*),[48] he recommended an "algebraical notation," with which he neatly calculated different "crosses" of "white" and "negro" blood ("a third cross clears the blood") in a telling comparison to a merino ram and a country ewe ("It is understood in natural history that a forth cross of one race with another gives an issue equivalent for all sensible purposes to the original blood"). Conveniently, as can be added from today's perspective, his own children with his slave Sally Hemings were therefore considered to be white. "But observe," he concluded immediately following his unflattering parallel to the breeding of merinoes, "that this does not re-establish freedom, which depends on the condition of the mother." After emancipation, however, her offspring "becomes a free *white* man, and a citizen of the United States to all intents and purposes."

Our fathers' blood

Writers from the African American archive leading up to James McCune Smith found much in Jefferson's approach that could be used to illustrate problems in American culture at large. In many ways, Jefferson's particular form of "blood magic and blood thinking"[49]—an account of (national) character that united a universalist reliance on sympathy to a particularist emphasis on blood purity that became more virulent as the nineteenth century progressed—turned out to mark the borders of the dominant national discourse of his period. Meanwhile, of course, it was far from defining the limits of the historically thinkable in general. From early on, African Americans developed alternative approaches, approaches that not only refuted Jeffersonian thinking or extended its scope, but that were in many cases based on, and shaped, different categories.[50]

A prominent contemporaneous example by a writer who directly referred to the Declaration of Independence is Lemuel Haynes's essay *Liberty Further Extended: Or Free Thoughts on the Illegality of Slavery* (c. 1776). Together with its epigraph from the Declaration's second paragraph, the title of this early essay, which was not published until 1983, seems to suggest a stable frame of reference. Yet instead of keeping the premise of the sympathy-based, progressively evolving "liberty and independency" of American character intact, Haynes's *Liberty* gradually shifts from a first person plural to a division between the speaker and an audience composed of revolutionary "Englishmen," which is eventually narrowed down to "all such as are Concearn'd in the practise of *Slave-keeping*."⁵¹ The initial collectivity in the Preface is defined by a combination of republican discourse ("son[s] of freedom" fighting "Tyrony") and Calvinist introspection:

> But while we are Engaged in the important struggle, it cannot Be tho't impertinent for us to turn one Eye into our own Breast, for a little moment, and See, whether thro' some inadvertency, or a self-contracted Spirit, we Do not find the monster Lurking in our own Bosom.⁵²

In the essay's conclusion, however, what had been a common struggle based on shared "moral causes" of a first person plural has come to be identified with the second person plural of the revolutionary slaveholders alone:

> Sirs, the important Caus in which you are Engag'd in is of a[n] Exelent nature, 'tis ornamental to your Characters, and will, undoubtedly, immortalize your names thro' the Latest posterity. And it is pleasing to Behold that patriottick Zeal which fire's your Breast; But it is Strange that you Should want the Least Stimulation to further Expressions of so noble a Spirit.⁵³

Haynes's rhetorical detachment from the "important Caus," with whose language of natural rights he identified in the text and for which he had been fighting, is paralleled by the division of the slaveholders' "Characters" into "pleasing" and "Strange" aspects. His reason for this split—a lack of "congruity amidst your Conduct" in white Americans' continuation of slavery and the slave trade—is expressed by powerful imagery related to blood. Reflecting a life-long concern,⁵⁴ Haynes's creative preoccupation with blood had begun even before his argument in *Liberty* in a patriotic ballad titled "The Battle of Lexington" that was rediscovered along with the essay more than two centuries later. In it, a "bleeding" Liberty that was further "Sealed" with blood emerges as a central theme, with blood running in "great Effusion" on all sides, but shed especially by "Sons of Freedom" encountering an "inhuman" foe "thirsting" for blood.⁵⁵ Although chattel slavery is not directly criticized, the final stanza may obliquely refer to it in its use of the widespread argument that "Sin is the Cause of all our Woe." Most importantly, Haynes openly presents himself as a self-taught "young Mollato" in the poem's subtitle. Thus, his identification with the Calvinist-republican synthesis of a unified patriotic character also comments on the nature of the blood spilled in the revolutionary war. By inscribing himself into a New England history when *"Our* Fathers Blood did freely flow" and referring to the British as *"our* native kin,"[56] he claims membership in the family, turning the revolutionary bloodshed, consistently with his biographical experience as a New England minuteman, into a transracial struggle for liberty.

Drunk with human blood

This tendency is radicalized in *Liberty Further Extended*, which fuses the bloodshed of the revolutionary war with that of the slave trade and slavery. Countering the twin Jeffersonian arguments of an exclusively "English" blood community in the American Revolution and the essentially bloodless nature of the slave trade and slavery, Haynes begins by evoking a "great Effusion of Blood" as a universal characteristic of all fights for *"Liberty & freedom"* (understood as "the subject of many millions Concern").[57] As the following paragraphs make clear, his conception of blood relies more strongly on scriptural associations than does the Jeffersonian one. Possibly inspired by the version in Anthony Benezet's *Some Historical Account of Guinea* (1771),[58] Haynes quotes part of a verse from Acts 17:26: "It hath pleased god to *make of one Blood all nations of men, for to dwell upon the face of the Earth.*" At least since Samuel Sewall's ambivalent use in *The Selling of Joseph* (1700), versions of this verse were prominent in antislavery writing and would remain important in African American discussions of race in a long nineteenth century by writers as diverse as Olaudah Equiano, Hosea Easton, William Wells Brown, McCune Smith, and Pauline Hopkins. Haynes's rendering clearly stressed the verse's universalist implications (omitting, for instance, the "bounds" God has appointed to the "habitation" of nations toward the end of Acts 17:26). Thus, it prepared the ground for an impassioned attack on slavery and the slave trade that also reads like a specific attack on the complete omission of "Affrican-Blood" in the Declaration of Independence:

> O! what an Emens Deal of Affrican-Blood hath Been Shed by the inhuman Cruelty of Englishmen! that reside in a Christian Land! ... O ye that have made yourselves Drunk with human Blood! altho' you may go with impunity in this Life, yet God will hear the Crys of that innocent Blood, which crys from the Sea, and from the Ground against you, Like the Blood of Abel, more pealfull [?] than thunder,

vengence! vengence! What will you Do in that Day when God shall make inquisision for Blood? he will make you Drink the phials of his indignation which Like a potable Stream shall Be poured out without the Least mixture of mercy; Believe it, Sirs, their shall not a Drop of Blood, which you have Spilt unjustly, Be Lost in forgetfulness. But it Shall Bleed afresh, and testify against you, in the Day when God shall Deal with Sinners.[59]

As John Saillant has demonstrated, Haynes went beyond arguments by antislavery writers such as Benezet in "his characteristically expansive manner of superimposing scriptural allusions and expressions of emotion onto the white abolitionists' barer sentences and sparser references."[60] In this passage, Saillant shows, apart from the story of Abel's blood crying from the ground (Genesis 4:10), which Benezet had taken over from Bartolomé de las Casas, Haynes evoked not only a long Old Testament tradition linking blood and revenge, but also passages from chapters 16 and 17 of Revelation, in which rivers are turned into blood (Revelation 16:4), the murderers of saints and prophets are given blood to drink (Revelation 16:6), and the harlot is "drunken with the blood of the saints, and with the blood of the martyrs of Jesus" (Revelation 17:6).[61]

The dense, over-signified conception of blood in Haynes's *Liberty*—fusing the historical bloodshed of the Atlantic slave trade and the revolutionary wars with the scriptural blood of Abel, of all nations, of the innocent, of saints, and of God's punishments in the end times—provided a rich arsenal for much later African American writing. While Haynes used it to stress the universal nature and significance of blood, he specified character in his construction of a morally split patriotic cause. As a consequence, *Liberty Further Extended* can also be read, in part, as a structural inversion rather than simple "extension" of the Declaration of Independence, which had conversely stressed the universal foundations of American character while specifying its conception of blood. In a move that would be taken up, for instance, by Prince Hall's *Charge* of 1797, Haynes preempted the exceptionalism of later Jeffersonian rhetoric by making visible the universal claims that could be tied to "African-Blood." Until the slave trade and slavery were abolished, he showed, all rhetorical efforts to direct the flow of Virgilian or scriptural "rivers of blood" to a place outside of the United States would remain utterly implausible.

In his concern with blood, Haynes's most immediate heir in the nineteenth century may have been David Walker. In his famous *Appeal, in Four Articles; ... to the Coloured Citizens of the World, but in Particular, and Very Expressly, to Those of the United States of America* (1829), Walker takes up several of Haynes's scriptural references to blood—in particular, the themes of Abel's blood crying "from the ground," the manifold ties between blood and revenge, and the image of blood drinking—while transferring them into new contexts. Thus, Walker stresses the legal and economic rather than mainly apocalyptic dimension of white Americans having "sweet times on our blood and groans" and getting "so fat on our blood and groans," while African Americans "enriched" the soil with their "blood and tears," constantly in danger of being driven "from our property and homes ... earned with our blood."[62] Except for the changing priorities from transatlantic to class and race conflict, however, Walker largely concurs with Haynes in stressing the universal nature of blood, which Walker tends to couple either with another universal bodily fluid (sweat or tears), with the pre-linguistic sound of "groans," or with abstract terms such as "oppression."[63]

Walker's repetitive stress on the universality of blood is opposed, as it had been in Haynes's *Liberty*, to a specification of character, as when Walker asks his readers sarcastically:

> Do you know that Mr. Jefferson was one of as great characters as ever lived among the whites? See his writings for the world, and public labours for the United States of America. Do you believe that the assertions of such a man, will pass away into oblivion unobserved by this people and the world?[64]

In contrast to Haynes's *Liberty* or Daniel Coker's *A Dialogue between a Virginian and an African Minister* (1810), Walker's reflections on Jefferson and Jefferson's writings in Article I of the *Appeal* were composed at a historical moment when, as Walker was glad to remind his audience, this supposedly "great character" was "gone to answer at the bar of God, for the deeds done in his body while living."[65] As Gene Jarrett has recently argued, from this historical vantage point Walker tends to streamline his historical foe Jefferson with his contemporary foes among the colonizationists, depicting them as "racist bedfellows."[66] Walker's sharp criticism of Jefferson's racist utterances in Query XIV of *Notes on Virginia* thus becomes a representative criticism of the racism he saw ingrained in American culture at large. Jefferson's "great character" emerges as both a cause and a symptom of this racism: a cause, as his writings and relative prestige were able to influence and support other Americans in their racist views; and a symptom, as his "great character" had remained largely undamaged by the publication of Query XIV. Walker thus exploits various connotations of *character*—person, name, written expression, and historical reputation—in an argument that condenses his attack on a "set of blood-thirsty beings" in the character of a man whose rhetoric had tried to extract all blood from his black slaves: "the very learned and penetrating Mr. Jefferson."[67]

Fusing and selling blood

Despite his emphasis on the representativeness of Jefferson's views, Walker was writing in an intellectual environment that had changed considerably since the time of *Notes on the State of Virginia*, not least in its medical assessment of blood. While blood-letting was slowly going out of fashion, nineteenth-century physicians began to experiment with interhuman blood transfusion. In the period following a partly successful transfusion experiment in 1818 and before the discovery of different blood types in 1907, doctors were inclined to profess an exaggerated faith in the universal characteristics of blood.[68] At the same time, of course, many contemporaries of the nineteenth century also became more radical in attributing particular "racial" or national characteristics to blood. The tensions between the two poles produced a rich American literature concerned not so much with medical transfusion as with metaphorical forms of blood "mixture" and "amalgamation." Jefferson's and Haynes's eighteenth-century logic of blood-letting—transitively "bleeding" citizens and sinners for the sake of the nation and salvation—was losing ground to a growing nineteenth-century interest in questions concerning the racial "fusion" of blood.

Unfortunately for Jefferson's long-term reputation, his character lent itself beautifully to discussions of these new questions as well. Despite his ostensible praise for Jefferson, between the lines of his *Dialogue* Daniel Coker may already have included his case in the discussion of the "undeniable truth" of the "mixture of blood" on "gentleman's seats that abound with slaves."[69] However, it remained for James McCune Smith's character sketch "The Black News-Vender" (1852) and his essay "On the Fourteenth Query of Jefferson's Notes on Virginia" (1859) as well as William Wells Brown's novel *Clotel* (1853) to spell out the full implications of Jefferson's "mixed blood" descendants. Brown's novel, to begin with the best-known text among the three, opens its oft-discussed transition from historical to fictional writing by presenting its theme that "the blood of the first American Statesmen coursed through the veins of the slave of the South" as a quotation from a historical figure, the Virginia slaveholder John Randolph.[70]

Blending questions of blood and genre, the novel goes on to depict a generational development from Jefferson's historical character to the semifictional ones of his daughters, Clotel and Althesa, and his granddaughters, Ellen, Jane, and Mary. In the process, it repeatedly juxtaposes this "mixed blood" saga of Jefferson's historical-fictional slave family to passages from the Declaration of Independence and *Notes on Virginia*, as if to confirm McCune Smith's dictum that Jefferson's "descendants of mixed blood are to be found as widely scattered as his own writings."[71]

In his panorama of a racially mixed America, Brown integrates medical practices and arguments from different phases of the cultural history of blood. In his chapters on medicine (of which Brown himself would become a practitioner), the novel recounts a shocking instance of slaves being bled to death following anatomical experiments.[72] On the other hand, it also presents a newspaper account detailing an instance of interracial blood transfusion to make possible a marriage in New Orleans.[73] Traditional eighteenth- and early nineteenth-century conceptions of blood and national character may find their clearest articulation in the didactic monologues of the liberator-martyr Georgiana (ironically named, or so it would seem in this context, after Jefferson's foil in the Declaration of Independence). Arguing against colonization, Georgiana mentions Crispus Attucks and evokes the trope of an American soil "enriched" with African Americans' "blood and tears."[74] Meanwhile, the novel sharply criticizes the extreme commodification implied in tying blood to forms of "enrichment," most clearly in its key scene of the sale of Clotel:

> This was a Southern auction, at which the bones, muscles, sinews, blood, and nerves of a young lady of sixteen were sold for five hundred dollars; her moral character for two hundred; her improved intellect for one hundred, her Christianity for three hundred; and her chastity and virtue for four hundred more.[75]

Given the potential anti-humanism of dividing human beings into physical, mental, and moral aspects—a procedure performed, to some degree, by Jefferson's Query XIV—and even attaching monetary value to each, it is no wonder that the husband of Jefferson's second fictional daughter Althesa can argue, in a chapter entitled "A True Democrat": "Our nation is losing its character. The loss of a firm national character, or the degradation of a nation's honor, is the inevitable prelude to her destruction."[76]

The right development of character

In an interesting final twist of the novel, the reunion of Clotel's daughter Mary with her lover George takes place on a European graveyard, perhaps an image for the novel's dead (fictional or historical) characters, evoking, in addition, the gradually disappearing characters on tombstones, a famous image for fading ideas in the mind.[77] James McCune Smith was likewise intrigued by this setting. Three years after the appearance of *Clotel*, he published in *Frederick Douglass' Paper* an essay titled "The New Pen and Old Graveyards" that may help illuminate his approach to blood and character in his criticism of Jeffersonian attitudes. The essay begins as a mock obituary of Smith's old pen that "has written so many Communipaws," asking whether after its sale to a jeweler "it will 'go to pot' with other old gold and mix its ultimate particles with what will be changed into a beautiful finger ring, or maybe a lady's ear drop, or infant's bracelet clasp, or a flaming breast pin!"[78] Following this reflection on the vanity of his gold pen and its possible afterlife, the essay abruptly shifts to a plain Quaker graveyard, where Communipaw reflects on the "very distinct traits of human character" exemplified by modest Quaker burial rites, on the one hand, and the pyramids of ancient Egypt, on the other. In the end, he criticizes both cultures, despite their antislavery history and their potential for race pride, for their lack of "imagination" and "faith." He only finds his spirits lifting upon entering St Paul's cemetery with tombstone inscriptions detailing the names and lives of the deceased. What had been a cryptic connection to the essay's first part gradually comes to light here. In contrast to the Episcopal dead buried at St Paul's, Quakers and ancient Egyptians alike did not rely on letters on their way to the next life, a criticism of "people without characters" reminiscent of, for instance, Daniel Defoe's analogous one of Quakers and Egyptians in *An Essay upon Literature* (1726).[79] Smith's gold pen (alluding, perhaps, to the different conception of his friend Frederick Douglass's "gold" of character[80]), like the tombstones at St Paul's, is the means through which character, for Smith, gains its true significance: as an imaginary as well as faithful form of writing able to reflect on, and shape, its own moral status in this world and the next.

Thus, Smith both continued and moved to new literary grounds the relationship between blood and character developed by his African American contemporaries and forebears, from Haynes's partial inversion of the Jeffersonian argument onwards. Smith's conception of *character*, which can only be sketched here, may be described as a highly individualized and deeply literary one. With occasional departures that can be understood as mainly metaphorical,[81] his conception of *blood* was that of a well-trained physician and "thoroughgoing environmentalist"[82] who stressed its universal nature, for instance in his "Civilization" essay (1859). The "right development of character *here* where God has placed us," as Smith explained in his essay "Human Brotherhood and the Meaning of Communipaw" (1852), could therefore be consistent with "mingling" blood until differences of skin color would disappear in a better American future.[83]

In "On the Fourteenth Query of Thomas Jefferson's Notes on Virginia," Smith accordingly criticizes the Jeffersonian concern with blood purity in scientific, moral, and literary terms. Putting the term "admixture" of blood in quotation marks, he presents a long list of dated and mistaken scientific assumptions, ridiculing their lack of empirical grounding ("Microscopic science has exploded this idea") as well as their triviality.[84] Further, Smith casts Jefferson, who prided himself on his faith in human progress, as an essentially unprogressive thinker: "The question asked by Mr. Jefferson in his fourteenth query, would never have been propounded had he been acquainted with the philosophy of human progress."[85] What Jefferson did not see, Smith argues referring to two theorists of character who became important in the nineteenth century, Wilhelm von Humboldt and John Stuart Mill, was that it was the "diversity of character and culture," not their "characterless" universality, that was the source of progress.[86]

James McCune Smith's ultimate criticism of Jefferson is a literary one. Opposing his own clear structure to that of *Notes* ("so mixed and so confused"), he ironically turns the tables on Jefferson's literary criticism of Phillis Wheatley and Ignatius Sancho.[87] He also detects a "stain" in a form of literary influence that wrongly relies on universalizing character: "In fact, the only stain upon the literary merit of de Tocqueville's great work consists in this: he gives Mr. Jefferson's views as if they were de Tocqueville's views, and gives them in Mr. Jefferson's own words."[88] Whereas Jefferson had warned about "stains" on the pure English blood of Virginia slaveholders, and Brown had criticized slavery as a "stain" on "America's otherwise fair escutcheon,"[89] for Smith a "stain" could result simply from this: Jefferson's words.[90]

Notes

1. James McCune Smith, "'Heads of the Colored People,' Done with a Whitewash Brush: The Black News-Vender," in *The Works of James McCune Smith: Black*

Intellectual and Abolitionist, ed. John Stauffer (Oxford University Press, 2006), 190–4. On this text, I have consulted Stauffer's "Introduction" (at 187–90); Bruce Dain, *A Hideous Monster of the Mind: American Race Theory in the Early Republic* (Cambridge, MA: Harvard University Press, 2002), 239–42; John Stauffer, *The Black Hearts of Men: Radical Abolitionists and the Transformation of Race* (Cambridge, MA: Harvard University Press, 2002), 218–24.

2. Smith, "Black News-Vender," 191–2.
3. See Deidre S. Lynch, *The Economy of Character: Novels, Market Culture, and the Business of Inner Meaning* (University of Chicago Press, 1998), ch. 1, esp. 30–5; 40–1.
4. Carla L. Peterson, "Untangling Genealogy's Tangled Skeins: Alexander Crummell, James McCune Smith, and Nineteenth-Century Black Literary Traditions," in *A Companion to American Literary Studies*, ed. Caroline F. Levander and Robert S. Levine (Malden, MA: Blackwell, 2011), 500–16, esp. 511–15.
5. Smith, "Black News-Vender," 190.
6. Ibid., 191.
7. See also Smith, "Reforms Are Mere Acts of Intellection," in *Works of James McCune Smith*, 172–4, at 173–4 (for his "stamp" and "mark" metaphor).
8. Apart from the context of nineteenth-century phrenology as a background for Smith's title, there is also an eighteenth-century context of engraved portrait "heads." See Lynch, *Economy of Character*, 34–5.
9. Thomas Jefferson to John Adams (8 April 1816), in *Thomas Jefferson: Writings*, ed. Merrill D. Peterson (New York: Library of America, 1984), 1381–4, at 1382.
10. David Hume, "Of National Characters," in *Essays and Treatises on Several Subjects* (London, 1758), 119–29, at 119. I have discussed the Scottish background of the Declaration of Independence at greater length in "Character and Cosmopolitanism in the Scottish-American Enlightenment," in *Character, Self, and Sociability in the Scottish Enlightenment*, ed. Susan Manning and Thomas Ahnert (Basingstoke: Palgrave Macmillan, 2011), 207–24.
11. Hume, "Of National Characters," 121. On Enlightenment views of national character in Britain, see Colin Kidd, "Constitutions and Character in the Eighteenth-century British World," in *From Republican Policy to National Community. Reconsiderations of Enlightenment Political Thought*, ed. Paschalis Kitromilides (Oxford: Voltaire Foundation, 2003), 40–61.
12. Hume, "Of National Characters," 128.
13. On the example of horse races, see Werner Conze and Antje Sommer, "Rasse," in *Geschichtliche Grundbegriffe*, ed. Otto Brunner, Werner Conze, and Reinhart Koselleck, 7 vols (Stuttgart: Klett, 1984), V: 135–78, at 139–40.
14. Hume, "Of National Characters," 124. On the similarities between Hume's and Diderot's accounts of the enforced homogeneity of French national characteristics, see Jerrold Seigel, *The Idea of the Self: Thought and Experience in Western Europe since the Seventeenth Century* (Cambridge University Press, 2005), 204–5.
15. Susan Manning, *Poetics of Character: Transatlantic Encounters 1700–1900* (Cambridge University Press, 2013), esp. chs 5 and 2, at 129; 59.
16. Hume, "Of National Characters," 122.
17. See John Immerwahr, "Hume's Revised Racism," *Journal of the History of Ideas* 53.3 (July–September 1992), 481–6. Hume's revised formulation in the posthumous 1777 edition targeted blacks exclusively while, arguably, downplaying polygenesis in omitting the phrase "other species of men."

18. Hume, "Of National Characters," 125 n.
19. See *Early Responses to Hume's Moral, Literary and Political Writings*, ed. James Fieser, 10 vols (Bristol: Thoemmes Press, 1999), 2: 255–9; 343–8.
20. Silvia Sebastiani, "National Characters and Race: A Scottish Enlightenment Debate," in *Character, Self, and Sociability*, ed. Manning and Ahnert, 187–205, at 191.
21. Hume, "Of National Characters," 119–20, 121 n.
22. Manning, *Poetics of Character*, 26.
23. Thomas Jefferson, "A Declaration by the Representatives of the United States of America, in General Congress Assembled" (including his draft version), as inserted in his "Autobiography," in *Jefferson: Writings*, 19–24, at 22.
24. On "characterlessness" in the context of nationality, see Manning, *Poetics of Character*, esp. 59–62; 123–7.
25. Seigel, *Idea of the Self*, 150–1.
26. Jefferson, "Declaration," 22–3.
27. Frederick Douglass, *Narrative of the Life of Frederick Douglass, an American Slave, Written by Himself*, ed. William Andrews and William McFeely (New York: Norton, 1997), 15.
28. J. St John de Crèvecœur, Letter III, "What is an American?," *Letters from an American Farmer* (1782) (New York: Penguin, 1986), 69; 71.
29. On the problems of connecting sympathy to blood, see Christine Levecq, *Slavery and Sentiment: The Politics of Feeling in Black Atlantic Antislavery Writing, 1770–1850* (Hanover: University of New Hampshire Press, 2008), esp. ch. 4.
30. See Peter Onuf, *Jefferson's Empire: The Language of American Nationhood* (Charlottesville: University of Virginia Press, 2000), ch. 5.
31. Jefferson, "Declaration," 22.
32. Jefferson to William S. Smith (13 November 1787), in *Jefferson: Writings*, 910–12, at 911.
33. Friedrich Meinecke, *Cosmopolitanism and the National State* (1907), trans. Robert Kimber (Princeton University Press, 1970), 15. I have discussed Meinecke and Jefferson's shared interest in blood imagery in the "Introduction" of *Cosmopolitanism and Nationhood in the Age of Jefferson*, ed. Peter Nicolaisen and Hannah Spahn (Heidelberg: Winter, 2013), 3–22, at 7–8.
34. Jefferson to Jean-Nicolas Démeunier (26 June 1786), in *The Papers of Thomas Jefferson*, ed. J. Boyd et al., 39 vols to date (Princeton University Press, 1950–), 10: 63.
35. I have made this argument at greater length in my *Thomas Jefferson, Time, and History* (Charlottesville: University of Virginia Press, 2011), chs 5 and 6.
36. See, for example, Jefferson to Benjamin Austin (9 January 1816), in *Jefferson: Writings*, 1370. On Jefferson's use of the Sibylline prophecy, see Peter Onuf and Nicholas Onuf, *Nations, Markets, and War. Modern History and the American Civil War* (Charlottesville: University of Virginia Press, 2006), 343–52.
37. John Locke, "Of Slavery," in Book II of *Two Treatises of Government*, ed. Peter Laslett (Cambridge University Press, 1964), 302. On Jefferson's use, see Onuf, *Jefferson's Empire*, ch. 5; Onuf, "Domesticating the Captive Nation: Thomas Jefferson and the Problem of Slavery," in *Jefferson, Lincoln, and Wilson: The American Dilemma of Race and Democracy*, ed. John

Milton Cooper and Thomas Knock (Charlottesville: University of Virginia Press, 2010).
38. Thomas Jefferson, "Query XVIII," *Notes on the State of Virginia*, ed. William Peden (1954; Chapel Hill: University of North Carolina Press, 1982), 162–3, at 163.
39. For such a use of *family* as *household*, see, for example, Jefferson's letter to his daughter Martha (13 October 1805), in *The Family Letters of Thomas Jefferson*, ed. Edwin M. Betts and James A. Bear (Charlottesville: University of Virginia Press, 1986), 279.
40. Jefferson, *Notes*, 138.
41. See Dain, *Hideous Monster*, 38; 32.
42. Jefferson, *Notes*, 138.
43. Ibid., 143.
44. Jefferson, "Autobiography," in *Jefferson: Writings*, 92.
45. For his elaboration of the concept, see Jefferson to John Adams (28 October 1813), in *Jefferson: Writings*, 1304–10.
46. Douglas Egerton, *Death or Liberty: African Americans and Revolutionary America* (Oxford University Press, 2009), ch. 9, esp. 247.
47. Jefferson to Francis Gray (4 March 1815), in *The Writings of Thomas Jefferson*, ed. Andrew Lipscomb and Albert Bergh, 20 vols (Washington: Thomas Jefferson Memorial Association, 1903–4), 14: 268–70.
48. On the history of the term, see Elise Lemire, *"Miscegenation": Making Race in America* (Philadelphia: University of Pennsylvania Press, 2002), 4; 51.
49. Ralph Ellison, "The Little Man at Chehaw Station," in *The Collected Essays of Ralph Ellison*, ed. John Callahan (New York: Modern Library, 1995), 489–519, at 505.
50. On the problems of overemphasizing Jefferson as a source of one-directional influence on African American literature, see Dain, *Hideous Monster*, ch. 1, esp. 4–6, and Gene Jarrett, "'To Refute Mr. Jefferson's Arguments Respecting Us': Thomas Jefferson, David Walker, and the Politics of Early African American Literature," *Early American Literature* 46.2 (2011), 291–318, at 308.
51. Lemuel Haynes, "Liberty Further Extended," in *The Literatures of Colonial America: An Anthology*, ed. Susan Castillo and Ivy Schweitzer (Malden, MA: Blackwell, 2001), 573–80, at 579.
52. Ibid., 573.
53. Ibid., 580.
54. See John Saillant, *Black Puritan, Black Republican: The Life and Thought of Lemuel Haynes, 1753–1833* (Oxford University Press, 2003), chs 1 and 5.
55. Ruth Bogin, "'The Battle of Lexington': A Patriotic Ballad by Lemuel Haynes," *William and Mary Quarterly* 42.4 (October 1985), 499–506. The quotations are from stanzas 12 and 14; 13; 24 and 25; 17.
56. Haynes, "Battle," stanzas 29 and 31, my emphasis.
57. Haynes, "Liberty," 574; 575.
58. Saillant, *Black Puritan*, 17–23.
59. Haynes, "Liberty," 576.
60. Saillant, *Black Puritan*, 22.
61. Ibid., 22–3.
62. David Walker, *Appeal, in Four Articles; Together with a Preamble, To the Coloured Citizens of the World, but in Particular, ... of the United States of America*, ed. Charles Wiltse (New York: Hill and Wang, 1965), at 36 n.; 65; 13, 65, and 68; 65.

63. Ibid., for example 54 n.; 46.
64. Ibid., 15.
65. Ibid., 14.
66. Jarrett, "To Refute," 310.
67. Walker, *Appeal*, 16; 14.
68. See Guy Williams, *The Age of Miracles: Medicine and Surgery in the Nineteenth Century* (London: Constable, 1981), chs 1 and 10; Geoffrey Sanborn, "Mother's Milk: Frances Harper and the Circulation of Blood," *English Literary History* 72.3 (Fall 2005), 691–715; Jules Law, *The Social Life of Fluids: Blood, Milk, and Water in the Victorian Novel* (Ithaca, NY: Cornell University Press, 2010), esp. 3–6; 86–7.
69. Coker, "Dialogue between a Virginian and an African Minister," in *Pamphlets of Protest: An Anthology of Early African American Protest Literature, 1790–1860*, ed. Richard Newman (New York: Routledge, 2001), 53–65, at 60–1.
70. William Wells Brown, *Clotel, or, The President's Daughter*, ed. M. Giulia Fabi (London: Penguin, 2004), 43.
71. Smith, "Black News-Vender," 191.
72. Brown, *Clotel*, 102–3.
73. Ibid., 154. A similar incident in New Orleans is related, for example, by Eliza Potter in *A Hairdresser's Experience in High Life*, ed. Xiomara Santamarina (Chapel Hill: University of North Carolina Press, 2009), 104.
74. Brown, *Clotel*, 134.
75. Ibid., 50.
76. Ibid., 153.
77. See John Locke, *An Essay Concerning Human Understanding* (London: Everyman, 1961), II.10, 84–5.
78. Smith, "The New Pen and Old Graveyards," in *Works of James McCune Smith*, 155–8, at 156.
79. On Defoe's "panegyric to the lettered Englishman" and his criticism of "people without characters," see Lynch, *Economy of Character*, 32–3.
80. Douglass's conception of the "gold" of character in "What Are the Colored People Doing for Themselves?" is analyzed in James B. Salazar, *Bodies of Reform: The Rhetoric of Character in Gilded Age America* (New York University Press, 2010), 174–7.
81. See, for example, Smith's introduction to Douglass's *My Bondage and My Freedom*, 17: here he connects aspects of Douglass's character to his "negro blood," and describes the "mixed race" of the ancient Egyptians as the result of "Negro blood circling around the throne."
82. Mia Bay, *The White Image in the Black Mind: African-American Ideas on White People, 1830–1925* (Oxford University Press, 2000), 58–63, at 61.
83. Smith, "Human Brotherhood and the Meaning of Communipaw," in *Works of James McCune Smith*, 91–3, at 92. See Peterson, "Untangling," 511–12.
84. Smith, "On the Fourteenth Query of Thomas Jefferson's Notes on Virginia," in *Works of James McCune Smith*, 264–81, at 270; 278.
85. Ibid., 280.
86. Ibid.
87. Ibid., 266.
88. Ibid.
89. Brown, *Clotel*, 209.
90. Smith comments on the significance of words later in the essay, quoting Mirabeau's "Words are things." See Smith, "On the Fourteenth Query," 279.

Part III
Medicalizing the Political Body

8
Flowing or Pumping? The Blood of the Body Politic in Burton, Harvey, and Hobbes

Robert Appelbaum

One of the most important events in the cultural history of blood was the discovery by William Harvey of the circulation of blood, which was first publicized in print in 1626. It would take several centuries for medical science to understand the chemical composition of blood, and how, by virtue of this composition, blood actually served the life-functions of the body. The discovery of oxygen in the 1770s was one of the keys. But already, in 1626, William Harvey was publicizing the fact that blood circulated in the bodies of mammals and other animals, that the heart operated as a kind of pump, and that blood moved through the body both out of the heart and back into it, delivering nutrients and "spirit" to the rest of the body. Harvey thus solved a technical problem in Galenic physiology of which people had been aware at least since the time of Michael Servetus, who published speculations about pulmonary circulation in 1553.[1] But Harvey's achievement was not only an empirical finding; it was also a demonstration of the utility of the experimental method in natural philosophy; and it signaled a revolution in the terms by which the body and the mind were to be understood.

How this happened, how the scientific revolution got under way, how it was sustained by a mixture of intellectual adjustments, theorizations, experimental findings, social reorganizations, political and religious controversies, and so forth—that is a perennial subject in the history of ideas, and I have nothing to add to it.[2] My subject is the cultural history of blood. And so, in the study of this history I want to underscore how Harvey signaled a moment of change in that history. For Harvey initiated a decisive shift in the metaphorical value of the idea of blood. Blood had been something that flowed in the body, not only in the physical sense that moved outward from the center of the body to its perimeters, but also in the sense that it expressed relations and affinities, moving outward from a central source toward peripheries dependent upon it. It had been tied, in its flowing, ebbing, and absorption in the body (absorbed, because it didn't circulate, and had to end up somewhere), to a microcosm of correspondences. It was a humor, a quality of life, a force, a generator and conduit of moods, pleasures, desires, creative powers, and spirits; it was both a resource of material life and a cousin to the divine. It was vulnerable too, vulnerable to corruption, obstruction, dilution, thickening, under-heating and overheating; it was vulnerable the way the flesh was vulnerable to the powers of life and death and good and evil. But it was also a quality that reflected the centripetal movement and value of the cosmos.

Harvey's theory, however, proved decisive in undermining the whole apparatus of Galenic medicine and the correspondences associated with it. Not just the blood and the heart but all the parts and functions of the human body were on the way to being absorbed into a mechanistic world picture in Harvey's lifetime (1578–1657). In fact, three of the major contributors to this new world picture in the realm of philosophy as well as physics, Gassendi, Descartes, and Hobbes, were familiar with Harvey's work, and developed reflections on the body in a similar vein, so to speak. (Descartes would even weigh in on the nature of the heart and Harvey's discovery of the circulation of the blood in the fifth chapter of his *Discourse on Method* [1637] as well as other writings.)[3] By 1660 Harvey's theory of circulation was widely accepted in Britain and France,[4] even if how the theory fitted or ought to be fitted into medical practice was little understood, and by 1800 this process of coming to understand the blood as a mechanism among other mechanisms in the body was complete.

What I want to point out in this essay is how, before the chemical nature of blood was understood, the switch to an understanding of blood as a mechanical functioning could also affect a switch to an understanding of political life. This may seem a bit far-fetched. The principles of justice, one might say, have nothing to do with whether blood is a flowing humor or a pumping solution of chemicals. But in the realm of cultural history, convergences between scientific understanding and other forms of understanding, or between literal and figurative meanings of basic concepts, can be key to understanding how historical processes work. And there is at least one clear example of how Harvey's theory of the blood impacted political theory: the political philosophy of Thomas Hobbes. For in Hobbes, who was a friend of Harvey and publicly expressed admiration for the latter's work, we find that old commonplace of Western thought, the body politic, transformed into an "artificial man" on a model meant to recall Harvey's experiments.[5] And we come upon an original thought: "Bloud [is] the mony of the commonwealth."[6] This expression had in various forms been used before, at least since the late Middle Ages.[7] But now it signified a new idea, for blood was a material that circulated, not something that flowed and expressed relations and affinities, and blood therefore performed a different kind of function, whether in the body or the body politic. It did not only sustain a body, but also preserved its surplus, allowing it to distribute resources into the future. So too, then, did money. The political significance of this new idea is huge, as I will show in what follows. It is a big part of Hobbes's revolution in political theory, where a sovereign government is at once consensual and mandatory, at once representational and absolutist. And it depends on a convergence between medical thinking and political thinking—a convergence that was actually insisted on by Hobbes himself.

I am not trying to propose a simply linear account here between theories of the blood and theories of the state, or even in the development of theories of the blood. What converged in one part of the history of thought diverged in another. There are some ways in which Hobbes is *not* consistent with Harvey. There are some ways in which Harvey is perhaps not consistent with himself. And it is important to keep in mind that even though Harvey's theories became widely accepted by the time of Harvey's death, Galenic science still ruled in many areas of medical practice, and would continue to do so for many years after.[8] Even after the dethronement of Galenic science, Galenic therapeutics—purging, bloodletting, cupping and the like—continued to dominate professional medicine. But I would like, on the model of what historians call the "micro-history," to present here a bit of a micro-textual-history. For I can show by way of four texts (with some supplementation by other texts by the authors in question, as well as macro-historical background information) that the discovery of the circulation of the blood had a decisive impact on how the operations of the body politic could be understood. This impact was not mandatory. What is in question is a horizon of possibility, a new field of what it was possible to say and write. But its historical significance is clear. The central texts are William Harvey's *De Motu Cordis* (1626) and *De Circulatione Sanguinis* (1649), which I cite here in their 1653 translations, printed and distributed probably with the approval of Harvey. In addition, published in its first edition a few years before *De Motu Cordis*, there is Robert Burton's *Anatomy of Melancholy*, which after its initial 1621 edition would be revised and expanded four times up to Burton's death in 1640, and then in a fifth posthumous edition of 1651.[9] And then there is Hobbes's *Leviathan* (1651), an elaboration and re-systemization of thoughts that Hobbes had published in 1640 (*The Elements of Law*) and 1642 (*De Cive*). Harvey, Burton, and Hobbes—rough contemporaries, and descendants of the same social class, the minor provincial gentry—had similar likes and dislikes when it came to politics. They were monarchists, but they were also open-minded humanists with an affection for the cause of the common good. Yet when he thinks about the body politic, Burton persists in a Galenic idea of the body; whereas Hobbes adopts the mechanistic model. And that affects what it is possible for each of them to say about the body politic, and especially about its economic life, what both call its "growth" and "nutrition." The movement of the blood in the natural body, whether it flows, as in the Galenic model, or circulates, as in the mechanistic model, provides a key figure for the movement of life in the commonwealth too.

* * *

As for Harvey himself, we have little information about his political opinions.[10] He was a personal physician to both James I and Charles I and thrived as a member of the Royal College of Physicians. He did not flourish either publicly or privately during the Commonwealth, but unlike his two landowning brothers, who were heavily fined by the Parliament once it was in power, he seems to have been let alone, and he died in 1657 a wealthy man. So Harvey's politics are uncertain. But in 1964 Christopher Hill published a breakthrough article which showed that changes in the way Harvey framed his thoughts about the heart and the blood between 1626 and 1651 indicated that Harvey sympathized, if not with Parliamentarian politics, at least with the subversion of traditional hierarchies upon which royal authority had till then been based.[11]

In 1628, Harvey dedicated *De Motu Cordis* to Charles I, his patron, comparing the king to the heart and both the king and the heart to the sun:

> The heart of creatures is the foundation of life, the prime of all, the sun of their microcosm, on which all vegetation does depend, from whence all vigor and strength does flow. Likewise the King is the foundation of his kingdoms, and the sun of his microcosm, the heart of his commonwealth, from whence all power and mercy proceeds.[12]

Later in the text, dropping (but perhaps still implying) the allusion to the king, Harvey expanded by the following:

> So the heart is the beginning of life, the Sun of the Microcosm, as proportionably the Sun deserves to be called the heart of the world, by whose virtue and pulsation, the blood is moved, perfected, made vegetable, and is defended from corruption and mattering; and this familiar household-god doth his duty to the whole body, by nourishing, cherishing, and vegetating, being the foundation of life and author of all.[13]

These ideas are at once, as Hill notes, ancient and modern. The imagery is modern, Copernican, for the sun is the center of the universe. But much of the language is ancient, including the description of the heart as a foundation and an author. Because the sun-like heart is the source of the circulation of the blood, not only by impelling the blood through the body but also by having it made "perfected" and "vegetable," the body ends up as a hierarchical if concentric system. According to Hill, this accords perfectly to absolutism, where the monarch is just such a source of life as the heart is, even if the identity between the body and the body politic is only analogical, and not directly ideological.

But in the work Harvey published in 1649, and saw translated into English and published in 1653, the heart is no longer the foundation and author of all. In fact it is the blood which acts as a kind of "author" and which in itself provides heat, vegetable life, and pulsion to the circulatory system. The heart responds, as a pump, to the prompting heat and the vitality of the blood. And for Hill this indicates an approach to the body which is non-hierarchical, corresponding to anti-hierarchical impulses of the political thought of the 1640s. Perhaps Harvey changed his mind; perhaps he was always inconsistent about the relation between the heart and the blood; perhaps he did not change his mind, but found the atmosphere of the 1640s more congenial to antihierarchical language. In any case, Harvey's conceptualization of the heart and the blood reflected a wider change in the mentality of the seventeenth century, where it became more and more acceptable to reject ancient scientific authority along with ancient doctrines of the natural hierarchies of social and political life. By 1649 Harvey had no trouble using language which trumpeted the power of the blood of all creatures, as it circulated through the body, and which demoted the heart into a machine that responded to this power.

More recently, John S. White would seem to have rejected this connection by emphasizing that, if you look at the notes for lectures Harvey gave in 1616, Harvey never actually changed his mind about the relation between the blood and the heart. The blood *always* came first in Harvey's mind. "For the blood is rather the author of the viscera then they of it, because blood is in being before the viscera," he quotes Harvey as having written in 1616. Moreover, Harvey also wrote for his lecture that "the soul is in the blood," and that "innate heat is the author of life."[14] White's argument actually does not refute Hill's position. For Harvey was obviously thinking about two different phenomena from the beginning. In one he saw, in the study of embryos, that before embryos developed organs, they were fluid and were comprised of or contained a substance he identified as blood. Harvey was intuiting the origin of life on the molecular, cellular, and genetic level before he had any vocabulary to describe it apart from referring to "blood," "soul," and "heat." In the other phenomenon he saw, within fully formed bodies, a movement of blood with a pulsion apparently originating in the heart. White says that Harvey was aware of this duality and addressed it through a consistent application of Aristotelian terms, dividing the features of the body and the roles of blood and the heart with respect to one another into different kinds of causes. But White is not convincing on this score. In 1616 Harvey believed that the blood *originated* first. In 1626, though he still probably believed that blood originated first, he wrote that the heart was the "author" of the movement and vitality of the blood. In 1649, however, Harvey developed the idea not only that the blood originated first, but that it continued to originate life, heat, and motion in the body by its own innate power, impelling the heart to act as a pump. So here is a major change in Harvey's language. In 1626 the heart was metaphorically a sun, an author, a king; in 1649 and later the heart was de-metaphorized; it was only a kind of mechanical device.

* * *

The importance of how the blood moves through the body shows up in the political and anatomical sections of Burton's *Anatomy of Melancholy*. One can quote either the first, 1621 edition or the last, posthumous 1651 edition and come up with the same results. So I will quote here, for the sake of simplicity, from a definitive modern version of the last edition.[15] And I will begin by pointing out that although there is much that is innovative in *The Anatomy of Melancholy*, from its literary form to its interest in post-Copernican astronomy, when it comes to the anatomy of the human body Burton is altogether conventional, his understanding of the matter based on books written in the Hippocratic–Galenic tradition, with little input from the new anatomical understanding that began with the work in Italy of Visalius.[16] Here is some of Burton's characterization of the "middle region" of the body, that is the chest:

> Of this Region the principall part is the Heart, which is the seat and fountain of life, of heat, of spirits, of pulse and respiration—the sun of our body, the king and sole commander of it: The seat and Organe of all passions and affections. *Primum vivens, ultimum moriens*, it lives first, dies last in all creatures: Of a pyramidicall forme, and not much unlike to a Pine apple; a part worthy of admiration, that can yeeld such variety of affections, by whose motion it is dilated or contracted, to stirre and command the humors in the body: As in sorrow, melancholy; in anger, choler; in joy, to send the blood outwardly; in sorrow, to call it in; mooving the Humors, as Horses do a Chariot ...
> [The left side of the heart] hath the form of a Cone, and is the seat of life: which as a Torch doth Oyle, drawes blood unto it, begetting of it spirits and fire; and as fire in a torch, so are spirits in the blood; and by that great Artery called Aorta, it sends vitall spirits over the body, and takes aire from the Lungs by that Artery which is called Venosa.[17]

Some of this language, of course, sounds a bit like Harvey in 1626. (Since Burton was already saying this in 1621 perhaps Harvey borrowed a bit from Burton.) The heart is a sun, a king, a commander. But most of the language does not sound like Harvey. For Burton, the heart sends the blood out and draws it in; the blood moves back and forth, it ebbs and flows, but it does not circulate. And new blood constantly has to be generated by the liver. More importantly, for Burton the heart is not only a generator of life, but a regulator of the passions, and it sends out to the body not only blood but all the four humors, responding to changes in mood or other affective factors. Burton is usually very nuanced and detailed when he discusses causes and effects; but here, speaking in principle of a strictly physical matter, he says that the heart commands (or at least "draws" about) the humoral substances of the body, sending them out or inducing them back in, reacting to changes in affect. Perhaps he means to say that the blood is the chariot that carries the other humors back and forth in the body, but he does not actually say that. The blood is one humor among four being moved through the body. All the humors seem equally to be subject to the impulses of the heart—an expression I use here literally, but that still has the figurative meaning that Burton is drawing upon. Perhaps Burton means to say that the heart by itself can generate affections, that the impulses of the heart are impulses of passion, but he does not actually say that either. If one tries to interpret Burton's anatomy of the body by modern mechanistic principles, one will inevitably be frustrated. If one tries at least to find nuance in the anatomy of the body, one will end up frustrated as well. But the human being as a whole is a complicated creature in Burton, with many different agencies, physical and mental, imaginary and real, endogenous and exogenous, and even if the heart is the king of the body and the blood its happiest humor—indeed the humor of happiness itself—there are many other factors that can make a person happy or sad, angry or phlegmatic, not to mention mentally ill or just plain crazy. If an obstruction in the blood can have an impact on well-being, so can a movement of the stars, the death of a parent, or theological anxiety.

But what of the body politic, then? Burton only uses the term, body politic, about five times, sometimes in the form of "politic body," and it is clear from his discussion of political society that the bodily metaphor only goes so far: just as the quality of the feeling self is an effect of a variety of endogenous and exogenous causes, so is the state. The kingdom, the province, the city-state, the association of city-states—for these are Burton's models—are politic bodies, but their success or failure is not only an embodied sort of condition. A kind of spiritual condition is vital to these "bodies" as well. Like human beings, states can suffer from "melancholy," a condition which is rooted in the flow and balance and of humors in the body, but which also translates into morbidities of affect, activity, mentality, and conscience. As he "anatomizes" the general condition of melancholy in the world, with a particular interest in individuals, so, in a section of the Preface to the *Anatomy*, "Democritus Junior to the Reader," Burton "anatomizes" melancholy in the state. That is to say, he approaches the state as a physician would approach an individual: he delineates its symptoms, speculates about its causes, and prescribes preventions, remedies, and cures.

Rather than elaborating a theory of the state as a historical institution, as Hobbes would come to do, Burton explains the state according to the simple distinction between illness and health, or between the sanguine state and the melancholy one. States are one or the other. "For where you shall see the people civill, obedient to God and Princes, judicious, peaceable and quiet, rich, fortunate, and flourish, to live in peace, in unity and concord, a Countrey well tilled, many faire built and populous Citties," there you shall find, according to Burton, a kind of happiness or beatitude. "But whereas," he goes on,

> you shall see many discontents, common grievances, complaints, poverty, barbarisme, beggary, plagues, warres, rebellions, seditions, mutinies, contentions, Idlenesse, Riot, Epicurisme, the land lye untilled, waste, full of bogges, Fennes, Desarts, &c. Cities decayed, base and poore townes, villages depopulated, the people squalid, ugly, uncivill; that Kingdomes, that Country, must needs be discontent, melancholy, hath a sicke body, and need to bee reformed.[18]

Of course, most states are neither fully happy nor fully unhappy. There is something utopian about the notion that any state could be fully happy, and that is one of the reasons why Burton, who needs to discuss the state in absolute terms, goes on to discuss what he calls a "Utopia of mine owne."[19] But for the complexities of Burton's not quite abstract notion of the modern state, it is clear that Burton emphasizes four qualities that make a state exist in an acceptable condition: concord, prosperity, industry, and urbanity.

Each of these qualities could be subdivided into others. Concord, for example, depends upon a mixture of consent and obedience. And as Burton goes on to say, that means that in a well-run commonwealth, if the subjects obey it is in large part because the leaders lead, putting the interest of the common good above private considerations. Similarly, if the body politic is industrious, that is in large part because, although inequality of property and income is inevitable, and Burton prefers a society which experiences the "splendor and magnificence" that only a hereditary nobility can furnish, everyone can be confident of reasonably profiting from his or her own labor, and individuals may be advanced in society according to their merit.

A principle of utilitarianism is embedded in Burton's political ideas. That which maximizes happiness for the maximum number of people is good, and ought to be put into practice. But the mechanism to achieve a maximally utilitarian state is not mercenary. People are meant to achieve the goals of a utilitarian state primarily because the goals are good in themselves. A finely built city may help its inhabitants act more civilly, and an intensely farmed land ("I will not have barren acre in all my Territories, not so much as the tops of mountains"[20]) may produce more food, but both a fine city and a cultivated countryside are achievements in themselves. They express a will toward concord, prosperity, industry, and urbanity, a will toward what Burton cites Aristotle as calling the "common good."[21] And they do not only *will* them; they *exemplify* them; they *are* them. The common good is at once an achieved condition and a condition of achievement.

This is perhaps an aporia of infinite regression. And the aporia is perhaps another reason why Burton's theory of the state inclines not only toward utilitarianism but also toward utopianism. Happiness creates happiness, and vice versa. There is no *outside* of happiness from which happiness may stem. But in the present context it may be added that the aporia may also be due to the fundamental metaphor upon which it is based: that is, that the state is a body, and the body is Galenic. Every "member" or "part" of Burton's Galenic body has to do two things at once: it has to cooperate with the other members and it has to initiate its own welfare. Cooperation is both consensual and deferential. And welfare is a matter both of a part of the body using well what it has been supplied with by other parts and of making sure that it gets what it needs, uses what it needs according to its proper functioning, and contributes what it is supposed to contribute to the other members. So, for example, fishing around for a way to reduce to order the "divers and confused" condition of melancholy in that divers and confused "Species" called man, Burton at one point ventures that there are three kinds in all, a "Head melancholy" that proceeds from the brain, a "whole body" melancholy when the "whole temperature" suffers from that condition, and a "Hypocondriacall, or windie melancholy," which arises "from the Bowels, Liver, Spleene, or Membrane."[22] Even this division is not exhaustive, for there are also love melancholy as well as religious melancholy, and the three zones of melancholy affect and combine with one another and are affected by exogenous causes as well. But here is the analogical root of the aporia of happiness. Whether in an embodied person or the body of a state, every part has to play its role, both independently and cooperatively. Every part has to be vigilant for its own health, as well as for the health of the parts it cooperates with, and there is no part in which something cannot go wrong, bringing about an illness suffered by the whole. One of the contributing factors toward illness, along with the "divers and confused" nature of the subject of illness, is that even as the parts attempt to play their roles, they are constantly using themselves up, consuming and exhausting their own nutrients; they continually have to replenish and remake themselves. Nothing circulates. And at any moment an obstruction or a deficiency—a lack of warmth or fluidity—can turn a part of the body ill, and make the whole of the body melancholy.

* * *

Against the metaphoricity and circularity characteristic of the thought of Burton and the dozens of humanists he borrowed from, Thomas Hobbes would wage a special kind of war. For in the first place, though the analogy between the human body and the body politic remains, both the one and the other are usually reduced by Hobbes to mechanics. "Life is but a motion of Limbs," he states in the opening paragraph of Leviathan; "the beginning whereof is in some principall part within; why may we not say, that all Automata (Engines that move themselves by springs and wheeles as doth a watch) have an artificiall life? For what is the Heart, but a Spring; and the Nerves, but so many Strings; and the Joynts, but so many Wheeles, giving motion to the whole Body."[23] On the one hand, following Descartes Hobbes asserts that the body is a machine; on the other hand, in an idea original to himself, Hobbes asserts that the body politic is also a machine, but with a difference, for the human body is a machine made by nature, whereas the body politic is a machine made by art. In fact, the body politic is an "artificial animal."

When Hobbes compares the heart to a "spring" he goes much further than Harvey in constructing a mechanistic perspective. For such a comparison never appears in Harvey, who was interested in the "innate heat" of the blood and the body, a vitalistic principle. Yet Hobbes took inspiration from Harvey. In the Epistle Dedicatory to his *Elements of Philosophy*, Hobbes compared Harvey to Copernicus and Galileo. He claimed that "the science of man's body" was "first discovered" by Harvey.[24] And as Hobbes tried to go on to claim a place as a first inventor of the modern science of "civil philosophy," so he also, in all of his major theoretical works, claimed that to understand the civil society one first had to understand what he calls, in *Leviathan*, "the Matter thereof, and the Artificer; both which is Man."

So Hobbes was trying to do for political philosophy what Harvey had done for human biology, and he began by adopting what he took to be the new science of the human body as an indication, first, of the nature of the matter and author of civil society and, second, of a model of discovery and discourse with regard to the forms and ends of civil society. Hobbes took from Harvey and some other contemporaries the idea that the body was a kind of machine, the life of the machine being a kind of perpetual motion; he then took the further step of arguing that the state was also a kind of machine, constructed out of the "art" and in view of the needs of natural men.

But Hobbes would then go on to face the same kind of problems as Harvey when trying to account for the inherent motion and life of bodies, whether natural or artificial. For where did the vitality of a live organism come from? Looking at his natural and artificial men, Hobbes had to account for both voluntary and involuntary motions. "There be in Animals," as Hobbes puts it,

> two sorts of Motions peculiar to them: One called Vitall; begun in generation, and continued without interruption through their whole life; such as are the Course of the Bloud, the Pulse, the Breathing, the Concoctions, Nutrition, Excretion, &c; to which Motions there needs no help of Imagination: The other in Animal Motion, otherwise called Voluntary Motion; as to go, to speak, to move any of our limbes, in such manner as is first fancied in our minds.[25]

Natural and artificial animals have both a "vital" or involuntary life and an "animal" or voluntary life. Moreover, there would seem to be behind the vitality of involuntary motion something like a soul, just as behind the voluntary animal motion there seems to be mind and imagination. Furthermore, principles of appetite, desire, will, and rational deliberation also lay behind motion in some sense. Whether these principles are innate and immanent or in some way disembodied and extrinsic (Hobbes always preferring the first sort of explanation, motion coming from "one principall part therein"), they are categorically different from "matter." "Nothing can move itself," Hobbes wrote in an unpublished work. "Whatever moves is moved."[26] And yet the body seems to be moving itself, at least so far as involuntary motion is concerned, from the moment of its "generation"; and so does the artificial body of the state.[27]

The state, at least, is "authored." That is, the state for Hobbes is always a deliberate creation of men, who are acting out of self-interest, transferring the right of government from themselves to the artificial "body" that they are making, "the sovereign" of the commonwealth. If the body politic must get its motion from something apart from itself, that motion comes from the original creative act of the commonwealth's authors. Whatever the nature of the first movement of the ever-moving body of a natural creature, in Hobbes the first movement of the ever-moving body of the state is the covenant, when men bind themselves together by contract to the authority of the sovereign. And the crucial point is that once they are contracted, once they have transferred their rights of self-government to the artificial being of the sovereign, there is no going back. In a complicated but definitive statement, Hobbes says the following: "Because the Right of bearing the Person of them all, is given to him they make Soveraigne, by Covenant onely of one to another, and not of him to any of them; there can happen no breach of Covenant on the part of the Soveraigne; and consequently none of his Subjects, by any pretence of forfeiture, can be freed from his Subjection." So long as the state continues to function as a state, sovereignty is permanent, and so is absolute subjection to sovereignty. There is only one kind of case where the covenant can be broken: "The Obligation of Subjects to the Soveraign is understood to last as long, and no longer, than the power lasteth, by which he is able to protect them."[28] That is, if the power of sovereignty expires, which is above all the power to protect the subjects of the state and preserve the common good, then the body politic has in effect expired, and subjects are no longer obliged to it. When France conquers Burgundy, Burgundy as an independent sovereignty is no more, Burgundians have become French, and owe their allegiance to the king of France. When the empire of Rome dissolves in anarchy, the people are no longer Romans and owe no allegiance to anyone calling himself the emperor of Rome. In fact, they revert, according to Hobbes, to the state of nature.

What does any of this have to do with the circulation of the blood? The fact is, the Hobbesian commonwealth has to operate as a kind of *imperfect* machine. Things can go wrong. The machine is subject, no less than in Burton's analysis, to a variety of "diseases." But its health does not have to be renewed in each and every part, every day, in a complex organization of cooperation and self-preservation. The commonwealth is meant to operate as an automata. Once set in motion, it is meant to stay in motion through perpetuity. Hence the concept of circulation becomes important. For though there will be many voluntary motions within the commonwealth, as men are by nature egoistic and ambitious, not to mention troubled by fantasy, the stability of the state depends upon its involuntary motions.

It depends on these involuntary motions, moreover, into an indeterminate future. This is a subject where Hobbes differs from a traditionalist like Burton: he not only sees that people must act providently, and husband their resources, but that acting providently is the very principle of government. "The finall Cause, End, or Designe of men, (who naturally love Liberty, and Dominion over others,) in the introduction of that restraint upon themselves, (in which wee see them live in Commonwealths,) is the foresight of their own preservation, and of a more contented life thereby."[29] Men covenant themselves to a sovereign, even to the point of sacrificing part of their natures, out of concern about the future. That is why, from the point of view of *voluntary* action, *terror* plays such a large role in Hobbes's Leviathan. In the sovereign state, men obey first of all out of a fear of the consequences—the future—of disobeying. The fear of punishment, universally experienced, is what allows the state to preserve itself against voluntary internal disruptions. And as for *involuntary* action, here Hobbes departs from Burton by thinking of the commonwealth as a self-enclosed automatic system, which provides for the "more contented life" into the future to which he refers. There is something of a paradox in this, but the paradox is crucial to Hobbes's theory, and would remain crucial to all thinkers who would follow Hobbes's lead and think of the state as a liberal system.[30] *For the sake of the future*, that is, out of "foresight" with regard to "preservation" and a "more contented life," the commonwealth is constructed as an enclosed and automatic system, as if it had no future but the present. Hobbes is constructing what is supposed to be a *universal* theory of the state; so what is essentially true of one state is true of them all, at all times, and all states are, unless they decline into disease and disorder, enclosed systems. And the systems are enclosed so far as they set perpetual limits to themselves, with a view toward sustaining themselves as they are indefinitely into the future.

Leaving aside the principles and institutions of law, Hobbes then looks at the political economy of the commonwealth. But he adopts what may seem to be a peculiar language to describe economy. He talks, in chapter 24 of *Leviathan*, about the "nutrition and procreation of a commonwealth." If the state is an automata, it is also a living body. It needs to "procreate": the original contract is not historically renewed with each successive regime, but rather recreated from one generation to the next by a kind of insemination. And if the state is a living body, it also needs to be nourished. This nourishment is the economy: that is, in Hobbes's words, "the Plenty, and Distribution of Materials conducing to Life: In Concoction, or Preparation; and (when concocted) in the Conveyance of it, by convenient conduits, to the Publique use."[31]

If the state is a machine, in other words, it is a machine more on the model of the clinical human body than on the model of the automatic clock.[32] The body politic is a mixed metaphor, in this respect. And however automatic it is supposed to be, it has to be fed. In fact, it not only has to be fed, it has to digest and distribute its food, and it has to do so generously, fairly, conveniently, efficiently, and providentially. The body politic has to continually take care of itself and for itself, for the sake of both the present and the future. And so, because he is committed to the metaphor of nutrition in order to explain the economy of the state, Hobbes falls back on language which predates Harvey's "discovery" of the science of the body, and which resembles the language of a traditionalist like Burton. For Harvey has almost nothing to say about "concoction" or "nutrition"; it is out of his range of ideas about the blood, the heart, and the body; it is a subject, I believe, that he deliberately ignores, because he is unable to explain it.

Hobbes has no time, however, for the study of concoction as a process of metaphoricity and well-being. Concoction is not a regulator of moods or of spiritual conditions for Hobbes, and when he talks about the natural body that is man, Hobbes skips the subject altogether. Concoction is a concept that applies only to the artificial body of civil society, so far as the argument of *Leviathan* is concerned. It is a kind of essentialization of substances, as much on the order of alchemical distillation as on the model of Galenic digestion. It is "the reducing of all commodities, which are not presently consumed, but reserved for Nourishment in time to come, to some thing of equal value, and withall so portably, as not to hinder the motion of men from place to place; to the end a man may have in what place soever, such Nourishment as the place affordeth." In other words, concoction is the alchemical or digestive transformation of commodities into money.

That is how money becomes the blood of the commonwealth. By means of the conversion of surpluses into gold, silver, or other forms of money, Hobbes writes in the paragraph with the marginal heading "Mony the Bloud of the Commonwealth,"

> all commodities, Moveable, and Immoveable, are made to accompany a man, to all places of his resort, within and without the place of his ordinary residence; and the same passeth from Man to Man, within the Common-wealth; and goes round about, Nourishing (as it passeth) every part thereof; In so much as this Concoction, is as it were the Sanguification of the Common-wealth: For naturall Bloud is in like manner made of the fruits of the Earth; and circulating, nourisheth by the way, every Member of the Body of Man.[33]

Hobbes is especially interested in how public funds are to be collected and distributed. By 1651 Hobbes would especially seem to be concerned with the fact that one of the reasons England devolved into civil war, and why the Royal party was unable to win it, was that there was not enough money in the national treasury. But the main point for purposes of this essay is simply that, in order to understand political economy, Hobbes requires an analogy between money and blood, and between the circulation of the one and the circulation of the other. He continues with the analogy elsewhere in his discussion of the

Leviathan. When a commonwealth does not save enough money for future needs, and distribute its savings effectively, the result is a kind of "Distemper," like an "Ague," where blood gets stuck in the veins, and the body overheats itself to get the heart pumping more fiercely and get the blood unstuck.[34] When the money in a commonwealth gets controlled by too few people, and hoarded in one part of the body politic at the expense of other parts, another "disease" results, much like "pleurisie," where blood gathered overabundantly "into the Membrane of the breast, breedeth there an Inflammation, accompanied with a Fever, and painfull stitches."[35] Money has to circulate continually in Hobbes's commonwealth, just as blood has to do in Hobbes's human body. Otherwise obstructions and inflammations result, both of which cause illness. And otherwise, too, parts of the body that require resources are unable to access them.

It is noteworthy that Hobbes never talks about diseases of circulation as causes of something like "unhappiness." The metaphors of the state, even if they are sometimes mixed, and automatic machines get confounded with vital bodies, are strictly physical. The state is no longer like a human body in the sense that it is subject of multiple agencies, in correspondence with elemental, cosmological, and spiritual principles, where concepts like happiness correlate with concepts like perfection, and microcosms adjoin macrocosms. Men may aim for happiness in their private lives—they aim to be "contented" as Hobbes often puts it. But the state is just itself, serving the ends of individuals, and it cannot be "contented." It is not emotional. However, the state needs its blood. It needs its blood to be pumped—not to flow but to be pumped—efficiently, fairly, and continuously. But the blood of the state is only money. Thanks to Harvey's disenchantment of the human body, and thanks no doubt to many other historical developments, material and scientific, political and linguistic, history is well along the road not only to a mechanistic world picture, but also to capitalism and that discipline of thought, common as much to Adam Smith as to Karl Marx, that would come to be known as "political economy."

Blood, in Hobbes, is that substance which, in a living body, may continually be replenished but never be exhausted. Until the body itself deceases, the blood never loses what Harvey would have called its vitality. And this is the lesson that Hobbes draws from the discovery of the circulation of the blood. We do not only have to live for today, everyday renewing the life blood of society, in peril of dying. Nor do we have to deny ourselves in the present, hoarding our wealth, or concentrating it in the hands of the few, out of fear of shortages in the future, or out of contempt for the needs of the many. For we have sovereignty and therefore security. And we also have money, the essence of commodities, drawn out of commodities as if by "concoction." Sovereignty is not the "author" of wealth or of anything else, according to Hobbes. It is not the center of the human universe, a heart of human creation or a fountain from which life may flow. Men are always the authors of civil society, and the authors of it equally. But having formed themselves into a civil society, men provide for themselves within an involuntary system, one of whose essential characteristics is the circulation of money. And money is the blood of the commonwealth.

To trace the significance of this switch in the understanding of bodies and body politics would require another study, one which took into account the future of both economics and political science. Certainly, economic thought since Adam Smith and *The Wealth of Nations* (1776) has been much occupied with the circulation of money, and the problem of what is now called "liquidity." A circulation metaphor, complete with wealth as a kind of blood, moving about in a blood vessel, dominates Smith's criticism of England's overreliance on its colonies:

> A small stop in that great blood-vessel, which has been artificially swelled beyond its natural dimensions, and through which an unnatural proportion of the industry and commerce of the country has been forced to circulate, is very likely to bring on the most dangerous disorders upon the whole body politic. The expectation of a rupture with the colonies, accordingly, has struck the people of Great Britain with more terror than they ever felt for a Spanish armada, or a French invasion.[36]

Just as in Hobbes, the body politic is subject to "disorders" like pleurisy, where too much blood gets forced into a single conduit. To think of money or wealth as a kind of "blood" is to think of a body politic and a political economy that can only thrive in condition of the free, open, and diverse circulation of wealth. The concentration of wealth in a few hands, and the reliance on the production of wealth by a limited source of output and trade, is always an "unnatural" danger. "Rupture," another metaphor with roots in the phenomenology of the body, is a danger so threatening as to evoke terror.

As Hobbes originally devises the idea, human society is both cut off from the natural world and identified with it. Human society is cut off from nature because it is just a machine, an "artificial" being: something made by men to suit their originally selfish but ultimately cooperative purposes. But human society can be imagined as a machine, it turns out, because nature itself is a machine, created by God for God's inscrutable purposes. In Hobbes's universe, blood is one of the more clever creations of God; and money is one of the most clever creations of Man. Without either, neither kind of machine can work. Animals would be unable to live: and modern societies, whose existence both mimics and defies the state of nature, would not be able to "live" either. That is to say, they would not, indefinitely and involuntarily, spring into motion.

Notes

1. Edwin Clarke, "Michael Servetus," *The British Medical Journal* 2.48 (1953), 934; Christodoulos Stefanadis, Marianna Karamanou, and George Androutsos, "Michael Servetus (1511–1553) and the Discovery of Pulmonary Circulation," *Hellenic Journal of Cardiology* 50.5 (2009), 373–8.
2. The most relevant study with regard to the present essay is Jonathan Sawday, *The Body Emblazoned: Dissection and the Human Body in Renaissance Culture* (London: Routledge, 1995). For a general overview, see Steven Shapin, *The Scientific Revolution* (University of Chicago Press, 1996).
3. Geoffrey Gorham, "Mind–Body Dualism and the Harvey–Descartes Controversy," *Journal of the History of Ideas* 55.2 (1994), 211–34.
4. See for example Steven Shapin and Simon Schaffer, *Leviathan and the Air-Pump: Hobbes, Boyle, and the Experimental Life* (Princeton University Press, 1985), 127.
5. On the history of the concept, see David George Hale, *The Body Politic: A Political Metaphor in Renaissance English Literature* (The Hague: Mouton, 1971).
6. Thomas Hobbes, *Leviathan*, ed. C. B. Macpherson (Harmondsworth: Penguin, 1968), 300.
7. Jerah Johnson, "The Money=Blood Metaphor, 1300–1800," *Journal of Finance* 21.1 (1966), 119–22. After Hobbes, the expression became a commonplace of physiocratic and classical economic theory. See Christine Desan, "From Blood to Profit: Making Money in the Practice and Imagery of Early America," *The Journal of Policy History* 20.1 (2008), 26–46, for its emergence in American discourse. For other accounts of Hobbes and the blood metaphor see Paul P. Christensen, "Hobbes and the Physiological Origins of Economic Science," *History of Political Economy* 21.4 (1989), 689–709; and Deborah Valenze, *The Social Life of Money in the English Past* (Cambridge University Press, 2006), esp. 62–4.
8. For a succinct account, see Laurence I. Conrad et al., *The Western Medical Tradition: 800 BC to AD 1800* (Cambridge University Press, 1995), 325–62. For a sociological reading of this history, see Harold J. Cook, *The Decline of the Old Medical Regime in Stuart London* (Ithaca, NY: Cornell University Press, 1986).
9. For background on the text and its many editions, an invaluable resource is Angus Gowland, *The Worlds of Renaissance Melancholy: Robert Burton in Context* (Cambridge University Press, 2006).

10. On Harvey's life and work see Jerome J. Bylebyl, *William Harvey and His Age: The Professional and Social Context of the Discovery of the Circulation* (Baltimore: Johns Hopkins University Press, 1979); Geoffrey Keynes, *The Life of William Harvey* (Oxford: Clarendon Press, 1966); and Thomas Wright, *Circulation: William Harvey's Revolutionary Idea* (London: Chatto & Windus, 2012).
11. Christopher Hill, "William Harvey and the Idea of Monarchy," *Past & Present* 27 (1964), 54–72. Citation of Harvey on 54. Also see a response to Hill, Gweneth Whitteridge, "William Harvey: A Royalist and No Parliamentarian," *Past & Present* 30 (1965), 104–9, and Hill's reply, "William Harvey (No Parliamentarian, No Heretic) and the Idea of Monarchy," *Past & Present* 31 (1965), 97–103.
12. 1653 English translation of *De Motu Cordis*, in William Harvey, *The Anatomical Exercises*, ed. Geoffrey Keynes (New York: Dover, 1995), vii.
13. Ibid., 59–60.
14. John S. White, "William Harvey and the Primacy of the Blood," *Annals of Science* 43 (1986), 239–55. The quotation from Harvey appears on 242, and it comes from William Harvey, *Anatomical Lectures*, ed. Gweneth Whitteridge (Edinburgh: E. & S. Livingstone, 1964), 127.
15. Robert Burton, *Anatomy of Melancholy*, ed. Thomas C. Faulkner, Nicholas K. Kiessling, and Rhonda L. Blair (Oxford University Press, 1989).
16. The same point is made in J. B. Bamborough, Introduction, in ibid., xxi.
17. Burton, *Anatomy of Melancholy*, 146.
18. Ibid., 67.
19. I discuss Burton's utopia in the context of the utopian tradition in Robert Appelbaum, *Literature and Utopian Politics in Seventeenth-Century England* (Cambridge University Press, 2002), 81–8. The readings in Gowland, *Worlds of Renaissance Melancholy*, and Adam H. Kitzes, *The Politics of Melancholy from Spenser to Milton* (New York: Routledge, 2006) are very different from my own, partly because they don't take the utopian tradition seriously—or jocoseriously either.
20. Burton, *Anatomy of Melancholy*, 88.
21. Ibid., 67.
22. Ibid., 169
23. Hobbes, *Leviathan*, 81. And see Thomas A. Spragens, Jr, *The Politics of Motion: The World of Thomas Hobbes* (London: Croom Helm, 1973).
24. Thomas Hobbes, *Elements of Philosophy*, in *The English Works of Thomas Hobbes of Malmesbury*, ed. W. Molesworth, 11 vols (London: Bohn, 1839–45), 1: viii–ix.
25. Hobbes, *Leviathan*, 118.
26. Thomas Hobbes, *Critique du Mundu de Thomas White*. Cited in Jürgen Overhoff, *Hobbes's Theory of the Will: Ideological Reasons and Historical Circumstances* (Lanham, MD: Rowman and Littlefield, 2000), 22–3. Overhoff's extended discussion of this issue is worth consulting. Especially of interest in the present context is his discussion of the relation between Hobbes and Harvey, 25–8.
27. With voluntary motion Hobbes perhaps has even more difficulty. Christensen ("Hobbes and Physiological Origins," 699–701) and William Sacksteder ("Speaking About the Mind: 'Endeavour' in Hobbes," *Philosophical Forum* 11

(1979), 65–79, esp. 74–5) believe that Hobbes solves the problem through the notion of "endeavor," which involves the initiation of activity in relation to the world and its stimuli. This notion of the voluntary "endeavor" in the natural body obviously has analogues in the artificial body of the state, or in human action generally.
28. Hobbes, *Leviathan*, 230.
29. Ibid., 223.
30. Hobbes's relation to liberalism and classical economic theory is a subject of some controversy, but I am inclined to the view that Hobbes's relation to liberalism and classical economic theory is direct. The most influential (but also still controversial) analysis of Hobbes's economic theory is still C. B. Macpherson's neo-Marxist analysis, *The Political Theory of Possessive Individualism, Hobbes to Locke* (Oxford: Clarendon Press, 1962). As Taylor points out in a very helpful summary of the subject, there has been a lot of confusion between understanding the economic implications of Hobbes's theory and the actual theory of economics that Hobbes develops. Quentin Taylor, "Thomas Hobbes, Political Economist: His Changing Historical Fortunes," *The Independent Review* 14.3 (2010), 415–33. I find myself in agreement with Taylor about the genuinely proto-liberal tendencies in Hobbes's economic thought, and even closer in agreement with Thea Vinnicombe, "Thomas Hobbes and the Displacement of Political Philosophy," *International Journal of Social Economics* 32.8 (2005), 667–81. Vinnicombe sees Hobbes's politics as being *essentially* economic.
31. Hobbes, *Leviathan*, 294–5.
32. The body, according to Sacksteder, signifies "the ruling or organizing part of the complex whole" ("Speaking about the Mind," 66). But the ruling of the whole has at least two incompatible models: the rule over the automata and the rule over the living creature, and Hobbes appeals at various points to both.
33. Hobbes, *Leviathan*, 300.
34. Ibid., 373.
35. Ibid., 374.
36. Adam Smith, *An Inquiry into the Nature and Causes of the Wealth of Nations*. Project Gutenberg E-book. Book Four, Chapter Seven, www.gutenberg.org/files/3300/3300-h/3300-h.htm.

9
Linnaeus and the Four Corners of the World

Staffan Müller-Wille

Many accounts of the history of the race concept place the naturalist Carl Linnaeus (1707–78), and his *Systema Naturae* (1735), at the beginning of modern concepts of race, in contrast to older notions that did not yet reduce race to physical traits, but presented it as the outcome of an inextricable entanglement of blood, soil, and customs.[1] In the slim, 11-page folio *Systema Naturae* (1735) that laid the foundations for the 22-year-old Swedish medical student's future claim to fame, "man (*Homo*)" was presented as part of the animal kingdom in a two-page tabular arrangement of classes, orders, and genera (Figure 9.1). Placing humans among the class of four-footed animals (*Quadrupedia*)—animals possessing a hairy body (*corpus hirsutum*), four feet (*pedes quatuor*), as well as viviparous and breastfeeding females (*feminae viviparae, lactiferae*)—and, within that class, among the order of the "human-shaped" (*Anthropomorpha*)—alongside the apes (*Simia*), and the sloth (*Bradypus*)—Linnaeus cleverly defined the genus *Homo* not by some presumably universal morphological or physiological feature, but by the human capacity for self-knowledge. What is interesting about this definition is that it addresses the reader by citing the famous dictum "Know thyself" (*Nosce te ipsum*), and then proceeds to split up the genus *Homo* into four distinct groups: the white European, the red American, the tawny Asian, and the black African.[2] In a single stroke, Linnaeus thus produced a universal scheme of naturalized human difference while at the same time highlighting that such a classification is the supreme product of human self-reflection. "Know thyself," Linnaeus suggests by typographic alignment, translates into "Distinguish thyself," and "race"—if that is what he was talking about here, a question, as we will see, that is not so easy to decide—hence turns out to have been conceived from its very

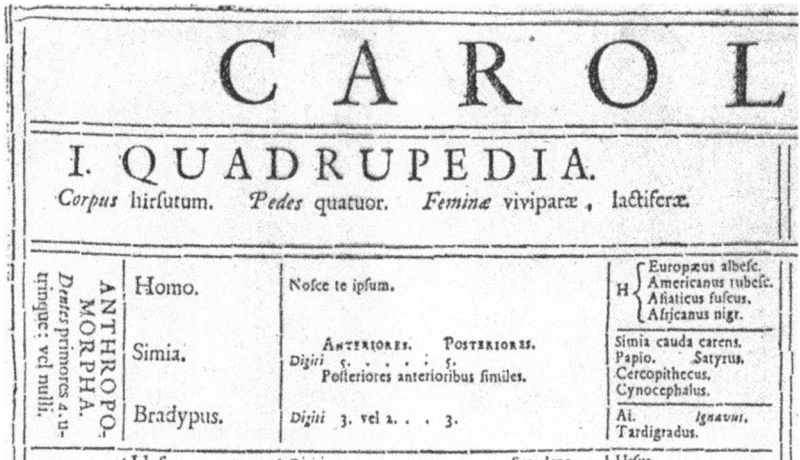

Figure 9.1 "Man's" place in the system of nature. Detail from Carl Linnaeus, *Systema Naturae, sive Regna Tria Naturae Systematice Proposita per Classes, Ordines, Genera, & Species* (Leiden: Theodor Haak, 1735)

beginning as a Janus-faced concept, facing nature on the one hand, and facing culture as reflection on nature on the other.

Despite its significance for the history of anthropology, there only exists one detailed and systematic study of Linnaeus's original writings on human races, published in Swedish in 1975 by Gunnar Broberg as part of a book on Linnaeus's general philosophy of nature and anthropological outlook.[3] As far as I know, Broberg's exhaustive and careful analysis of the original sources (including manuscripts) has had no reception in the Anglophone literature on the history of the race concept, which therefore continues to be riddled by the widespread misconception that Linnaeus was a staunch essentialist, and presented human races as distinct types. In fact, as we will see, Linnaeus shared contemporary views that skin-color—the chief criterion of distinction employed in the *Systema Naturae*—was largely a product of climate, and hence as variable as other "accidental" bodily characteristics of humans, such as stature or weight.

The significance that Linnaeus's classification of four human "varieties" (as he himself called them) would gain can therefore not be reduced to the fact that it pre-empted the racial typologies of the nineteenth century. Something else must have attracted Linnaeus himself, and eventually his readers—among them enlightenment luminaries such as Georges Buffon and Immanuel Kant—to the seductively simple scheme of four races distinguished by skin color. In this chapter, I am going to try to reveal, by a close rereading of relevant sources, that it was not the dubious value of race as a representation of actual, clear-cut difference that made it attractive to eighteenth-century naturalists. In fact, as I already indicated and will show in detail in the first section of this chapter, Linnaeus did not believe that such differences existed. And yet—as I will argue in the second section by turning to some of the possible sources on which Linnaeus relied—there was something unique and unprecedented about the way in which Linnaeus presented human diversity in 1735, namely the very abstract way in which it correlated physical characteristics with global distribution over the four continents. Section three will place this within the context of Linnaeus's general fascination with the four continents, and will argue that, rather than serving as a representation of human diversity, the distinction of four different varieties of humans served Linnaeus as a tool to orient himself on a global scale, and to guide him in the further collection of factoids about humans, resulting in a highly idiosyncratic association of the four races with medical temperaments, political inclinations, and psychological and cultural dispositions. This explains, as I will demonstrate in the final section of this chapter, why race played a very minor role only in Linnaeus's physiological and medical speculations about the human body. While an element of struggle comes to the fore in these speculations by portraying the body as being composed of two fundamental, antagonistic substances, this struggle is one between the sexes. Even in his proposals to interpret the diversity of life as the outcome of repeated hybridizations, Linnaeus did not build on the apparently obvious example of interracial mixing among humans, in stark contrast to Buffon.

For the general theme of this volume, this means that "race" in the eighteenth century was not straightforwardly connected with conceptions of bodily constitution. Race as a category was still in the making, and meshed with a variety of medical and philosophical ideas which upon closer inspection turn the category into a much more fluid one than a more superficial reading would suggest. While Linnaeus believed that classification provided the royal road toward truth, he did not necessarily believe that classifications should always and everywhere result in the distinction of stable types, nor that they should and would always refer to some underlying essence. Heredity, environment, and culture remained inextricably entangled in Linnaeus's conception of human variation. And yet, the net result of Linnaeus's deployment of the category was a set of geopolitical stereotypes on which later anthropological writers relied as a matter of course.

Subspecies, races, or varieties?

Linnaeus, as far as I am aware, never used the term race (Swedish *ras*), neither with reference to humans, nor with reference to other organisms. In Latin, he used the word *varietas* (variety) to designate different groups within one and the same species, in Swedish the word *slag*, a term introduced from the language of gardeners and breeders.[4] The reason for this is simple. The word had not reached the Swedish language yet; according to the Swedish Academy's dictionary, it appears first in print in 1765, in a translation of Henry Fielding's *The History of Tom Jones, a Foundling* (originally published in 1749). Whether Linnaeus, who at this point had passed the height of his career, would have accepted the term as an adequate neologism into his own taxonomic language is a matter of speculation.

The question whether Linnaeus would have referred to the four groups of humans he distinguished in *Systema Naturae* as "races" is nevertheless relevant. It has become quite common to read Linnaeus's classification as if it distinguished subspecies, and hence stable types.[5] This is reflected in more specialist literature by rendering the names of the four groups that Linnaeus distinguished—in line with a taxonomic custom that was established in the nineteenth century—as trinomials: *Homo sapiens europaeus, Homo sapiens americanus*, etc. A particularly prominent example is Phillip R. Sloan's essay "The Gaze of Natural History," which contrasts Linnaeus's anthropology with that of his contemporary Georges Buffon, who favored a view of human races as relatively fluid spatio-temporal entities and rejected abstract universals as the ones seemingly proposed by Linnaeus's classification of humans.[6] On the other hand, however, it is a well-known fact—which Sloan also acknowledges—that Linnaeus believed that all variation within a species was caused by local, environmental factors.[7] The *Systema Naturae* of 1735, and its subsequent editions, do not provide any clue to resolve the question whether Linnaeus thought of races as stable (sub)species or as environmental varieties. In these works, he never addressed this question explicitly. The way in which he presented the fourfold classification of humans in the *Systema Naturae* of 1735 might suggest a status of different species, but then no other animal genus is resolved into its constituent species.

To clarify the taxonomic rank of the four human races within Linnaeus's taxonomy of the animal kingdom, one has to turn to an unlikely source. In Linnaeus's botanical work, the distinction of varieties from species played an important role, since it was Linnaeus's great ambition to reduce the number of species—and species names—within botany.⁸ To achieve this, Linnaeus made a strong distinction between traits whose formation is determined by intrinsic "laws of generation" and which therefore remain "constant" across all members of a species, and traits that vary within a species due to "accidental" factors such as soil or climate.⁹ In *Critica Botanica*, a work detailing the rules and conventions according to which plant names should be formed, Linnaeus discussed the distinction at great length, and this is the only occasion on which he entered a lengthy discussion on the significance of physical differences among humans.

This discussion relates to a difficulty that the distinction of species and varieties encountered, namely the fact that certain varieties continue to transmit their distinctive character, even if external conditions change. The example Linnaeus adduced in this context—alongside the "variety of seeds that gardeners sell"—was human skin color. "Who would deny that the Ethiopian is of the same species as our people (*ac nos homines*)," Linnaeus asks rhetorically, only to add: "And yet the Ethiopian produces black children on our soil (*nigros infantes in nostra terra*)."¹⁰ A very clear distance makes itself felt here in the use of the first person plural ("our people" could also be rendered as "us humans"); but the insistence that this distance does not indicate a species difference is equally clear, and repeated with great force in another passage from *Critica Botanica* that is worth quoting at length:

> Certainly, if each trait would equally constitute a new species, there would be no wiser and accurate Botanists among mortals than those FLOWER-LOVERS, who each year point out to the curious, among tulips, primroses, anemones, daffodils and hyacinths alone, some thousand species as yet unknown to Botanists, and hence [claimed to be] new ones. But the Omnipotent Builder abstained from the work of creation on the seventh day, so that there are no new creations with each day, but a continued multiplication of things already created. He created one human, as the Holy Scripture teaches; but if the slightest trait [difference] was sufficient, there would easily stick out thousands of different species of man: they display, namely, white, red, black and grey hair; white, rosy, tawny and black faces; straight, stubby, crooked, flattened, and aquiline noses; among them we find giants and pygmies, fat and skinny people, erect, humpy, brittle, and lame people etc. etc. But who with a sane mind would be so frivolous as to call these distinct species? You see, therefore we assume certain characters, and query deceptive ones, which lead astray and do not change the thing.¹¹

The inclusion of skin color with other highly variable physical characteristics, including deformations, leaves little doubt that Linnaeus did not believe that this trait pointed to any essential difference, and that he also did not believe that it allowed for the formation of discrete categories. It may well be that aligning complexion with other highly variable traits in humans was motivated by Linnaeus's belief in scripture, as Broberg has surmised.[12] But he was surely also acknowledging the simple, empirical fact that skin color is indeed highly variable. Linnaeus actually acknowledged this fact in the 1735 edition of *Systema Naturae* by the choice of color terms; none of these terms states a clear-cut color, but rather a hue or coloring: Europeans are said to be "whitish (*albesc[ens]*)," not white; Americans "reddish (rubesc[ens])," not red; Asians "tawny," not yellow; and Africans "blackish (nigr[iculus])," not black.

If anything, this lets Linnaeus's scheme of four human varieties appear even stranger than to begin with. Apparently, it was not meant to present the reader with some kind of image, or representation, of what the (human) world is actually like. It must have had some additional function. In order to approach this function, it is worthwhile to contrast Linnaeus's classificatory schemes with some of its potential sources, in order to see more clearly what it is, exactly, that marks it as the beginning of something new.

Linnaeus's sources

Linnaeus was never explicit about the sources for his anthropological knowledge. Neither the first, nor the tenth, nor the twelfth edition of *Systema Naturae*—the latter two substantially revised and augmented versions of the former—cite any authorities on the classification of mankind. It rather seems that Linnaeus remained exceptionally uninformed about matters of race throughout his long career. In the treatise *Sponsalia Plantarum* (1746), which dealt with organic reproduction in general, and plant sexuality in particular, all that can be found on this matter, for example, is a citation of an account by the seventeenth-century Danish physician Thomas Bartholin (1616–80) about an "Ethiopian" slave and a Danish maidservant in Copenhagen who had a male child "whose whole body was due to the mother, except the penis which by its black color showed his paternal kind (*paternum genus*)."[13] This was only three years before George-Louis Leclerc, Comte de Buffon (1707–88), produced his more than 150-page chapter on "varieties within the human species (*variétés dans l'espèce humaine*)" which was based on an extensive review of existing travel literature.[14] Even later, Linnaeus would prefer to ask his French correspondents—Bernard de Jussieu (1699–1777) in particular, who was serving under Buffon as *demonstrateur des plantes*—what Buffon was up to, rather than reading the French original.[15]

It is nevertheless possible to speculate about some of the sources that may have been available to Linnaeus, if only to contrast them with his own curious division of mankind of 1735. There is first of all the chapter on the "Inhabitants of Brazil" from Georg Marcgrave's *Historia Naturalis Brasiliae* (1648). The book was in the possession of the Uppsala professor of theology and oriental languages Olof Celsius (1670–1756) with whom Linnaeus lodged as a student, and whose extensive botanical library he studied assiduously.[16] Marcgrave's account of the inhabitants of Brazil is remarkable in several respects; first, it notes with a modicum of surprise that the Portuguese, Dutch, German, French, and English are collectively referred to as "Europeans" in Brazil;[17] second, it proposes that the "mixture of various nations (*nationum*)" happening in Brazil had led to the emergence of "five distinct kinds of people." What follows is one of the earliest accounts of a classification system known as *las castas*, which tried to get a grip on *mestizaje* through an elaborate terminology designating its various products: "Who is born from a European father," wrote Marcgrave, "and a Brazilian mother is named *Mameluco*"; "[who is] born from a European father and an Ethiopian mother is called *Mulatto*."[18] Again, skin color plays a role in this system—Marcgrave mentions, for example, the birth of twins from an "Ethiopian woman (*Aethiopissa*)," one of which was "white," the other "black" (*unum album, alterum nigrum*). But it is not highlighted as a universal criterion of distinction; quite the contrary, as the example of the twins shows. Marcgrave's description rather places emphasis on the singular and local character of race mixture. In contrast, Linnaeus's classification clearly was meant to be global and exhaustive, effectively correlating his four human varieties with the four continents then known.

A second likely source that Linnaeus may have drawn upon is an obscure pamphlet produced by the composer and mathematician Harald Johannson Vallerius (1646–1716) in 1705 in the form of an academic dissertation at Uppsala University, the university that Linnaeus studied medicine at from 1727 to 1731. Under the title "About the Various External Appearance of Men," it reproduced the argument of François Bernier's (1625–88) well-known essay "New Division of the Earth According to the Different Species or Races that Inhabit it," adapting it to the purposes of the home-grown ideology of *Göticism* (Gothicism).[19] Like Bernier, Vallerius began with an overview of the various kinds of people that inhabit our planet, only to embark on a long-drawn-out argument aiming to show that the most beautiful women are *götiskt*, that is, Swedish. The chart he presents of human variation is rather odd: according to Vallerius, there are "Ethiopians" who are "black" (*nigri*); lapps and samojeds who are "tawny" (*fusci*); Italians, Spaniards, and French whom Vallerius curiously describes as "ashgrey" (*cinericio colore*)"; and, finally, "White Ethiopians" (*Leucoaethiopes*), who again, as the name indicates, include some inhabitants of Africa, but mainly those of Germany and its "neighboring countries."[20] Like Linnaeus 30 years later, Vallerius used skin color as a chief criterion, and there are similarities down to the color terms used. There is a striking difference also, however. Unlike Vallerius—and Bernier, who mentions the "Lapps" (*Lappons*) as a separate "species" of humans[21]—Linnaeus's classification does not make reference to smaller, marginal populations. His classification seems to be the product of an urge to establish a fourfold, symmetric division of humankind. The four varieties are presented as inhabiting the globe in equal parts, thus excluding polarities like metropolitan vs peripheral, natural vs monstrous, domestic vs exotic, or, for that matter, beautiful vs ugly.

There is a third likely source for Linnaeus's classification of mankind. In the notebooks he kept as a student, there is a drawing of a bat that closely resembles a plate from Richard Bradley's (1688–1732) *A Philosophical Account of the Works of Nature* (1721), of which Linnaeus possessed a copy.[22] Bradley's book was a remarkably materialistic presentation of the "scale of life," arguing, for example, that the difference in "capacity and understanding" between apes and humans "proceeds from the various Frames of those Parts which furnish the Brain with nourishing Juices."[23] According to Bradley,

> we find five Sorts of Men: the White Men which are *Europeans*, that have *Beards*; and a sort of *White Men* in *America* (as I am told) that only differ from us in having no *Beards*. The third sort are the *Molatoes*, which have their *Skins* almost of a *Copper* Colour, *small Eyes*, and *strait black Hair*. The fourth Kind are the *Blacks*, which have *strait black Hair*: And the fifth are the *Blacks of Guiney* whose *Hair* is *curl'd*, like the *Wool* of a *Sheep*.[24]

Although Bradley distinguishes five, rather than four, "sorts of Men," and although he includes hair color and form, as well as eye shape, as additional criteria, the similarities with Linnaeus's scheme are striking; both classifications make reference to physical characteristics, and both propose a global and symmetric division of mankind. It is all the more remarkable that Bradley as well does not cite any sources, and also refuses to draw any conclusions. Like Linnaeus 14 years later, he presents his classificatory scheme *ad hoc*, with no apparent context.[25] Its function must therefore have been different from simply synthesizing what was supposedly known already. In order to see what that function might have been, I will turn to a feature of Linnaeus's classification of man that is often overlooked: its close correlation with the geographic division of four continents.

Orientation and accretion of facts

There are many signs that Linnaeus was fascinated from early on with the four continents. In a commonplace book he kept as a student at Lund University, one finds a table that associates various drinks with the four continents: Asia is associated with tea (*Theè*), Africa with coffee (*Coffi*), America with chocolate (*Chocolaten*), and Europe with beer (*Cerevisia*).[26] The journal from his Lapland journey in 1732 contains a famous passage in which he describes his first visit to the highlands of this northern region and how the abundance of unknown species caused him to wonder "whether I was in Asia or Africa, as the soil, the situation, and all the plants were unknown to me."[27] A final example may suffice. One of Linnaeus's first botanical publications—the *Hortus Cliffortianus*, a folio volume published two years after the *Systema Naturae*, and consisting of a lush catalogue of the exceptionally rich botanical collections of the merchant banker and former director of the Dutch East India Company George Clifford (1685–1760)—contained a frontispiece which showed Europa at the center, surrounded by three figures to the left impersonating the three continents Asia, Africa, and America, each of them presenting a plant to her, and a male figure to the right caught in the act of removing a cloak from her head, and bearing an unmistakable resemblance to Linnaeus himself (see Figure 9.2). The preface to this volume has a long section that lists plant species characteristic of each continent, and highlights Linnaeus's own descent from Northern Europe.[28]

What these documents suggest is that the four continents served Linnaeus as a kind of geographic grid that helped him to orient himself on a global scale (or, for that matter, to express disorientation). If we apply this to his distinction of four human varieties, it becomes clear that this distinction was not so much the result of a careful synthesis of previously established facts, but rather a deliberate and arbitrary

Figure 9.2 Frontispiece of Carl Linnaeus, *Hortus Cliffortianus* (Amsterdam, 1738)

projection to support the future accumulation of facts. That this is indeed so becomes clear once one follows Linnaeus's treatment of human diversity through the various editions of *Systema Naturae*, and also attends to the handwritten annotations that peppered his personal copies of these editions (Figures 9.3 and 9.4). The tenth edition, published in 1758, saw the first substantial expansion of the classification of 1735. Again, it lists four main "varieties" of the human species, numbered consecutively by Greek letters. Skin color remains the first mark of distinction, although the color terms have altered to red (*rufus*), white (*albus*), pale yellow (*luridus*), and black (*niger*), indicating both a hardening and, in the case of *luridus*, a more judgmental distinction.[29]

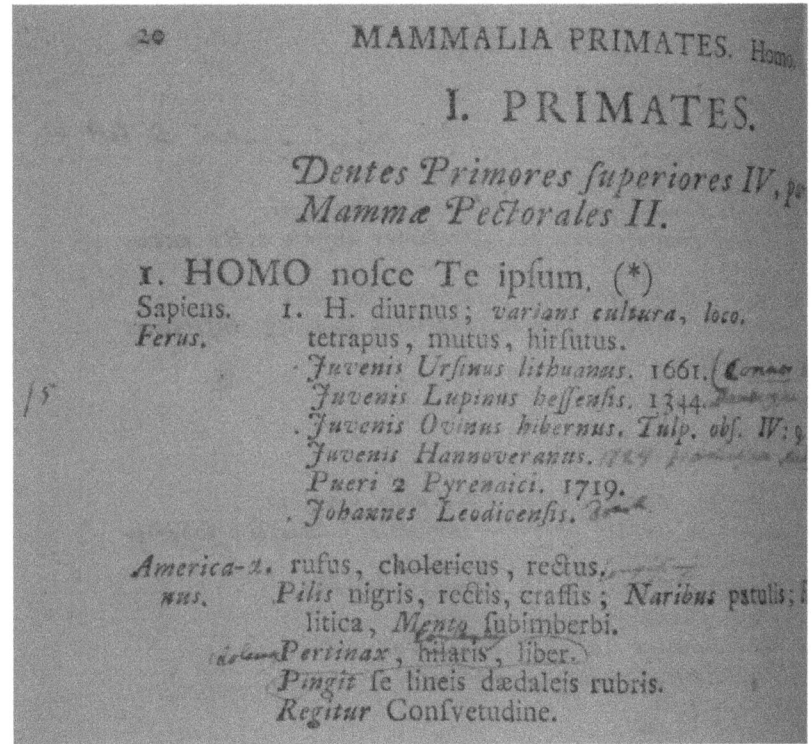

Figure 9.3 Entry for "Homo Sapiens" in Linnaeus's personal, annotated copy of the tenth edition of *Systema Naturae* (1758). Linnean Society London, Library and Archives, Linnaean Collections, *Systema Naturae*, 10th edn (1758), Sign. BL.16

In addition, Linnaeus associated a range of other characteristics with his four human varieties, arranging them in five lines: the first line describes skin color, medical temperament, and body posture; the second line adds further physical characteristics pertaining to hair color and form, eye color, and distinctive facial traits; the third line refers to behavior; and the final two lines to manner of clothing and political constitution respectively.[30] Many of these characterizations relied on nascent racial stereotypes—Africans, for example, are said to be governed by *arbitrio*, which can be translated as caprice or dominion, that is mastery by others—yet the corrections and additions in Linnaeus's personal copies also make clear that the classification was fluid. The notes

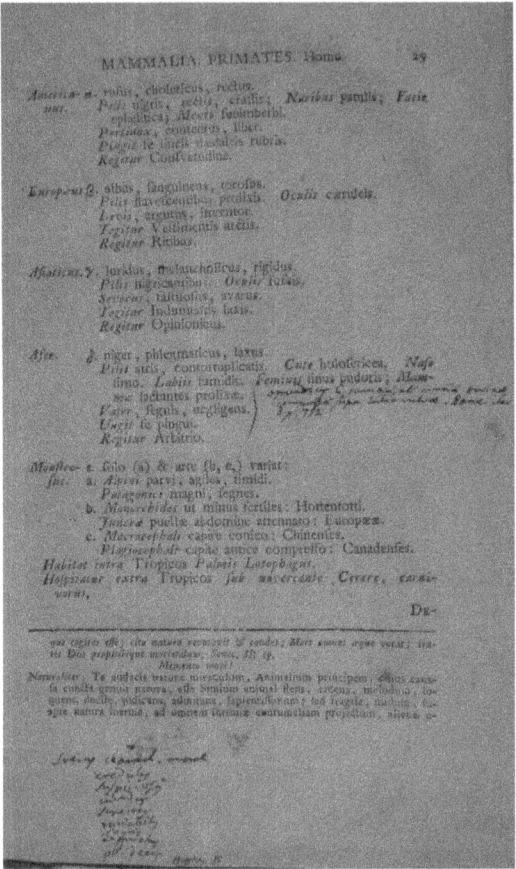

Figure 9.4 Page with descriptions of human varieties in Linnaeus's personal, annotated copy of the twelfth edition of *Systema Naturae* (1766). Linnean Society London, Library and Archives, Linnaean Collections, *Systema Naturae*, 12th edn (1758), Sign. BL.21

in his personal copy of the tenth edition, for example, indicate that Linnaeus wanted to change the characterization of "Americans" from "cheerful" (*hilaris*) to "content" (*contentus*), and contemplated moving the medical temperaments to the line dedicated to behavioral traits (see Figure 9.3).[31] Other annotations in the twelfth edition include a short definition of the "moral character of the Swede"—"credulous, distrustful, jealous, conceited, fickle, dull, fidgety, compliant"—and hence an attempt at a finer-grained differentiation within the category of "white Europeans" (see Figure 9.4).[32]

Two features of this new classification scheme fathoming human variation deserve highlighting and further comment. First, Linnaeus rearranged the order of the four varieties. It is not the "white Europeans" anymore that occupy the top position, as in all previous editions of *Systema Naturae*, but the "Americans," echoing ideas of the noble savage that particularly come to the fore in the behavioral and political traits assigned to the latter: "unyielding, content, free" (*pertinax, contentus, liber*) as well as "governed by customary right" (*consuetudo*), rather than laws (Europeans), opinions (Asians), or caprice (Africans). Second, the traits are arranged in five lines by their increasing "distance" from the body: traits in the first line refer to bodily constitution as gauged by complexion, temperament, and composure; the second line singles out characteristic facial features; the third what we would call "behavioral" traits; the fourth takes up apparel—with Americans "painting [themselves] with red streaks" and Africans "smearing [themselves] with fat," while "Europeans" and "Asians" wear clothes, the former tight, the latter wide clothes; and the fifth spells out the presumed social and political constitution of the four varieties. The impression that this arrangement is meant to progress from internal, and hence more constant, to more variable, external properties is confirmed by the fact that in his annotations to the tenth edition, Linnaeus experimented with a different arrangement that would place apparel before behavior (Figure 9.3). Further confirmation is provided by the addition of a fifth, "monstrous" human variety (*Monstrosus*) which includes a variety of groups clearly shaped by external conditions: natural conditions as in the case of the *Alpini*, that is, humans living at high altitudes, which Linnaeus believed to be "small, agile, and timid" (*parvi, agiles, timidi*); cultural conditions as in the case of "slender girls with constricted waists" to be found in Europe (*Junceae puellae abdomine attentuato: Europaeae*).

As Mary Floyd-Wilson has remarked about Linnaeus's late classifications of humankind, one can clearly discern in them "the residual matter of early modern geohumoralism," that is, the idea that climate and medical temperaments, external conditions, and inner constitution are causally contiguous, and hence mirror each other in the shaping of human differences, whether physical, behavioral, or cultural. Floyd-Wilson also notes, however, that Linnaeus, like many of his contemporaries and predecessors in the early modern period, performs a radical reevaluation of this relationship between medical temperaments and climates. Phlegmatic temperament, most notably, is now associated with a black complexion and a "hot" climate, whereas sanguine temperament is associated with whiteness and the North, in stark contrast to ancient and medieval lore. Just like Francis Bacon and Thomas Browne before him, Floyd-Wilson observes, Linnaeus loosened the "tie between skin colour and humoral disposition," thus allowing for a radical "restructuring of geohumoral theory."[33] This leads to a surprising conclusion, however. The classification of human diversity by skin color that Linnaeus introduced in 1735 was not only used for the accretion of new facts about, and increasing entrenchment of, presupposed racial categories. At the same time, its abstract and *ad hoc* nature enabled fundamental inversions in the received framework of associating constitutions and climes.

Marrow and bark: the struggle of the sexes

Geohumoral theory was not only restructured by Linnaeus because he decided to realign it with his fourfold classification of mankind. More fundamentally, he embedded it in a physiological theory which relegated the four bodily humors, and hence the four medical temperaments, to the status of mere epiphenomena of more fundamental substances and forces. In speculations that grew more and more elaborate the older he became, Linnaeus assumed that all physiological processes were due to the antagonistic interaction of two fundamental substances, the marrow (*medulla*) which had a capacity for uninhibited growth, and the bark (*cortex*) which contained and structured this growth. In addition, he assumed that these two substances were passed on from one generation to the next along paternal and maternal lines respectively: the marrow came from the mother, and the cortex from the father.[34] The distinction of the two substances clearly reflects Linnaeus's ideas of male and female roles in the economy—he identified females as largely responsible for the drain of bullion through the consumption of luxury goods from abroad[35]—but also his own growing exhaustion with trying to tame the diversity of nature through his taxonomic enterprise.[36]

Linnaeus's medulla-cortex theory was highly idiosyncratic—combining elements of iatromechanism's understanding of bodies as hydraulic machines, with a curious brand of vitalism—but is relevant for understanding his race concept for two reasons. First, it provides a strong indication that Linnaeus, in his later career, began to think of living nature as being constituted and shaped by an underlying struggle between antagonistic forces—and ideas of a "struggle for life" would become one of the central elements of scientific racism.[37] Second, it provided him with an explanation for the origin of diversity that went beyond climatic degeneration and thus could account for the formation of essential, rather than merely accidental, difference. Assuming that even widely different life forms were able to hybridize, Linnaeus developed the view in later works that God had only created a few forms in the beginning and that subsequently new species arose through hybridization, and hence through the combination of different cortical substances with the medulla of the original form. In *Fundamentum Fructificationis*, a late essay published as a dissertation in 1762, Linnaeus explained unique features of the North-American and African flora on this basis, citing strong winds at the Cape of Good Hope as a possible mechanism that may have led to the particularly pronounced proliferation of unusual species in this region.[38]

But did Linnaeus ever apply this theory to explain human diversity? Curiously, there is no sign that he ever tried to do so. What we do find in terms of explanations of human diversity are hints at accounts that rely on migration and subsequent climatic degeneration. In an undated zoological manuscript Linnaeus jotted down, for example, that humans enjoy a "rich and blessed immaterial soul" (*Anima immateriali beata dives*), form a "single species" (*Species unica*), and "roam about" (*peregrinat*), even to places like Nicobar and Ambon Island.[39] In other words, what unites humans has no immediate relation to the body, and all differences among humans have thus come about accidentally. With the possession of a rational soul, humans essentially remained part of the divine order for Linnaeus, even if they were hardly distinguishable from their next of kin in the animal kingdom—"Man's cousins," as the Swedish version of a text by Linnaeus on primates was entitled—and even if they could be subjected to classification just as any other animal species could.[40]

Conclusion

Race is tied up with metaphors of blood; talk of bloodlines, the mixing of blood, or the "one drop of blood" rule provides abundant evidence. The connection goes back to the late medieval period, when ancient conceptions of "noble blood" were revived in the context of animal breeding and transposed to debates around nobility.[41] The connection is tenuous nevertheless, as the example of Linnaeus that I have analyzed in this chapter clearly demonstrates. Race as we know it, while clearly rooted in the racist preconceptions that colonial encounters precipitated, did not simply grow out of the ancient entwinement of the microcosm of bodily humors and the macrocosm of climates and regions. Quite the contrary. With Paul Feyerabend, one might want to claim that Linnaeus engaged in an exercise of "counter-induction" when setting up his racial classification according to skin color and later associating it with the four medical temperaments, and hence the balance of body humors. This classification dissociated physical traits from both bodily constitution *and* natural environment, only to open an entirely new space of phenomena that would form the subject of speculations about the contingent relationship of organic bodies and their "natural places" in theories of inheritance and, eventually, evolution.[42] Immanuel Kant, in particular, would have no qualms in filling the explanatory gap that Linnaeus had left.[43]

Linnaeus's color scheme became, as Renato Mazzolini recently pointed out, an "integral part of all subsequent classifications of the late eighteenth and the first half of the nineteenth century." It did so, as Mazzolini argues, on the basis of a careful bibliometric analysis, not because skin color was associated with bodily constitution, but because it quite literally had turned out to be a "skindeep" phenomenon only, located in the so-called Malpighian layer of the skin, and hence was freed up to define a European "somatic identity mainly constructed on political-social relationships."[44] It cannot be emphasized enough how fanciful Linnaeus's color scheme actually is, if judged in terms of the humoral doctrine: white is red (sanguine), black is white (phlegmatic), yellow is black (melancholic), and red is yellow (choleric). The fact that it sticks with us to this day only demonstrates how overwhelmingly powerful the discourse was that took hold within the conceptual space thus freed up.

Notes

1. Stephen J. Gould, *The Mismeasure of Man*, 2nd edn (New York: W. W. Norton, 1996), 66; see C. Loring Brace, *Race Is a Four-Letter Word: The Genesis of the Concept* (Oxford University Press, 2005), 17–36, for a more recent version of the standard account.
2. Carl Linnaeus, *Systema Naturae* (Amsterdam: Schouten, 1935), unpag. [p. 10].
3. Gunnar Broberg, *Homo sapiens L.: Studier i Carl von Linnés naturuppfattning och människolära* (Uppsala: Almquist & Wiksell, 1975), ch. 5.
4. Carl Linnaeus, "Rön om växters plantering grundat på naturen," *Kungliga Svenska Vetenskaps-Akademiens Handlingar* 1 (1739), 5–24.
5. See, for example, Jonathan Marks, *Human Biodiversity: Genes, Race, and History* (New Brunswick: Aldine Transaction, 1995), 50.
6. Phillip R. Sloan, "The Gaze of Natural History," in *Inventing Human Science: Eighteenth-Century Domains*, ed. Christopher Fox, Roy Porter, and Robert Wokler (Berkeley: University of California Press, 1995), 112–51, 128. Presenting Linnaeus's distinction as a series of trinomials goes back at least to Stephen Jay Gould's *Mismeasure of Man*, 66, and probably has its origin in an English translation of the first part of the thirteenth, posthumous edition

of *Systema Naturae* that was published in 1792; see Carl Linnaeus, *The Animal Kingdom, or Zoological System*, ed. Johann Friedrich Gmelin, trans. Robert Kerr (London and Edinburgh: A. Strahan, T. Cadell, and W. Creech, 1792), 45. As Kerr stated quite openly in the full title of the publication, this edition contained "numerous additions from more recent zoological writers."

7. Sloan, "Gaze of Natural History," 121.
8. Carl Linnaeus, *Genera Plantarum* (Leiden: Wishoff, 1737), "Ratio operis," aph. 8 [unpag.]. For a translation of this important methodological text, see Staffan Müller-Wille and Karen Reeds, "A Translation of Carl Linnaeus' Introduction to *Genera Plantarum* (1737)," *Studies in History and Philosophy of the Biological and Biomedical Sciences* 38 (2007), 563–72.
9. Linnaeus, *Genera Plantarum*, "Ratio operis," aph. 5; see Staffan Müller-Wille, "Collection and Collation: Theory and Practice of Linnaean Botany," *Studies in History and Philosophy of the Biological and Biomedical Sciences*, 38 (2007), 541–62.
10. Carl Linnaeus, *Critica Botanica* (Leiden: Wishoff, 1737), 255. Linnaeus knew of many cases of "constant varieties" among plants, and seems to have shared the widespread conviction that the environment has effects on organisms that will only recede after many generations upon transplantation; see John Ramsbottom, "Linnaeus and the Species Concept," *Proceedings of the Linnean Society London* 150 (1938), 192–219. Conversely, he believed that exotic plants, even from warmer regions of the globe, could be acclimatized to Swedish conditions; see Lisbet Koerner, "Linnaeus's Floral Transplants," *Representations* 47 (1994), 144–69.
11. Linnaeus, *Critica Botanica*, 152–3.
12. Broberg, *Homo sapiens L.*, 228.
13. Carl Linnaeus, *Sponsalia Plantarum* (Stockholm: Salvius, 1746), 26.
14. Georges-Louis Leclerc, Comte de Buffon, *Histoire naturelle, générale et particuliére* (Paris: Imprimerie Royale, 1749), 3: 371–530.
15. Carl Linnaeus to Bernard de Jussieu, 25 March 1752, *The Linnaean Correspondence*, http://linnaeus.c18.net, letter L1387.
16. A catalogue of Celsius's botanical library has been preserved which lists Marcgrave's work; see "Catalogus Bibliothecae Botanicae ... Olavo Celsio, Bibliotheca haec Regia suo aeve emit d. XV. Novemb. MDCCXXXVIII," Uppsala University Library, Donationskataloger över tryckta böcker m.m. A-J, Bibl. Arkiv K 52:1. On Linnaeus and Celsius, see Wilfrid Blunt, *The Compleat Naturalist: A Life of Linnaeus* (London: Collins, 1971), 30–6.
17. Willem Piso and Georg Marcgrave, *Historia Natvralis Brasiliae* (Amsterdam: Elzevir, 1648), 268: "In genere autem vocant omnes Europaeos."
18. Ibid.: "Denique ob misturam variorum nationum, aliae quinque distinctae hominum species haec reperiuntur." On the *castas*-system, which was only really popularized in Europe through the writings of Buffon and Cornelis de Pauw (1739–99) in the 1770s, see Renato G. Mazzolini, "Las Castas: Inter-Racial Crossing and Social Structure (1770–1835)," in *Heredity Produced: At the Crossroads of Biology, Politics and Culture, 1500–1870*, ed. Staffan Müller-Wille and Hans-Jörg Rheinberger (Cambridge, MA: MIT Press, 2007), 349–73.
19. Harald Vallerius, *De Varia Hominum Forma Externa* (Uppsala: Werner, 1705); François Bernier, "Nouvelle division de la terre par les différentes espèces ou races d'hommes qui l'habitent," *Journal de Sçavans*, 24 April (1684),

133–40. As far as I can see, there is no evidence that Linnaeus ever read Bernier's essay.
20. Quoted from Broberg, *Homo sapiens L.*, 221. The "white Ethiopians," as Broberg explains, go back to Pliny's account of black albinos.
21. Bernier, "Nouvelle division," 136.
22. Carl Linnaeus, "Manuscripta Medica," vol. I, Linnean Society Library and Archives, Linnaean Collections, Box LM Gen, Folder LINN PAT GEN 2, f. 83v. The plate from which Linnaeus copied the bat can be found in Richard Bradley, *A Philosophical Account of the Works of Nature* (London: Mears, 1721), 88, pl. xiii, fig. ii. For a reproduction and discussion of Linnaeus's drawing, see Isabelle Charmantier, "Carl Linnaeus and the Visual Representation of Nature," *Historical Studies in the Natural Sciences* 41.4 (2011), 365–404, 380, fig. 5.
23. Bradley, *Philosophical Account*, 95.
24. Ibid., 169.
25. It is easy to see, however, that Bradley's contribution stands in the tradition of naturalizing human diversity, and treating it as a question of natural history, rather than theology, which began in Britain with John Locke; see David Carey, *Locke, Shaftesbury, and Hutcheson: Contesting Diversity in the Enlightenment and Beyond* (Cambridge University Press, 2006), 15–23.
26. Carl Linnaeus, "Manuscripta Medica (1727–1730)," Linnean Society Library and Archives, Linnaean Collection, Manuscripts, vol. I, f. 38v.
27. Carl Linnaeus, *Iter Lapponicum*, ed. Thomas M. Fries. Skrifter af Carl von Linné, vol. 5 (Uppsala: Almqvist and Wiksells, 1913), 106.
28. For a detailed analysis of the frontispiece to *Hortus Cliffortianus*, see Gunnar Broberg, "Naturen på bild: Anteckningar och Linneanska exempel," *Lychnos* 1979–80 (1980), 231–56.
29. On the change from *fuscus* to *luridus*, see Michael Keevak, *Becoming Yellow: A Short History of Racial Thinking* (Princeton University Press, 2011), 51–7.
30. Carl Linnaeus, *Systema Naturae*, 10th edn, 3 vols (Stockholm: Salvius, 1758), 1: 20–2.
31. Carl Linnaeus, *Systema Naturae*, 10th edn, 3 vols (Stockholm: Salvius, 1758), Linnean Society London, Linnaean Collections, Library, BL.16, vol. 1, 20–2. The change in the characterization of "Americans" happened with the twelfth edition (see citation in note 32), the medical temperaments remained in the first line, however.
32. Carl Linnaeus, *Systema Naturae*, 12th edn, 3 vols (Stockholm: Salvius, 1766–68), Linnean Society London, Linnaean Collections, Library, BL.21, vol. 1, 29. The regional fauna that Linnaeus produced for Sweden contains a classification of his home country's population into four varieties: "Goths" (*Gothi*), "Finns" (*Fennones*), "Lapps" (*Lappones*), and "Various mixtures of the preceding" (*Varii & mixti ex praecedentibus*); see Carl Linnaeus, *Fauna Suecica* (Stockholm: Salvius, 1746), 1.
33. Mary Floyd-Wilson, *English Ethnicity and Race in Early Modern Drama* (Cambridge University Press, 2003), 86. In the same way, "Creole physicians found ways to adapt the wide and permissive Hippocratic landscape to their New World circumstances"; Carlos López Beltrán, "Hippocratic Bodies, Temperament and Castas in Spanish America (1570–1820)," *Journal of Spanish Cultural Studies* 8 (2007), 253–89, 276–7.

34. For a detailed discussion, see Peter F. Stevens and Steven P. Cullen, "Linnaeus, the Cortex-Medulla Theory, and the Key to His Understanding of Plant Form and Natural Relationships," *Journal of the Arnold Arboretum* 71 (1990), 179–220.
35. Lisbet Koerner, *Linnaeus: Nature and Nation* (Cambridge, MA: Harvard University Press, 1999).
36. On the older Linnaeus's disenchantment with natural history, see Elis Malmeström, *Carl von Linnés religiösa åskådning* (Stockholm: Svenska Kyrkans Diakonistyrelse Bokförlag, 1926).
37. Michel Foucault, *"Society Must Be Defended": Lectures at the Collège de France, 1975–1976* (London: Picador, 2003).
38. Carl Linnaeus, *Fundamentum Fructificationis* (Uppsala: No publisher, 1762). The proposition that the environment of Africa, in particular, fosters the production of new species echoes ancient ideas; see Harvey M. Feinberg and Joseph B. Solodow, "Out of Africa," *Journal of African History* 43 (2002), 255–61.
39. Carl Linnaeus, "Zoologia," Linnean Society London, Linnaean Collections, Manuscripta Medica I, Folder "Pertinet ad Linnaei Manuscr. Med."
40. On Linnaeus's classification of man, see Gunnar Broberg, "Linnaeus's Classifications of Man," in *Linnaeus: The Man and His Work*, ed. Tore Frängsmyr (Berkeley: University of California Press, 1983). Eric Voegelin has made the general point that the kind of body–soul dualism that we find exemplified in Linnaeus tended to prevent the full naturalization of human difference; see Eric Voegelin, *Die Rassenidee in der Geistesgeschichte von Ray bis Carus* (Berlin: Junker und Dünnhaupt Verlag, 1933).
41. Maaike van der Lugt and Charles de Miramon, "Introduction," in *L'hérédité entre Moyen Age et époque moderne*, ed. Maaike van der Lugt and Charles de Miramon (Florence: SISMEL—Edizioni del Galluzzo, 2008), 3–40.
42. Staffan Müller-Wille and Hans-Jörg Rheinberger, *A Cultural History of Heredity* (University of Chicago Press, 2012), 59.
43. On Kant's theory, see Raphaël Lagier, *Les races humaines selon Kant* (Paris: Presses Universitaires de France, 2004).
44. Renato Mazzolini, "Skin Color and the Origin of Physical Anthropology," in *Reproduction, Race, and Gender in Philosophy and the Early Life Sciences*, ed. Susanne Lettow (New York: SUNY Press, 2014), 131–61, 151.

10
"Who Got Bloody?": The Cultural Meanings of Blood during the Civil War and Reconstruction

James Downs

Blood still stains the report that a federal doctor in Charleston, South Carolina sent to officials in Washington, DC, about the outbreak of smallpox in the winter of 1865. Smallpox, he reported, had infected 1568 freedpeople throughout the state from Orangeburg to the Sea Islands. The bloodstain on the report remains one of the few visual markers that illustrates the suffering and illness that the war engendered for newly emancipated slaves.

The doctor's claim that smallpox infected 1568 freedpeople provided federal officials in Washington with all the information they needed.[1] The federal government agreed to establish hospitals and to employ physicians only in an effort to segregate the population of sick freed slaves from those who could work. The hospitals, in turn, were to simply quarantine not treat the infected.[2] Consequently, the statistic was one of the few ways that former bondspeople's suffering became recorded during the postwar period. Only through numbers, documented in the service of organizing a labor force, did freedpeople's suffering and mortality become registered in the public record.

The fact that blood had somehow dripped onto the report about the mortality in South Carolina and stained it was purely accidental. This document is one of the few occurrences in which freedpeople's blood, both literally and figuratively, manifested itself in the official correspondence of federal doctors serving in the Reconstruction South. In all the accounting of the number of freedpeople who died during the Civil War and Reconstruction, there is no reference in the public record to their bloody and wounded condition. Instead, throughout the thousands of documents that detail freedpeople's health during this period, doctors and federal officials defined black people as "dirty," "destitute," and "dependent." This lexicon that developed had more to do with federal efforts to evaluate freedpeople's health in terms of their labor power, rather than in terms of their actual medical condition.[3]

Freedpeople were sick and suffering during the Civil War and Reconstruction—over one million freed slaves required federal assistance in response to the epic outbreak of epidemics, the chilling rates of starvation, and the massive dislocation that accompanied emancipation—but the official federal and military records rarely describe formerly enslaved people as wounded, sick, and bloody. In fact the historiography often renders freedpeople during this period as triumphant—accentuating the degree to which freedpeople exerted their agency. These interpretations leave little rhetorical room to illustrate how the Civil War era also left many black soldiers and freedpeople infected, bruised, and bloody. The Civil War was the largest biological crisis of the nineteenth century; more soldiers died from disease than from battle. Once freed, formerly enslaved people entered into a world plagued by dysentery, typhus, and other camp diseases. By the winter of 1865, a smallpox epidemic that began in Washington, DC, moved into the Lower South and Mississippi Valley, infecting thousands of newly emancipated slaves. Freedpeople lacked shelter, clean clothing, inoculations, vaccinations, and even the spaces to quarantine those infected. Consequently, the smallpox outbreak exploded into a major epidemic and took the lives of roughly 60,000 freed slaves.[4]

Further, both the nineteenth-century press and contemporary historians often narrate the emancipation of over four million enslaved people in the United States as a celebratory turning point, often overlooking the bloodshed that led to their liberation. A comment made by James L. Smith, a formerly enslaved man, challenges this interpretation of emancipation as simply a triumphant narrative in which freedpeople danced in the streets and sang alleluia. He writes, "Some were violently beaten, or rudely scourged; many were deliberately shot down in open day, on the public streets; others were way-laid and cruelly butchered, and some ... The streets have been drenched with their blood."[5] His is one of the few voices that reveals how many white Southerners violently contested emancipation, which led to the streets being soaked with formerly enslaved people's blood.

Despite the inordinate mortality and suffering that freedpeople experienced during the Civil War and Reconstruction, very few references describing their bloody condition appear in the voluminous documents from the era. According to the unspoken rhetorical logic of the Civil War era, to depict formerly enslaved people as bloody would be to acknowledge their suffering and in so doing suggest that their suffering connoted patriotism. During the Civil War and Reconstruction, blood only became visible when it was processed through a political register—such as nationalism, honor, or even abolition. Blood or even its adjective, bloody, rarely carried medical meaning in newspapers, correspondence, federal records, and diaries; more often it functioned as a metaphor that excluded black people's experience. According to the cultural norms of nineteenth-century America, only injured and wounded white soldiers could be depicted as bloody. Their suffering represented the aftershocks of combat, and, more broadly, the corporeality of the nation, and the sacrifices made in the name of national honor and patriotic valor. In short, when white people bled, their blood could be framed within a particular cultural context; whereas, when black people bled, it lacked the cultural register to be seen. Black people's blood during the Civil War functioned like the proverbial tree in the woods that fell and did not make a sound; without a cultural metaphor, it could not be seen.

Throughout the Civil War era (and earlier), references to racial identity, color, and blood appeared interchangeably and often synonymously in census records, newspapers, and other forms of print culture.[6] In the few instances when white Americans referred to freedpeople's blood it was often in terms of some formulation referring to a biological racial identity. White observers created categories of racial classification—mulattoes, quadroons, and octoroons—that attempted to measure the quantity of "black blood" in those that appeared to have mixed ancestry. For example, in a letter to the Secretary of the Treasury during the Civil War, Edward Pierce, a Treasury Department Special Agent for South Carolina, refers to blood as a system of racial classification when he writes of visiting a school in Beaufort, South Carolina where "One sees there those of pure African blood, and others ranging from the lighter shades, and among them brunettes of the fairest features."[7]

During the Civil War and Reconstruction, the reference to blood among white—not black—people connoted a diverse range of meanings from suffering and death to the vitality of the nation.[8] In other instances, nineteenth-century white American writers reserved the use of the term blood to evoke a range of cultural narratives that applied to white people's experience. Not until after the Civil War, toward the end of the nineteenth century, when an increasing number of black writers entered into the literary marketplace did references to black people's blood as a metaphor enter the historical record.

In the early years of the Civil War, military officials and physicians deployed throughout the South often used the term blood in order to convey the unexpected suffering and death of white soldiers—despite the enlistment of black men in the Union Army in July 1862. Many Army leaders in both the North and South predicted that the war would end quickly. As a result, they did not develop the necessary health precautions in order to address the massive mortality, disease outbreaks, and suffering that erupted from a prolonged war, nor did they employ an adequate number of doctors and other medical staff to respond to the suffering and deadly conditions that the battles produced. Two main points illustrate this predicament: first, the Surgeon General's office had developed as a sinecure for retired army officials not as an active agency that was prepared, let alone equipped, to deal with the medical catastrophes wrought by war. Second, many Northerners and Southerners did not take seriously the severity of military fighting; in fact, at one of the earliest battles waged between the Union and the Confederacy, the Battle of Bull Run fought on 21 July 1861 near the city of Manassas in Virginia, both Northerners and Southerners swarmed the battle fields armed with picnic lunches, hoping to be entertained by the spectacle of combat. When the firing of cannonballs, the shooting of muskets, and the slashing of swords dismembered and killed many soldiers, these spectators, along with many Union white soldiers, fled the scene, giving rise to the term "skedaddle" to describe their flight. Due to the Surgeon General's Office, the military, and the government's utter lack of preparedness to deal with the outcome of this battle, white soldiers with missing limbs were left to suffer on the field, some caught on fences, others left among the trampled remnants of the earlier festivities, and many of the dead were left unburied. The public health crisis that the Battle of Bull Run engendered impelled a group of New York reformers to establish the US Sanitary Commission, a wartime agency that transported maimed and dead bodies and developed a set of health procedures to be distributed among the various regiments.[9]

Many nineteenth-century Americans had a romantic notion of war and did not anticipate the large-scale suffering that the war would cause. Throughout the war, soldiers, military officials, and the press's continued use of the term blood points to the ways in which the war disrupted romantic notions of combat and informed American readers about the brutalities of war. Blood signified the unexpected trauma of war, which explains why observers who documented such cases often wrote about blood with detail and narrative power, offering their readers visual images that captured the prolonged suffering that the war produced. In a letter to his mother penned in December 1863 from Camp Stoneman, Union soldier John Parris Sheahan evoked an image of blood to depict the "horrors of war." "My dear mother I have seen many sad sights—and some of which if you had seen them would have frozen your very brain you know nothing of the horrors of war at home only think of men strewn for miles over the field wounded and some torn in pieces some crying to God for help others for water, and some were beging [sic] to be taken from pools of their own blood such are the sights on the field of strife. May it soon be over is my earnest prayer."[10] Sheahan's reference to "pools of their own blood" produces a visual image that reveals the "horrors of war." The physical realities of men strewn for miles likely produced a number of additional images in his mother's mind—from vultures picking at the bodies to the stench of this immeasurable human carnage. Yet, Sheahan does not evoke any of these ghastly images; instead he hones in on the image of blood to convey the pervasive mortality of the battlefield, thus revealing the extent to which blood became the nineteenth-century metaphor to describe the brutalities of war.

References to blood in correspondence and diaries continued to eviscerate the valor and vaunted mythology often associated with nineteenth-century American notions of war and instead produced a cultural narrative in which the wounded and corporeal body became the emblem of modern warfare. The American Civil War, unlike the Mexican War, the War of 1812, or even the War for Independence, ranked as the first modern war due to advancements in the technology of warfare. The invention of rapid-fire weapons, Gatling machine guns, and submarines armed with torpedoes raised the stakes in terms of military engagement and introduced new ways for soldiers' bodies to be maimed, dismembered, and injured. References to blood reflected these technological innovations, and pointed to the extent to which nineteenth-century Americans attempted to depict the violence produced by these new forms of battle. Evoking images of blood became a way to capture the violence of this new warfare and to highlight its devastation. John Sneden recounted in his diary, "The shells from the gunboats crashed through the trees at the rate of 2 per minute; while the ambulances, with their blood dripping burdens wound down the centre roads and across the marshy ground to the hospital at the Old house by the River."[11] By stating "at the rate of 2 per minute," Sneden attempted to assign a particular logic to the gunfire sounds that he heard; his reference to blood points to the effects of this new military technology. "Burdens" refers not to one wounded soldier, but to many. The advances in war technology increased the number of wounded, and blood serves as a metaphor to depict the massive and incalculable death toll.

In a letter to the *Baltimore Sun*, Dr Henry Janes similarly conjured an image of blood in order to document the countless injuries and deaths caused by technological advances when he wrote, "a terrible, crippling, smashing invasion of the sacred machine, splitting bones like green twigs and extravasating blood in a vast volume of tissue about the path of the projectile."[12] A Union nurse summoned up yet another image of blood as a preface to describe the bodies of Union soldiers torn apart by the new wartime machinery when she wrote, "Astride a tree sat a bloody horror, with head and limbs severed by shells, and birds having banqueted on it, while the tattered uniform, stained with gore, fluttered dismally in the summer air."[13]

Union white soldiers and doctors consistently invoked the term blood or bloody in order to portray the suffering that white troops endured. These writers sought to depict the actual conditions of war and to debunk the otherwise romantic notion that many Americans harbored about the valor of war. Thomas Ellis, a Union physician, visited a hospital camp for Union soldiers who had fought in the Battle of Fair Oaks, which was part of the Peninsula campaign, in 1862. "I found over 300, many of them in a dying condition; and all of them more or less mutilated, and still enveloped in their filthy and blood-stained clothing, as they were found on the battle-field. Some of these had been attended to by the surgeons, but by far the greatest number were sent down to the station before receiving any surgical care."[14] Ellis further noted that there was no medical officer on duty and he then detailed his struggle to telegraph information about the troops to a supervising medical official at Fortress Monroe, a central Union barracks in northern Virginia. Ellis described the soldiers' clothing as "blood-stained" in an effort to amplify the picture of the mutilated and dying bodies and to offer a more visual narrative of the suffering they endured. In another passage, he returned to the image of blood to underscore the abysmal conditions of the troops. To magnify how circumstances worsened, he explained that the men were left "lying on the battle-field uncared for two or three days," and he described the blood as "congealed." "Nor could this be avoided," he wrote, "as the ground was alternately in the possession of the Union and rebel troops; and their condition, on arrival at the White House, was filthy in the extreme, their wounds alive with maggots, their clothing saturated and stiff with congealed blood and dried mud."[15]

No one had anticipated that the outbreak of the war would lead to maggots gnawing at the wounds of injured, bloody soldiers. Although nineteenth-century Americans likely realized that death and injury were unavoidable costs of war (even creating, as historian Drew Faust claims, a number of cultural narratives about how to respond to death), the scale, magnitude, and utter gore that the war produced challenged Americans' conceptions of the realities of war.[16] Despite being a trained surgeon, Ellis was haunted by the images of the slaughtered soldiers' bodies left untreated with "congealed blood." Of the many medical descriptions he had available to him as a doctor to describe the soldiers' medical condition, he chose to focus on the state of their blood as the most alarming aspect. Congealed blood emphasized their suffering as well as their vulnerability. Just as they could not protect themselves from the infestation of maggots, they could not clean their clothing from the blood that congealed on it.

Invoking images of blood revealed that battles could not be simply contained to a field nor could they be orchestrated like a chess match in which military supervisors could plot each movement. Blood instead posited a new cultural narrative about war; it emphasized the aftermath of the battle, and in so doing shifted the definition of war away from one that glorified military engagement to the aftershocks of combat—to the pain, trauma, and the unexpected discontents that could not be contained or coherently rationalized. Thomas T. Ellis explained the march of soldiers to Union gunboats after a battle as a "painful and hopeless pilgrimage" that "left many a drop of blood on the sandy track as they filed through brook and wood and over hill and dale."[17] He further added, "many poor wounded fellows sat on the tail of the ambulances, their blood-dripping feet dangling behind."[18] The actual blood of the war likely traumatized him, which explains why he continues to evoke it throughout his text. For Ellis, blood functions as a form of melancholia. Ellis certainly mourns the loss of the soldiers who died or were injured but his repeated use of blood as both an adjective and noun throughout his writings suggests an almost internalized form of suffering, one which he could not transcend even as a physician trained to work with blood and injury. Blood became for Ellis a symbol of both the physical suffering that the war produced and the deep emotional impact that the war had on him.

In addition to illustrating the trauma and the discontents of battle, blood served other multiple forces—it not only symbolized an injured and maimed nation, but also entered political discussions as a metaphor to resist claims of national reconciliation. In the immediate aftermath of the war, Radical Republicans propagated the notion of "waving the bloody shirt," a political slogan that reminded voters of the struggles of the war and those who died for the Union cause. As historian David W. Blight has perceptively noted, "Yankee retribution never had a more vehement voice than [Thaddeus] Stevens, and no one ever waved the 'bloody shirt' with greater zeal." Blight quoted Stevens, "Do not, I pray you, admit those who have slaughtered half a million of our countrymen until their clothes are dried, and until they are reclad. I do not wish to sit side by side with men whose garments smell of the blood of my kindred."[19]

Many writers of the period also invoked blood as a personification to dramatize the otherwise abstract quality of the nation. Blood gave the abstract concept of the nation corporeality. As Alfred Lewis Castleman noted, "Our Generals seem to have forgotten that this is *the people's* war, not their's; that it is waged at the cost of the treasure and of the best blood of the nation, not to promote the ambitious views of individuals or parties but to protect the people's right to Government."[20] Benjamin Franklin Palmer also turned to images of blood in an effort to personify the sacrifices of a nation made in the name of liberty. "The small appropriation, I thought, might become the nucleus of a National fund which would confer incalculable blessings on an army of gallant men," Palmer wrote, "who, to save the Nation from dismemberment, have consecrated the soil of an empire with blood, and sacrificed a thousand—yes, ten times a thousand—of the noblest hands that ever brought so precious tribute to the altar of Constitutional Liberty."[21]

* * *

During the Civil War and Reconstruction, many newly freedpeople also suffered from the brutalities of war, but the white soldiers and officers who documented their condition did not refer to them as bloody, which would have suggested that their bloodshed contributed to the corporality of the nation. Instead, the unwillingness to define them as bloody implies that their suffering could have been avoided, and even if it could not be, it lacked cultural significance.

References to black women as victims of the war in the historical record are sparing and confirm white officials' cultural devaluation of black blood. Unlike the representation of white soldiers as victims, their suffering does not puncture the mythologies of romantic war or even offer a realistic account of the effects of the new wartime technologies.[22] Black women suffered but to no rhetorical or patriotic end as white soldiers did—proving that even in death, white soldiers mattered.

A rare utterance of a black woman as a fatality appears in one of the accounts of the Battle of Fort Pillow. When black Union soldiers raised a white flag to indicate their surrender, the Confederate troops ignored their plea and continued their assault. A soldier who documented an eyewitness account of the war noted, "He saw a black woman, who was wounded during the action, shot through the head and killed, by one of the rebels."[23] Despite the fact that being shot likely caused her to bleed, the writer does not describe her as bloody in the same way that white gunshot victims appear as bloody throughout the historical record—proving that the term bloody carried cultural significance that negated black people's suffering and death. The acknowledgment of her death fails to articulate it as politically significant or symbolic.

Similarly to the antebellum era, references to blood, at times, served as a euphemism to obfuscate slaveholders raping enslaved women. Many white nineteenth-century Americans avoided discussing rape in such direct terms, not only because of Victorian convention but also because they refused to acknowledge the violence that it engendered. References to "blood" and "veins" became a way to understate and neutralize this phenomenon. In the same letter in which he wrote about visiting the school in Beaufort, during the Civil War, Edward Pierce referred to blood as evidence of the raping of enslaved women by white slaveholders. Yet, in it he vacillates between recognition of sexual violence and the enslaved woman's (improbable) complicity. "The women, it is said, are easily persuaded by white men—a facility readily accounted for by the power of the master over them ... whose solicitation was equivalent to a command, and against which the husband or father was powerless to protect ... Yet often the dishonor is felt, and the woman, on whose several children her master's features are impressed, through his veins his blood flows, has sadly confessed it with an instinctive blush."[24]

While Pierce explains that the enslaved women remained powerless and lacked the protection of the men of their community to safeguard them against their masters' alleged persuasions, the reference to blood appears both in Pierce's effort to explain the slaveholder as father of the enslaved woman's children, which illustrates the nineteenth-century phenomenon of "miscegenation," and in the subtle reference to the enslaved woman's "instinctive blush." As literary critic Geoffrey Sanborn has smartly noted, nineteenth-century Americans made distinctions between "blush" and "flush." In the nineteenth century, blushing was understood to be a "distinctly feminine and white manifestation of sensibility, indicating that one was highly attuned to social and moral codes of behavior."[25] Flushing, by contrast, is "a cardiovascular event, a rise in the rate and pressure of the blood's circulation. It is prompted not by one's thoughts about what others might be thinking, but by a sudden and overwhelming passion, like anger or joy or sexual arousal."[26] By saying that the enslaved women blushed, Pierce disavows the enslaved women's rape and instead frames it as a violation of a social code; conversely, he does not say she "flushes," which would suggest an arousal and by extension a complicity. The violence of the rape, the blood that oozed from her skin, due to the physical confrontation, evaporates from the narrative. Instead, the only way that her blood enters the historical record is in relation to the slaveholder for violating the social code, and for producing mixed-race offspring.

Meanwhile, a military official at the Benton Barracks, Missouri reported to his commanding officer that one of the black soldier's wives who refused to go with the military from Missouri to Kentucky was threatened that she would be jailed and her children would be sold if she did not follow the Union command. The official then reports that this enslaved woman, who we only know as Martha Glover, was "most cruelly" whipped "with a leather strap from Buggy Harness, and this when she is pregnant and near confinement."[27] Yet, nothing in the document refers to her as bloody, bruised, or suffering; instead the officials represent her as disobedient, which discloses any depiction of her bleeding from the whipping. Furthermore, the representation of black people being whipped appeared throughout antebellum literature in an effort to expose the sheer violence of slavery and to advance the abolitionist movement. In most of these depictions, the enslaved person is often referred to as bloody. In telling the story of his Aunt Hester whom the slaveholder often whipped, Fredrick Douglass notes, "Now, you d—d b—h, I'll learn you how to disobey my orders!" and after rolling up his sleeves, he commenced to lay on the heavy cowskin, and soon the warm, red blood (amid heart-rending shrieks from her, and horrid oaths from him) came dripping to the floor."[28] For Douglass, blood refers to the literal consequence of the whipping as well as having a metaphorical meaning of suffering to advance the abolitionist movement. Yet in the reference to the black soldier's pregnant wife being whipped, there is no reference to her blood, as her suffering is not articulated to advance a political movement, to suggest sympathy, or to symbolize the campaign for nationhood. Without a political context to make her blood visible, her bleeding is invisible.

In general, references to black people's blood are not accentuated, elaborated, or dramatized in the way that white people's suffering and death appear in the sources. Instead, their bloody experience embroiders these sources but often in inconspicuous and complicated assertions. In an article published in the *L'Union* newspaper, located in New Orleans, at the time occupied by the North and then later reprinted in the *New York Weekly Times*, the writer does, in fact, report on the bloodshed of African American military prisoners. He writes, "Our correspondent from the Mississippi squadron states that rebel correspondence captured at Simmsport showed that 200 negroes were recently slain in cold blood at Franklin, La., by the rebel soldiers."[29] While the reference to black people killed in cold blood on the surface seems to evoke the corporeality of nationalism that appears throughout wartime writings about white soldiers, a closer inspection reveals that the writer seemed less interested in the massacre of black people and more concerned that Confederate soldiers violated wartime codes of honor and respect. The reporter describes how the rebels had raided villages in Maryland and Virginia, seizing black people and turning them into slaves. He writes, "those who are brutal enough to steal men and make slaves of them, care nothing whatever for the civilized world … we demand the Government and we urge upon our Generals everywhere not to let such things pass unnoticed."[30]

Blood in this formulation does not signify the sacrifices made in the name of the Union nor the suffering endured by enslaved people. Instead, the reference to blood points to the barbaric nature of the Confederacy. Enslaved people function in this article simply as the capital that both the North and the South desired and demanded. The reporter calls on the government and the generals to act not because these people have been butchered in "cold blood" but rather because their commodities have been stolen and the Confederacy has behaved in an "uncivilized" manner. By July 1863, when this article was published, the Emancipation Proclamation had been in effect for seven months and the Union Army had a deep investment in the labor power that freedpeople could offer the North.

This fact further reveals the extent to which the article had less to do with black people's blood and suffering and more to do with winning the war. Indeed, throughout the war, both the Union and the Confederacy considered enslaved people as assets that could bolster their economy and reinforce their armies. President Lincoln emancipated the slaves as a military strategy to deplete the Southern labor force and to increase the manpower of the Union Army by attracting enslaved people to Union lines where they would be employed as constructor workers, washerwomen, and cooks.

Shifting the focus away from the reporters' or even the government's investment in enslaved people as capital betrays a chilling reality of war that the metaphors of blood actually mystify. Hidden within this fact is the disturbing reality that innocent, unarmed enslaved people were killed during the war, a part of Civil War history that rarely if ever makes it into the historiography. The military prisoners are not black soldiers but civilians—black men, women, and children—but this fact is buried in the nomenclature of the period, namely in the use of the term "prisoners of war." Throughout the Civil War, "prisoners of war" referred to captured soldiers from the opposing side.

In the *L'Union* article, "prisoners of war" refers to black refugees. First, he states "at Port Hudson the rebels hung a negro soldier whom they captured." By referring to the man as a "negro soldier," he distinguishes him from the prisoners of war, who are the captured black civilians. Second, he writes that the rebel cavalry "have everywhere sent or carried off, not horses and cattle, not boots, drugs and provisions alone, but every black man, woman, and child upon whom they could lay hands—and as everyone knows, they have not carried them off as prisoners of war to be paroled and exchanged, but as chattels, to be sold into slavery."[31] Here, the reporter draws a clear connection between black civilians as prisoners by stating that "every black man, woman, and child" were not carried off as "prisoners of war" but as slaves.

Finding references to freedpeople's suffering and bloody condition has enormously important consequences for how we write and remember the history of the Civil War and Reconstruction. As the newspaper report reveals, the experience of kidnapping enslaved people and then killing them appears nowhere in the historiography. According to the logic of the source, the author mentions how enslaved people died in "cold blood," but because the reporter remains more invested in charging the Confederacy with war crimes, the actual death of these enslaved people, which is in fact noted, almost disappears by the end of the article; it simply fades from view.

What is at stake in this article is not that the Confederacy behaved in a way that was "uncivilized"—how does one even make such an assertion in war when both sides had raped and pillaged their way through their enemy's territories? The point lies in the fact that the term "cold blood" points to the kidnapping and killing of innocent black people. This should serve as the caption to the war's otherwise heroic images of armies in combat. Yet, because those in power reserved the term "blood" to refer only to the suffering and death of white soldiers, "cold blood" does not reveal much about the vulnerability of the innocent people slain but instead emphasizes the actions of the killer. The narrative thrust of the term "cold blood" shifts attention away from the dead to the culprits; it tells more about how the act of killing was done than about the actual blood of the 200 people who died. Nowhere again in the report are the victims mentioned; instead the article remains tightly focused on the Confederacy as the culprit of wrongdoings.

While much of this could be interpreted differently or even outwardly dismissed, the point of this analysis, nevertheless, seeks to highlight how references to black people as bloody appear in the public historical record, and how historians define fatalities during the war. In 2012, J. David Hacker, a demographer, published a study that claimed that the mortality of the Civil War exceeded 600,000, a figure that historians have universally accepted for over a century. Based on new analysis of census records, he claimed that roughly 750,000 people died during the war.[32]

While impressive, Hacker's estimate failed to consider the number of African American people who perished. Hacker counted black troops, but refused to count black civilians, arguing that they did not fit the definition of wartime casualties. White soldiers who died of disease during the war, not from battle, counted as casualties, but black people, freed from slavery who died in Union camps from the same camp diseases that killed white soldiers, were not calculated as part of the Civil War death toll.[33]

Part of this problem lies in the representation of whose blood the official historical record captured. Black civilians' blood was not recorded or documented because it did not symbolize the nation. Further, in the case reported in the *L'Union* newspaper, when enslaved people's experience could be recorded as part of the larger mortality induced by the war, the focus of this scenario turns instead to the Confederacy's barbaric nature. The call for the government and the military to stop the Confederacy only further obfuscates the death of those enslaved people and prevents their mortality from being counted.

The metaphor of blood during the war has important political import for how historians even approach the archives, interpret evidence, and, most of all, define suffering and mortality. Blood rarely provides direct material evidence—it often serves as a metaphor that attempts either to expose the realities of war or to personify the abstract concept of the nation. It does not offer insight into health conditions or provide an accurate depiction of wounds and injuries. Further, blood functions mostly as a metaphor that represented the wartime experience of white soldiers.

* * *

Writers, diarists, and nineteenth-century observers recorded white soldiers' wounds and suffering in order to highlight the unexpected consequence of violent combat and a prolonged war—freedpeople's injuries or suffering rarely enter into the historical record under these auspices. Further, when nineteenth-century Americans conjured images of blood as a metaphor for the corporeality of the nation, they imagined white people as its citizenry.

With access to black newspapers and other publishing outlets, however, poet, essayist, and novelist Frances Ellen Watkins Harper and other African American writers from the period framed their discussion of the nation, freedom, and the war around the image of blood, and in so doing shifted the discussion of blood away from one that only pertained to the white soldiers' wartime experience. In response to the news that a Union General had begun to emancipate enslaved people in 1861 before President Lincoln's famous Emancipation Proclamation in 1863, Harper wrote "And yet I am not uneasy about the result of this war. We may look upon it as God's controversy with the nation; His arising to plead by fire and blood the cause of His poor and needy people."[34] Here, Harper overturns a wartime cultural discourse that has otherwise neglected the wartime experience of freedpeople and uses the image of blood to argue that it constitutes God's reason for the war and the ending of slavery.

Harper continues to evoke images of blood in order to write African American experience into the national discourse about the war. Poetry became an alternative discourse for black writers, like Harper, to transcribe the contributions made by black people during the war. In a poem entitled "To the Cleveland Union-Savers," Harper condemns the Union Army for refusing to allow an escaped enslaved woman from entering Union camps. She writes, "And ye sent her back to torture … There is blood upon your city,—Dark and dismal is the stain."[35] In another poem about the bloodshed that African American soldiers who were part of the Massachusetts 54th endured, she writes, "To heal the wounded nation's life, And from the soil drenched with their blood."[36] Similarly to Harper, in his poems James Madison Bell documented how freedpeople's bloodshed contributed to the nation, "No! by the blood of freeman slain."[37]

Throughout such Civil War writings, African American writers highlighted black people's sacrifices and struggles, and challenged the prevailing cultural representations in which only white soldiers could appear bloody. The representation of white soldiers as bloody signaled that only their suffering mattered; their bloodshed epitomized the sacrifices made in the name of the nation. By referring to black people as bloody, Harper and Bell constitute their experience as part of the nation.

Long after the Civil War ended and the nation began to reel from the effects of the inordinate suffering and bloodshed, a number of African American writers continued to invoke images of blood in order to illustrate black people's contribution to the nation. Taking her cue from the deluge of war memoirs published at the turn of the twentieth century, Susie King Taylor published her autobiography that, like Harper, acknowledged how black people's bloodshed contributed to nationhood. "Justice we ask,—to be citizens of these United States, where so many of our people have shed their blood with their white comrades, that the stars and stripes should never be polluted."[38] Taylor, one of the only known African American women to publish a memoir of her wartime experience, described bloodshed to illustrate both black people's patriotism and to fuel the campaigns against Jim Crow-sanctioned violence. Like white wartime writers and antebellum abolitionist writers, Taylor draws on the metaphorical value of blood to expose the irony of Jim Crow violence—that black men "shed their blood with their white comrades" only to be victims of segregation and violence after the war.

Taylor further makes references to blood for political reasons when she attempts to chastise a younger generation of black people for not appreciating the sacrifices that black veterans made for freedom. "I look around now and see the comforts that our younger generation enjoy, and think of the blood that was shed to make these comforts possible for them, and see how little some of them appreciate the old soldiers. My heart burns within me, at this want of appreciation. There are only a few of them left now, so let us all, as the ranks close, take a deeper interest in them."[39] Despite chiding the younger generation, the rhetorical thrust of this passage lies in Taylor's efforts to use the metaphor of blood to emphasize black military service that had been unacknowledged during the war, and had continued to be unacknowledged in 1902 when her memoir was published.

Still other black writers of the postbellum era found access to publishing opportunities, and employed references to blood as a metaphor to personify nationhood, the Civil War, and slavery. In essence, they continued to develop cultural registers to make the black people's blood visible. Like Susie King Taylor, Sojourner Truth also used images of blood in order to convey how slavery was made in the name of nationalism in her autobiographical writings. "Our nerves and sinews, our tears and blood, have been sacrificed on the altar of this nation's avarice."[40] James Smith also narrated the history by drawing on images of blood. "Brought here from our mother country, we have bedewed the soil with our blood and tears."[41] The genre of biography enabled writers like Susie King Taylor, James Smith, and Sojourner Truth to craft highly stylized accounts of the war that articulated black people's contribution to nationalism.

* * *

While there was a rise of African American writers in the last quarter of the nineteenth century, who created images of blood to describe the

African American experience during the Civil War, the official wartime discourse of the period did little to capture the experience of black people. When images of black suffering appeared, it actually masked the real flesh and blood reality of their experience, serving more as an ideological pawn to indict their enemy than to account for their suffering. The representation of blood thus demarcated whose bodies warranted the sympathy of the nation and whose suffering could be ignored. In this formulation, Confederate and Union soldiers who are often seen as divided sharply by the Mason–Dixon line, in fact share the cultural emblem of patriotism marked by blood.

By defining soldiers' bodies as wounded, bloody, and even dead, these men's injuries and deaths became tabulated as part of the war's lauded, national statistics. Meanwhile, freed slaves, whose bodies were similarly wounded and infected by the war, remained both figuratively and literally removed from the war's calculations of casualty and mortality rates despite the publication of census data that proves the sharp decline in the black population during the war years.[42]

The Medical Division of the Freedmen's Bureau is the largest archive on black people's health during the Civil War and Reconstruction era, but in the thousands of letters, reports, and charts that constitute this archive, there are few references to black people's blood or freedpeople's condition as bloody. Yet the extant evidence, however, suggests that freedpeople did, in fact, bleed from the aftershocks of battle, epidemic outbreaks, and other violent episodes. The lack of references to blood in this archive indicates the extent to which the notion of blood functioned more as a cultural metaphor during the Civil War than as an indication of one's medical condition.

Furthermore, not referring to the former slaves' bodies as bloody enabled federal officials and doctors to ignore the deadly realities of war and emancipation. It also justified federal efforts to view black people only in terms of their labor power. Finally, and perhaps most importantly, by not including references to black people as bleeding or bloody, many white nineteenth-century chroniclers foreclosed any discussion about how the war that was intended to liberate bondspeople from chattel slavery led in fact to widespread suffering and death.

Notes

I am deeply grateful to historian Megan Kate Nelson who shared with me a key number of sources mentioned in this chapter and to Carla Peterson whose incisive comments compelled me to reorganize the essay and to further probe the archival record.

1. W. R. DeWitt, *Report of Sick and Wounded Refugees* [sic] *and Freedmen*, 30 December 1865, e. 2979, South Carolina, Box 38 (Chief Medical Weekly Reports), RG 105, National Archives.
2. This practice of creating hospitals, asylums, and almshouses in order to develop a labor force developed in the late eighteenth century in many American cities. See David J. Rothman, *The Discovery of the Asylum: Social Order and Disorder in the New Republic* (Boston: Little Brown, 1971).
3. Jim Downs, *Sick from Freedom: African-American Illness and Suffering during the Civil War and Reconstruction* (Oxford University Press, 2012).
4. Ibid., 95–119.
5. *Autobiography of James L. Smith, Including, Also, Reminiscences of Slave Life, Recollections of the War, Education of Freedmen, Causes of the Exodus, etc* (Norwich, CT: The Bulletin, 1881), 103.
6. "Proportion of Mulattoes in the Negro Population of the U.S.," *African Methodist Episcopal Church Review* 29.2 (October 1912). Patricia Morton, "From Invisible Man to 'New People': The Recent Discovery of American Mulattoes," *Phylon* 46.2 (2nd Qtr, 1985), 106–22. Martha Hodes, "Fractions and Fictions in the United States Census of 1890," in *Haunted by Empire: Geographies of Intimacy in North American History*, ed. Ann Stoler (Durham, NC and London: Duke University Press, 2006), 240–70. J. L. Hochschild and B. M. Powell, "Racial Reorganization and the United States Census 1850–1930: Mulattoes, Half-Breeds, Mixed Parentage, Hindoos, and the Mexican Race," *Studies in American Political Development* 22.1 (2008), 59–96. Jim Downs, "Her Life, My Past: Rosina Downs and the Proliferation of Racial Categories after the American Civil War," in *Storytelling, History, and the Postmodern South*, ed. Jason Phillips (Louisiana State University Press, 2013), 156–86.
7. Edward L. Pierce to Hon. Salmon P. Chase, 3 February 1862, vol. 19, no. 72a, Port Royal Correspondence, 5th Agency, RG 366 (Q-8), as quoted in Berlin et al., *Freedom: Series I, Volume III, The Wartime Genesis of Freed Labor: The Lower South* (Cambridge University Press, 1990), 133, 149.
8. On both the historical and contemporary ways that Americans have used the terms blood and race interchangeably, see Karen E. Fields and Barbara J. Fields, *Racecraft: The Soul of Inequality in American Life* (New York: Verso, 2012), 52–70.
9. For more on the history of the US Sanitary Commission, see Jeanie Attie, *Patriotic Toil: Northern Women and the American Civil War* (Ithaca, NY: Cornell University Press, 1998).
10. John Parris Sheahan Civil War Letters (1862–65), Folder 7, John Parris Sheahan Papers, Maine Historical Society.
11. Robert Knox Sneden Diary, vol. 3, 1862, 29 June–25 October (Mss5:1 Sn237: 1 v. 3), p. 226, Virginia Historical Society.
12. Henry Janes, MD, letter to editor of *Baltimore Sun* (27 October 1899), in vol. 6, "Gettysburg News Clippings," GNMP Library, as quoted in Gregory A. Coco, *A Strange and Blighted Land: Gettysburg, the Aftermath of a Battle* (Gettysburg, PA: Thomas Publications, 1995), 164.
13. Sophronia E. Bucklin, *In Hospital and Camp: A Woman's Record of Thrilling Incidents Among the Wounded in the Late War* (Philadelphia, PA: John E. Potter and Co., 1869), 187–8, as quoted in Coco, *A Strange and Blighted Land*, 20.

14. Thomas T. Ellis, *Diary of Thomas T. Ellis, 1862?*, in *Leaves from the Diary of an Army Surgeon: or, Incidents of Field, Camp, and Hospital Life* (New York: John Bradburn, 1863), 312.
15. Ibid., 66.
16. Drew Gilpin Faust, *This Republic of Suffering: Death and the American Civil War* (New York: Vintage, 2009).
17. Ellis, *Diary*, 123.
18. Ibid.
19. David W. Blight, *Race and Reunion: The Civil War in American Memory* (Cambridge, MA: Harvard University Press, 2001), 51.
20. Alfred L. Castleman, *The Army of the Potomac. Behind the Scenes. A Diary of Unwritten History; From the Organization of the Army to the Close of the Campaign in Virginia, about the first day of January, 1863* (Milwaukee: Strickland & Co., 1863), 203.
21. Benjaimn Franklin Palmer, "The Palmer Arm & Leg" (1862), 4, Francis A. Countway Library of Medicine, Harvard Medical School.
22. I found references to black women, but none to black men.
23. Captain Carl Adolf Lamberg to Lieut. Col. T. H. Harris, 27 April 1864, vol. 32, Union Battle Reports, ser. 729, War Records Office, RG 94 {HH-5}. Endorsements as quoted in Berlin et al., *Freedom, Series II: The Black Military Experience* (Cambridge University Press, 1982), 540–1.
24. Edward L. Pierce to Hon. Salmon P. Chase, 3 February 1862, vol. 19, #72a, Port Royal Correspondence, 5th Agency, RG 366 {Q–8}, as quoted in Berlin et al., *Freedom: Series I, Volume III, The Wartime Genesis of Freed Labor: The Lower South* (Cambridge University Press, 1990), 133, 149.
25. Geoffrey Sanborn, "Mother's Milk: Frances Harper and the Circulation of Blood," *English Literary History* 72.3 (Fall 2005), 694–5. Also, Thomas Jefferson notes that the blush illustrates the differences between the races. He writes, "Whether the black of the negro resides in the reticular membrane between the skin and scarf-skin, or in the scarf-skin itself; whether it proceeds from the colour of the blood, the colour of the bile, or from that of some other secretion, the difference is fixed in nature, and is as real as if its seat and cause were better known to us. And is this difference of no importance? Is it not the foundation of a greater or less share of beauty in the two races? Are not the fine mixtures of red and white, the expressions of every passion by greater or less suffusions of colour in the one, preferable to that eternal monotony, which reigns in the countenances, that immoveable veil of black which covers all the emotions of the other race?" Thomas Jefferson, *Notes On the State of Virginia* (CreateSpace Independent Publishing Platform; Ill edition, 2011), 165–6.
26. Sanborn, "Mother's Milk," 694
27. Brig. Genl. Wm. A. Pile to Maj. Genl Rosecrans, 23 February 1864, enclosed in Brig. Genl. Wm. A. Pile to Maj. O. D. Greene, 17 March 1864, P-197 1864, Letters Received, ser. 2593, Dept. of the MO, RG 393, Pt. 1 {C-160}, as quoted in Berlin et al, *Freedom*, 245–6
28. Frederick Douglass, *Narrative of the Life of Frederick Douglass, An American Slave, Written by Himself* (Boston: Anti-Slavery Office, 1845); reprint, edited with an introduction by David W. Blight (Boston: Bedford Books, 1993), 15–16.

29. "The Rebels and the Negro Question," *L'Union*, 14 July 1863, 2.
30. Ibid.
31. Ibid.
32. J. David Hacker, "A Census-Based Count of the Civil War Dead," *Civil War History* 57.4 (2011), 306–47.
33. For more on this debate, see Jim Downs, "Color Blindness in the Demographic Death Toll of the Civil War," 13 April 2012, OUP Blog; Guy Gugliotta, "New Estimate Raises Civil War Death Toll," *New York Times*, 2 April 2012.
34. Melba Joyce Boyd, *Discarded Legacy: Politics and Poetics in the Life of Frances E. W. Harper, 1825–1911* (Detroit: Wayne State University Press, 1994), 51.
35. *Liberator*, 8 March 1861, as quoted in Frances Smith Foster (ed.), *A Brighter Coming Day: A Frances Ellen Watkins Reader* (New York: The Feminist Press at CUNY, 1993), 93–4.
36. *Weekly Anglo-African*, 10 October 1863.
37. James Madison Bell, "Through Tennyson the Poet King from 'A Poem Entitled The Day and the War,'" in *The Columbia Book of Civil War Poetry*, ed. Richard Marius (New York: Columbia University Press, 1994), 193.
38. Susie King Taylor, *Reminiscences of My Life in Camp: An African American Woman's Civil War Memoir* (Athens: University of Georgia Press, 2006), 75–6.
39. Ibid., 51–2.
40. *Narrative of Sojourner Truth* (Boston: Published for the Author, 1875), 197.
41. *Autobiography of James L. Smith, Including, Also, Reminiscences of Slave Life, Recollections of the War, Education of Freedmen, Causes of the Exodus, etc* (Norwich, CT: The Bulletin, 1881), 135.
42. The 1860 census tallied 26,922,537 whites, 4,441,830 blacks (488,070 free blacks, 3,953,760 slaves). The 1870 census tallied 33,589,377 whites, 4,880,009 blacks. The white rate of growth was (33,589,377—26,922,537) / 26,922,537, about 24.76% over ten years. The black rate of growth was (4,880,009—4,441,830) / 4,441,830, about 9.86% over ten years. In spite of the enormous number of deaths from the war itself, the black rate of growth was less than half that of the whites. The rate of growth from 1840 to 1850 was about 28.82%, over ten years. The rate of growth from 1850 to 1860 was about 23.39%, over ten years. The white population grew even more rapidly over this period, but immigration was a substantial part of that increase. I am deeply indebted to statistician Gary Simon, Stern School of Business, New York University, for providing me with these important statistics. After finishing my book, *Sick from Freedom*, he decided to test my argument about the enormous mortality that black people endured during the war; his meticulous analysis of the census records substantiated my claims.

# 11
Colonial Transfusions: Cuban Bodies and Spanish Loyalty in the Nineteenth Century

David Sartorius

By the early nineteenth century, talk of blood continued to do a great deal of work in giving meaning to political and social life in Spain's American colonies. "Con sangre se hace azúcar"—"Sugar is made with blood": this was a saying popular among Cuban planters during the nineteenth century, and it grimly evoked the backbreaking labor and quotidian violence that made possible their extraordinary wealth.[1] Those planters were among many Cubans who undertook a major expansion of the island's sugar industry and enslaved African labor force just as antislavery sentiment gained momentum throughout the Atlantic world. By 1791, British abolitionists circulated pamphlets advocating the boycott of Caribbean sugar, "steeped in the blood of our fellow-creatures."[2] In August of that same year in St Domingue, the French colony less than 50 miles from Cuba's easternmost point, a slave named Boukman led a ceremony in which he and his conspirators sacrificed a black pig and drank its blood. That event initiated an island-wide revolt that ended slavery in far more radical and violent ways than sugar planters and officials, and even many abolitionists, had desired.[3] The Haitian Revolution (1791–1804) facilitated Cuba's global dominance in sugar production but at a bloody price, namely, the brutality elemental to slave societies. While inhabitants of mainland colonies of the Spanish empire declared independence after 1808, many Cubans continued to expect colonial rule to contain and regulate that violence. Spanish authority promised the stability necessary to avoid the turmoil that had shaken Haiti. Records of an 1811 town council meeting in Santiago de Cuba expressed fear about "the same force repeating itself on the island of Cuba, the same catastrophe that has covered in blood and ashes the greatest and most opulent of the Antilles."[4]

While references to blood spilled on battlefields and the bloody business of slavery echoed throughout and beyond the Atlantic world, the particular salience of blood discourse to the political logic of the Spanish empire became increasingly unstable with the onset of mainland independence. Defining common bonds among inhabitants of the colonies became as vital for the empire as it did for the anticolonial movements. Cuba, along with Puerto Rico and the Philippines, remained loyal to the Crown until 1898; yet invoking shared blood ties with Spaniards could not sustain popular allegiance for an island with so many Cubans with the "stain," or "defect," of black or African blood. Just as the politics of Spanish empire were changing, so, too, was the language through which social difference conditioned political inclusion. Blood had traditionally indexed both race and political allegiance—notions of community and exclusion expressed biologically and genealogically. This essay considers the cultural politics of blood not to understand the joined strength of those concepts but to understand how they came undone. Nobody stopped talking about blood in nineteenth-century Cuba, but it acquired declining importance in articulating the racial dimensions of political allegiance to the Spanish empire. At the end of the century, proponents of emerging sciences of the body, namely anthropology and biological and social formulations of Darwinism, renovated corporeal idioms of racial and political difference. And those disciplines absorbed some of what blood once indexed in order to define the uncivilized against the idealized subjects of colonial politics.

Beginning more or less in the 1870s, anthropological study attracted the attention of Cuban politicians and intellectuals who continued to biologize political affiliation. This field represented one of several tributaries flowing into an early twentieth-century racist social science that marked some Cubans—especially those of African descent—as unfit for citizenship.[5] Blood talk, which had long explained these linkages, now played a minimal role. Looking to the era of Iberian conquest and colonization, Roland Greene analyzes the cultural semantics of early modern blood discourse and identifies the instability of figurative and material references to blood. He suggests that a "conceptual envelope" contained the various meanings ascribed to concepts like blood: the envelope might be closed, fixing those meanings, or opened up to empty or fill its contents, especially "when one paradigm is in the process of giving way to another."[6] The history of nineteenth-century Cuba is a story of shifting paradigms as slavery, colonialism, capitalism, and racial ideology took dramatically new forms. Without minimizing entirely what Cubans preserved, added to, and emptied from the conceptual envelope of blood itself, I give attention to two other envelopes—race and political allegiance—and the seams through which blood trickled out of them.

Congealing loyalty

The idea that blood transmitted physical characteristics, social status, and character qualities across generations had a storied trajectory in the Iberian world. It carried legal and administrative weight, and over time the religious meanings of *limpieza de sangre* (purity of blood) in Spain gave way to linkages of purity with color, caste, and *calidad* (inherited qualities) that informed early modern ideas about race. In the Americas, people of African descent found that their "black blood" held a particularly stigmatized place in the hierarchies of race and *casta*.[7] But divisions of Spaniards and indigenous people into two legal republics allowed some indigenous people (ostensibly with no Jewish or Muslim ancestors) to claim their own purity of blood, which reinforced the political standing of native noble lines and of native communities more broadly. What had begun as a measure of religious orthodoxy had also become a metric linking blood to political allegiance. Such formulations could render loyalty to a political community an embodied trait. As María Elena Martínez notes of indigenous subjects as well as creole (native-born) elites in the eighteenth century, "their patriotism intensified and they began to imagine the merger of the two republics in reproductive and biological terms."[8]

As resistance to Spanish rule crested on the mainland, presumed affinities of blood did little to prevent "pure" Spaniards in the Americas from rebelling. A regent of the Real Audiencia (high court) in Havana opened the court's 1812 session with a speech warning of the unpredictability of blood to predict loyalty to Spain. Not only had the "vexations of individuals of races and castes" threatened the empire, but the unstable and "quixotic nobility of blue blood" had failed to impede white promoters of independence, freedom, and nationalism in Spanish America.[9] Moreover, calls for independence began attracting support from people of indigenous and African ancestry, forming a contingent and unequal union that transcended putative bloodlines.

It was during this crisis that Spaniards and Americans gathered in Cádiz from 1810 to 1812 to draft Spain's first written constitution. The racial terms of political inclusion figured prominently in the debates, with American delegates particularly worried that independence leaders might promise freedom to slaves and citizenship to free men of color.

And as anticolonial movements began to take military form, delegates recognized that the loyalty of racialized subordinates could just as easily maintain Spanish rule. These concerns motivated a speech by Quito delegate José María Lequerica, who argued that "the blood of colored men is red, and so is that of warriors, of healthy men: pure and noble blood."[10] Men of color, especially those equipped to combat insurgents, thus met an alternative blood requirement for allegiance. Indeed, the 1812 Constitution extended Spanish citizenship to adult indigenous men but left only a narrow window for free men of African descent to become citizens. Delegates expressed this distinction not in the language of blood but of ancestry and origin: indigenous people were natives and descendants of Spanish territory, whereas Africans in Spanish America could not claim native status.[11] The Cortes (parliament) delegates by no means negated blood as a powerful marker of political power, but its symbolism had fluctuated. Blood's universality could foster political unity as much as blood difference could reinforce stratification. Just as important, other terms and criteria for political membership, namely nativeness and foreignness, came to the fore.

The constitution, with its intricate racial logic, remained in effect only from 1812 to 1814 and 1820 to 1823, but sweeping transformations to the empire frustrated any attempt to return to "normal" colonial politics—including the politics of blood discourse. An unstable Spanish state and new republican American governments all struggled to regulate social difference. An 1805 Spanish decree about interracial marriages continued to mention purity of blood, and that language appeared in petitions parents filed to challenge their children's marriage choices. As newly independent Spanish American republics often dropped racial categories from offical record keeping, Cuban documents continued to use them well into the 1890s but rarely made mention of *limpieza de sangre* or being of mixed blood. Verena Stolke argued that "around the 1840s the notion of purity of blood ceases to be an issue," citing the abolition of the Inquisition in the 1830s, royal decrees in 1835 and 1836 ending purity of blood requirements for many government offices, and the complete abolition of those requirements in 1865. She adds that discrimination continued, "couched in other terms."[12] A cursory read of any novel from the nineteenth century will confirm that blood continued to mark racial difference and political allegiance in literary culture. And it never disappeared entirely from political discourse. When Cubans found themselves excluded anew from Spanish citizenship in 1837, José Antonio Saco, a leading intellectual and a shunned Cortes deputy, wondered why Cubans, "through whose veins also circulate

Spanish blood, [were] not deserving of even one liberal concession."[13] Somewhere in between state policies and fictional renderings were discussions of race and fidelilty that relied less and less on blood talk.

The shift became especially visible in popular expressions of allegiance. By the middle of the century, imperial upheaval had destabilized the institutions and practices by which Cubans could exercise political voice. Heightened censorship, the exclusion of Cubans from Spanish citizenship, and anxiety about slave rebellion suffocated public discourse; even the free-colored militias, in which free men of color had defended Spain since the sixteenth century, disappeared for a decade when the captain general, the island's highest-ranking official, temporarily abolished them after the 1844 La Escalera conspiracy.[14] References to blood in discussions of political inclusion took on new meanings and eschewed older ones. In 1862, one colonial administrator, Miguel Rodríguez Ferrer, made an elaborate appeal to Madrid for a louder Cuban voice in imperial politics. He wondered why Spain had so vigorously defended its empire during the independence wars if it was now brushing off appeals from the remaining loyal colonies: "The Nation, generously, splendidly, gave its blood and its gold for a problem that, for the most part, neither concerned nor interested it." Despite inheritance-based claims to unity, he avoided the common expression of biological connection: paternalistic ties that depicted a protective Spanish father-king in control of its colonial children. Spanish ancestry, he claimed, should generate fraternal ties, not a hierarchical relationship that dispensed concessions to subordinate colonial subjects.[15] Ancestry and fictive kinship took priority over blood in claims to political inclusion as officials such as Rodríguez Ferrer sought to renegotiate the colonial relationship in the interest of preserving Cubans' allegiance.

If there was one mode of blood talk that continued uninterrupted, it was wartime sacrifice: blood spilled on battlefields. And during the Ten Years' War (1868–78), the island's first armed struggle for independence, Cubans and Spaniards of all backgrounds sought to demonstrate their loyalty to Spain or to independence. Predictable appeals to Spanish "national integrity" at the outset of the war could drum up Cuban support while awaiting the arrival of Spanish troops. Among the newly raised, white-only volunteer units, the batallion in Cienfuegos won praise from the city council for protecting rural areas "with pride in its origins and the Spanish blood that circulates through its veins."[16] More complicated references spoke to the racism, the lack of integration, in both Spanish and rebel forces. Francisco de Camps y Feliú, a Spanish coronel, acquired the notebook of a rebel soldier killed in battle and reported to military superiors the soldier's recorded conversations within his *guerrilla*. Likely responding to a claim of white superiority, the soldier offered a radical take on the nature of *limpieza de sangre*: "The African says that his blood is as pure as the European's; and that being a free nation, he values his origins as much as the European."[17] Soldiers of color, both in free-colored militias and as auxiliaries in the Spanish Army, turned to blood discourse when making public demonstrations of their loyalty to Spain during and after the Ten Years' War. For example, the "good and loyal Spaniards, peninsular and islander," of the *bombero* (fireman) battalion in Santiago de Cuba sent a statement of their support for Spain in the early years of the Ten Years' War. The bomberos were units historically composed mostly of men of color, with white officers, and they blurred racial distinctions when citing the sacrifice "of their own blood" in defense of Spanish religion, customs, civilization, patriotism, and honor."[18] In a rare example of black mobility within the mostly segregated Spanish Army, a Dominican of African descent rose to the rank of general, but despite "his blood and loyalty," according to a war chronicler, a Havana theater denied "the loyal and respectable" Eusebio Puello of the privilege of sitting with his family in a theater box.[19]

The end of the war occasioned new, public opportunities to cite the blood spilled during war as a justification for political inclusion. African-descended Cubans after the war took advantage of the reforms (granted as a concession to rebels) legalizing associations and newspapers—an unprecedented expansion of the public sphere and a transformation of Spanish colonialism into a system of political parties. The Santiago bomberos informed the government that many of their comrades had died to preserve the integrity of the empire, "offering our blood in holocaust for its immaculate honor."[20] Whereas for much of the colonial period black blood indicated a stain that justified political exclusion, blood spilled in defense of Spain could denote political allegiance while escaping identifications with race. Black blood did not necessarily disqualify someone from being a loyal subject of the empire. But just as these writers and activists were developing a language of loyalty in which their blood marked national sacrifice rather than impurity, new languages of race and degeneration were working to redefine black inferiority as cultural, anthropological, and as an evolutionary stagnation.

Innovation and evolution

Perhaps the best-publicized evidence of blood's shifting significance in Cuba—and of new contents in its conceptual envelope—came in 1881, when Cuban doctor Carlos Finlay presented findings that identified the *Aedes aegypti* mosquito as the vector for transmitting yellow fever. Finlay demonstrated how the mosquito sucks infected human blood, the yellow fever virus moves through its own bloodstream and salivary glands, and when biting humans its saliva enters their bloodstream to cause infection. Finlay's epidemiological breakthrough brought international attention to the work of the Academia de Ciencias Médicas, Físicas, y Naturales de la Habana (Havana Academy of Medical, Physical, and Natural Sciences), established in 1868.[21] The idea that blood could pass along racial characteristics never fully dissipated, but blood's transmissive properties acquired new meanings backed by institutionalized scientific research. Cuban scientists had written about the "stain" of blood and its associations with degenerate behaviors, but they now focused less on genealogy than on the material substance itself: journal articles and university theses weighed the scientific reliability of bloodstains in the developing fields of forensic science and criminology.[22] Several decades after their initial development on the island, these new ideas would become instrumental in marking Cubans of African descent unfit for citizenship in an independent nation.

"Modern" scientists, of course, were perfectly capable of reproducing the idea of race irrespective of its associations with blood; other physical qualities could do the job, too.[23] Finlay had not discerned that yellow fever was endemic to Africa, which would have explained in part the relative immunity of African-descended Cubans to the virus. But he noticed differential rates of infection with "those of northern races" and considered "the preferences that the mosquito manifests toward certain races and individuals." He attributed the mosquito's "favoritism" not to blood or even color but to the "varying thickness of skin and the conditions that effect cutaneous circulation."[24] Finlay's comment obviously did not effect a wholesale paradigm shift from blood to skin thickness as the corporeal metonym for race. But the instability of blood discourse in Cuban politics and within the island's scientific community in the last quarter of the nineteenth century reveals that unresolved questions about race were finding their way to new deliberative venues.

The establishment of the Sociedad Antropológica de la Isla de Cuba (Anthropological Society of the Island of Cuba) in 1877 became the foremost association to consider ideas about anthropology, race, purity, mixture, and, in turn, their political salience. Most of the published research of various members (and of European scientists' anthropological studies) appeared in the *Revista de Cuba*, established also in 1877 by José Antonio Cortina, an orator who identified with liberal principles and spoke frequently about self-rule for Cuba.[25] In many of the *Revista*'s writings, and in the monthly meetings of the Sociedad Antropológica, race emerged as the foremost factor that stymied attempts to arrive at the universal definition of "man"—the gendered protagonist of human history—that anthropologists pursued. Racial differences expressed through blood discourse earlier in the colonial period now entered new scientific vocabularies—new conceptual envelopes—that maintained distinctions and hierarchies: a messy overlap between "The Old Orthodoxy and Modern Science," as the title of one *Revista de Cuba* essay argued.[26] Readers could take in full transcriptions of the Paris lectures of anthropologist Paul Broca, who advocated studying humans to learn about "racial determinism." Of particular interest to him were the relationships between physical characteristics and intellectual, moral, social, and civilizational capacities, and the relationships between "natural groups," such as races, and "accidental" or "historical groups," namely political collectivities such as tribes, populations, towns, and nations.[27] None of these groupings relied on blood metrics. With these articles, Cuba's reading public could find institutionally sanctioned proof of presumed civilizational differences between whites and blacks, backed by (specious) scientific evidence and interpreted through the lens of evolutionary thought. These ideas had determinative effects on the qualifications for political subjectivity in a system perceived to be on the march of progress.

While the anthropological society eschewed the idea of the blood's transmission of traits, its studies of other bodily phenomena actively reproduced bodily and behavioral associations with racial difference. The first winning essay in an annual competition for the best anthropological study went to a comparative craniological analysis of black and Chinese skulls.[28] A study of eye illnesses in 1878 revealed a tendency for conjunctivitis and corneal infections in Chinese Cubans, "exaggerated by poor medicine and opium abuse." Changing topics, it went on to identify infections of the nasal passage as more common among white than black Cubans. The report noted that both had wider noses than *peninsulares*, insinuating that nominally white, but probably mixed, *criollos* bore a telling resemblance to the Cubans of color with whom they lived in close proximity.[29] According to the article, inherited titles of nobility—long considered guarantors of loyalty to Spain—might transmit superior *calidad* to new generations, but good breeding and noble blood proved unreliable in the face of widespread intermingling among racialized populations.

Indeed, race mixture figured among the greatest concerns of sociedad members. Although their research drew heavily on non-Spanish subjects, early anthropologists in Cuba recognized the "primitive" elements on the island that compromised full membership in a modern scientific or political community. The sociedad's approach to race mixture acknowledged both its dangers and its promise. Like Brazil, a slave society whose intellectual elite "discarded" the European racist theory of the degeneracy of "mixed bloods," Cuba had so large a mixed population that anthropologists had to acknowledge it with the assumption that its ultimate end was the eradication of inferior races.[30] One discussion in the sociedad postulated that the extinction of blacks and mulattoes would only take one or two generations, and thus members criticized consanguineous marriage norms for unnecessarily trying to keep bloodlines pure. Whatever eradication looked like, the remaining population would never fully be pure. The "aristocracies and ruling castas" had not always been so distinguished, one member noted, and "fellow citizens" (*conciudadanos*) would realize at some point their common humble origins.[31]

If Northern European anthropologists elsewhere were to be believed, the situation was not so different than the one in Spain. The Iberian role in the transatlantic debate had less to do with its valorization of blood purity than with its imperial history. German scientist Pruner Bey, whose research attempted to link black blood to the moral deficiencies of Egyptians, told the Paris anthropological society (in a speech reprinted in the *Revista de Cuba*) that Spaniards had established the "auxiliary sciences" of anthropology centuries earlier in the course of exploring the world. Missionaries had catalogued populations and languages in the first century of American colonization, and no other nation in Europe "had been linked to so many different races."[32] Thus the inhabitants of Spanish territories contained "precious evidence" for the study of racial mixing, but sometimes because of the mongrelization or degeneration of European whiteness. Mixtures with other peoples demonstrated to some the superiority of the "Latin race" to extinguish or overcome the weaker races. And although the "first people that populated Spain came from Africa," the Iberian climate shaped racial characteristics that eventually differentiated black Africans from white Spaniards.[33] Climatological explanations of racial difference—variously a counterpoint and companion to the epistemological authority of blood—had circulated since at least the sixteenth century, but they acquired new political meaning in defending against Anglo-Saxon "ascendancy" and explaining Spain's imperial strength.[34]

Whereas Spanish anthropologists set themselves to the task of dismantling arguments about their mongrelization, they made quite similar arguments about Cubans. And in turn, some members of the Sociedad in Havana, much like Northern European writers, targeted Spain's own mixed past. They asserted that *peninsulares* could only be a pure race by forgetting "the mix with Berbers and Arabs, or those of the Canaries with Africans"; Vizcainos, José Montalvo posited, could be considered a race of their own.[35] If this first generation of dedicated anthropologists closely linked proximity to Europeans with degrees of civilization, they could not agree on its political and social valence. Like their European counterparts, they increasingly turned to Charles Darwin's writings, however inchoate their political stakes were in the 1870s, to develop a common vocabulary for considering the place of Cubans among the civilized peoples of the world. As early as 1861, Cuban scientist Felipe Poey recognized the potential for Darwinist interpretations to undermine the logic of colonial rule. Censors pressured him to recant much of a speech about "The Unity of the Human Species," in which he denounced evolutionist justifications for racism and slavery and explained racial difference not as a matter of blood but as a temporary climatological adaptation occurring under the general rubric of human unity. What concerned him—or what officials encouraged him to concern himself with—was the impression that he had invoked the principle of equality. Aside from a possible allusion to the equality of men's souls, he eventually explained, he had never intended to displease the captain general by threatening the political order.[36]

Two decades later, the bustling world of associations and clubs following the Ten Years' War enabled a far richer conversation about Darwinism and politics than the milieu in which Poey worked. Theories of evolution, both biological and political, reached vastly wider audiences. In an 1879 speech about Darwinism to the Liceo Artístico in the Havana suburb of Guanabacoa, for example, Antonio Mestre explained the history, if not the presence, of Spanish rule in Cuba as a matter of biology. "One consequence of that principle [of competition between races] is the historical event of the rapid extinction of the savage races of America and Australia," noted Mestre, signaling a key change from Darwinist invocations of "species." "Contact between the civilized man and the races that have remained in a rudimentary state diminished and extinguished those races." Not even the noble attempts of "the Bishop of Chiapas," Bartolomé de las Casas, and the 1542 New Laws (prohibiting indigenous slavery) could overcome biological destiny and the ultimate success of the "conquering race."[37] By this point, blood had largely ceased to explain much about race or civilization, at least within the elite community of scientifically minded Cubans who embraced

Darwinist principles to articulate the relationship between bodies, race, and politics.

The public dissemination of those ideas went beyond writings and speeches in clubs, associations, and journals. Francisco Calcagno, a doctor, writer, and bibliographer, produced one of the first popular reckonings with Darwinism when he published *En busca del eslabón* (Looking for the Link) in 1888. This novel was part of an impressive corpus of work. Calcagno was an active member of the Sociedad Antropológica, and at meetings would draw on reports of Henry Morton Stanley's exploits in Africa to consider in Darwinian terms the "physical, moral, and social" character of the "black race."[38] With *En busca del eslabón*, Calcagno made a more subtle argument about racial difference and mixture as he chronicled a group of explorers searching for the primate species that would definitively connect apes to humans. Coincidentally, they ran into Stanley himself on their journey, who initiated a lengthy argument with them when he claimed that a lack of political and social institutions justified the enslavement of Africans: "They do not recognize property rights; they have neither religion nor government, nor the authority of a head of family, because in reality, there is no family."[39] The wise Cuban expedition leader Don Sinónimo took on Stanley in great detail, defending the capacities of Africans to maintain such basic institutions. Don Sinónimo always had at his side a black servant named Procopio, a man prone to amusing capers but whose "atavism is nothing more than a caprice of nature." On their travels, an azagaya fruit fell on Procopio's head in Brazil, flattening his forehead; in Africa, Procopio tripped on tree roots, fell on his nose, and flattened it too. These accidents thus generated or accentuated his physical features, although one of the commission's findings at the end of the voyage was that "Africans belong to the human species."[40] A novel that presented Cuban readers with intricate descriptions of different peoples in different parts of the world, then, circled back to the theme of unity, informed by Calcagno's understanding of evolutionary theory.

Asserting that shared identity, whether in Calcagno's science fiction or in his public statements, served to lay a groundwork by which Cubans under Spanish rule might imagine a racially integrated post-emancipation society. If racial difference was but an accident, as Calcagno contended, and if Africans indeed possessed the basic tools of civilization, then their descendants' suitability as Spanish subjects or citizens might become more fathomable than during earlier periods of colonial rule in the Americas, when the stains of blood kept African-descended subjects at arm's reach from political personhood.

Body politics and bodies politic

How did the reach of these new ideas extend to the political subjectivity of the people whose putative inferiority rested on anthropological claims—an explanation once accomplished with references to blood? Certainly, doubting the evolutionary capacity of African-descended Cubans to civilize and embrace progress provided a language to rationalize their political exclusion—not coincidentally, the same justifying language deployed during the vast expansion of Northern European empires in Africa and elsewhere in the late nineteenth century. But civilization and progress often indexed moral, social, and cultural inequalities rather than explicitly political ones. In fact, many studies of "the black race" by Cuban anthropologists tended to express more concern with the cultural and laboring aptitude of African-descended people than with their capacities to be loyal subjects of the Spanish empire—or, in agentive contrast, rebellious insurgents. Sociedad member Agustín Reyes published a comparative study of African-born (*bozal*) and native-born (*criollo*) Cubans of color, for example, and argued that criollos had superior abilities and more refined physical features than bozales, as evidenced not only in their manual labor but in their achievements in skilled trades and their capacity to imitate white Cubans in dress, speech, and musical preferences: "the progress of civilized life."[41] The bodies of Cubans of African descent may have furnished evidence for all kinds of abilities and affinities, but for Reyes and many other anthropologists, institutional colonial politics did not figure prominently.

Yet the political significance of the activities of the Sociedad Antropológica becomes evident in both its composition and its ideology. First, a great many of the men developing and debating ideas about the study of race were themselves politicians. The founding committee included a dozen or so members of what became the Partido Liberal Autonomista, or Liberal Party, once the Spanish government sanctioned political parties on the island in 1879. In addition to José Antonio Cortina, the editor of the newly established *Revista de Cuba*, Liberal Party luminaries such as Antonio Govín y Torres, Enrique José Varona, and José Torralbas figured among the Society's first members and the most outspoken advocates of greater autonomy from Spain.[42] The Society often enjoyed direct patronage from the colonial government, as several of Cuba's captains general accepted the Society's invitations: Joaquín Jovellar served as the founding honorary president, and Arsenio Martínez Campos and Emilio Calleja paid visits.[43] These anthropologist-politicians successfully shaped public discourse in Cuba to a surprising degree, and with enough flexibility that references to biological metrics for civilization and barbarism could easily reinforce language often enlisted in the service of racist stereotyping.

Second, these men of politics and science often expressed both their scientific and political endeavors as a search for universal paradigms, and they found one in evolution. Darwinism offered new avenues of inquiry into the corporeal foundations of race, and it also framed aspirations to a forward-looking political system guided by evolutionary principles.[44] As a slave society on the wane and as a colony, Cuba became significant to a transatlantic Darwinian "geography of reading" that foregrounded questions of race and empire.[45] An article in the 1880 *Revista de Cuba* considered the shared trajectory of conservatives, democrats, and liberals as one driven by "the fundamental laws of sociological science." Politics was one with biology as part of a "total evolutionary unfolding," but it was the "evolutionist Liberal Party" that would serve as the beacon of progress. A minor Partido Democrático Asimilista (Assimilationist Democratic Party), the article claimed, "was born, to borrow a phrase from biological science, *out of sequence*" in part because it "bowed before the privilege of castas while swearing an oath to battle all privileges." Adherence to an enduring racial ideology, then, could cast a political agenda as unevolved. The unnamed author made no mention that Liberals who debated Darwinism at the Sociedad Antropológica bowed to racial hierarchies as well, arguing instead that the Liberal Party represented the "scientific, evolutionist, and positivist element" that, by observation and experience, would guide Cuban politics backed by the "evolutionary force of history."[46] If in the early modern Iberian world, race and blood purity "remained part of a grid of knowledge constituted not by scientific (biologistic) discourses but by religious ones," as María Elena Martínez argues, in the final decades of Spanish rule in Cuba scientific discourses (not primarily about blood) did now constitute a grid of knowledge through which race and political community found a common language.[47]

As with blood talk, an anthropological vocabulary could enable claims to political inclusion as well as instantiating hierarchies that justified exclusive membership in political institutions. Newspapers, party politics, and associational life after the Ten Years' War (a Spanish concession to rebel demands) provided the setting for new and public discussions about race and allegiance. Rodolfo de Lagardère, the most prominent pro-Spanish Cuban of African descent, was a leading, if controversial, figure among Havana's black intellectuals. He laid out his agenda at a lavish event that he organized in 1881 as head of the Casino Español de la Clase de Color (Spanish Club for the Class of Color), an auxiliary organization to the patriotic white-only *casinos españoles* that had organized volunteer troops during the war. He made a case for black citizenship and against racial discrimination in recognition of amply demonstrated loyalty to Spain. Lagardère's opening speech made a brief mention of blood but situated it in the context of how "modern science" affirmed the universality of "man" irrespective of color or locale. He also cited anthropology. His definition was unconventional—it included humoral theory, which dates to ancient Greece and Rome—such that to him, the discipline offered the "scientific dogma" that the blood and the other humors (especially black and yellow bile) of the *negro* were richer in carbon than that of the white, proving that "colors are nothing but mere accidents."[48] Religion also fortified claims to racial equality. Lagardère noted that St Paul's letter to the Galatians argued that Greeks, Gentiles, Romans, Jews, and servants/slaves (*siervos*) were all of "the same origin, the same, spirit, the same, blood, the same family." Christianity "established the holy dogma of equality" and as the "yellow," "African," and "caucasian" races emerged from the offspring of Adam and Eve. Science and religion did not have to oppose each other: "Adopt, if you please, the Adam of the Bible or the Adam of progressive evolution"—God or Darwin. Both emerged from a "fundamental law" that explained variety but identified common ancestors that set in motion evolutionary processes.[49] In terms of the racial politics of the Spanish empire, Lagardère acknowledged the many Cubans of African descent with "Spanish blood," but he stopped well short of making that blood the main or sole justification for citizenship. Instead, the willful loyalty of black and mulatto Cubans merited a stronger government commitment to education, combating discrimination, and other measures that would allow Cubans of color to improve—and evolve—into ideal subjects and citizens.

Popular (mis)perceptions and metaphorical representations of social difference and evolutionary politics took predictable embodied forms. Speeches and newspaper cartoons were replete with images of Cubans as children—usually children of color—slowly coming of age under Spain's tutelage. Full political adulthood was always a generation away. Questions arose as to how people of so-called inferior races could profess any kind of political allegiance if they were physically incapable of good judgment. A speech given in 1879 at the Liceo de Guanabacoa considered the question by drawing on recent analytical yields in the field of "evolutionary psychology." A comparison of the human brain to those of lemurs and monkeys revealed complex differences analogous to those between children and adults, uncivilized and civilized men, and "the men of inferior races with respect to those of superior races." Differences in spinal cords allegedly affected the expression of emotions and practice of logic.[50] Popular evolutionism held as much potential to denigrate Cubans, particularly those of color, as it did to justify the political rights that Lagardère championed. But it also ignored a paradox: evolutionary logic rested on the principle of mutability, and many of these authors intended it to fix in place the structured inequalities that they considered so vital to Cuban society.

Whether biological or political, evolution promised only slow change, and many Cubans by the late nineteenth century had grown weary of waiting to achieve political adulthood under Spanish rule. Piecemeal extensions of citizenship rights could be revoked as arbitrarily as they appeared. The frustrations of a sugar planter in eastern Cuba named Carlos Manuel de Céspedes led him to declare independence in 1868, intiating the Ten Years' War. Immediately—not gradually—the anticolonial insurgency had attracted support from slave and free people of color alike, and promises of citizenship and emancipation took concrete form in the unfolding of the war. When Céspedes wrote of "The Constitutional Evolution of Cuba," he actually referred to a quite radical shift in constitutionalist momentum from Spanish rule to a revolutionary national independence movement.[51] Once the war ended and Spain sanctioned political parties, rebels remained dissatisfied with the slow pace of political change. In 1886, the year of slavery's final demise, the mulatto rebel general Antonio Maceo chastised Cubans, especially Liberals, who "preferred parliament to arms" based on the "false promises" that "left them quiet in their evolution of reincorporation."[52] Pro-independence writer and activist Manuel Sanguily condemned the Liberal slogan "All through evolution, nothing through revolution" by noting that Darwin "was in favor of struggle."[53] Eventually, even Liberal politician Miguel Figueroa admitted that "our work and mission" was to fight tyranny, if not that of Spain then that of the rival Partido Unión Constitucional (Constitutional Union Party), "first by evolution, later, if it were necessary, by revolution."[54] Those terms benefited from their ambiguous referents: if gradual political advancement could take many forms, so, too, could radical change. And expressing political allegiance in evolutionary terms did not necessarily require coming clean about which Cubans counted as sufficiently civilized to be political subjects.

Forestalling bloody revolution by promoting bloodless evolution had become a leading rhetorical strategy in political discourse, but, as Figueroa's words suggest, the delicate balance between the two had weakened by the

1890s. Just a year before the beginning of the final war for independence (1895–98), activists of color, invoking their loyalty to Spain, were still negotiating with colonial officials to end racial discrimination. Captain General Emilio Calleja celebrated government circulars banning discrimination as evidence of "the natural consequence of the evolution that this society has made since slavery happily ended."[55] Evolution, not impurities of blood, guided his policies toward the African-descended population. In 1894, when he conceded to all men of color the right to use the honorific "Don" before their names, a local newspaper in Cienfuegos praised him for acknowledging how black Cubans "tend to improve themselves" over time and deserved the title, "which in early times was only given to nobles of blood (*nobles de la sangre*)."[56] Simultaneously, José Martí, the antiracist leader of the independence movement, eschewed talk of evolution but paired standard references to blood spilled for the cause with talk of "Cuban blood," the "free blood" of fellow citizens, and the "purest blood of the Cuban nation."[57]

Some Spanish writers and politicians offered an inclusive alternative. In the final years of Spanish rule, and in contrast to almost four centuries of exclusionary rhetoric, they began to invoke blood to signal all Cubans' inclusion in the Spanish empire and in the Spanish or Latin race, as they called it. Assertions of a common identity were borne in crisis: colonial officials and Spanish pundits hoped that Cubans would maintain their loyalty if they understood themselves to be elevated to equal status. So at the same time as the struggle against Cuban rebels prompted frequent accusations of inciting race war, there were also strategic attempts to look past racial distinctions in order to hail Cubans of all backgrounds as Spaniards. Since the 1860s and 1870s, according to Joshua Goode, Spanish intellectuals had considered national identity in terms of the Latin race, and that term now served politicians and intellectuals well when they naturalized Cubans' loyalty to Spain as an embodied affinity, conscious of "the value of racial mixing."[58] In an 1897 speech, the venerable Spanish politician Segismundo Moret wondered if "all are Spaniards, their *patria* is our *patria*, their blood is that of our race ... who dares to invoke the holy name of *patria*?"[59] Although he linked loyalty to blood, Moret, like others, cited it as an equalizing, not stratifying, characteristic. While the bodily metonym continued to hold political value, blood, in this case echoing José María Lequerica's pronouncement in 1810, had indeed become red: not black, white, stained, mixed, or otherwise divisive.

If articulations of a Latin race served as a last-ditch discursive effort on the part of Spaniards to hold together their crumbling empire, the concrete consequences of related expressions of Anglo-Saxon superiority served to accelerate its demise. Political cartoons in US newspapers during the run-up to US intervention in 1898 racialized Cubans and Spaniards alike; they could depict Cubans as monkeys at the same time as the brutal Spanish leader Valeriano Weyler appeared in a *Philadelphia Inquirer* cartoon as a caged ape, with a caption reading "an excellent addition to our museum."[60] Politicians, too, lumped together Cubans and Spaniards as uncivilized and unfit for modern government. Wisconsin senator John Spooner suggested that weaknesses in the insurgency could justify US force—for example, if Máximo Gómez, the leader of the Liberation Army, might not be able to subdue the Cubans, "for in them, remember, is the hot blood of the Spanish people."[61] Nominally a war against Spain to liberate Cuba from tyranny, US military action drew support from portrayals of the Latin race that deprived Spaniards and Cubans of the civilization that would qualify them for political equality.

Whether or not he was aware of these depictions, Captain General Ramón Blanco recognized the divergent political consequences of the idea of a common Latin race. Hot-blooded Cuban rebels, he forecasted, might become as much of a target of US aggression as the Spanish government. But their shared hot blood, he surmised, could also ground a politics of coalition. In May 1898, on the verge of the arrival of US troops, he wrote to Máximo Gómez, who led a multiracial army, to warn him that "we find ourselves before a foreigner of a different race." Not only did the prospect of US intervention jeopardize Spanish rule, it also threatened to "exterminate the Cuban people because of their Spanish blood." In suggesting a heretofore unfathomable alliance between forces that had battled each other for years, Blanco pleaded with the rebel general to set aside their differences temporarily, unite their armies in the town of Santa Clara, and mount a common defense against the United States. Blanco received a reply from Gómez that evinced the limits of blood discourse in the final months of the Spanish empire:

> You say that we belong to the same race and you invite me to fight against our foreign invader; but you are mistaken again because there are no differences of blood or of race. I only believe in one race: humanity; for me there are just good and bad nations. Spain, having been bad until now, and right now in Cuba the United States is fulfilling its obligation of humanity and civilization. From the swarthy Indian savage to the wise refined Englishman, a man, for me, deserves respect according to his honor and emotions, no

matter what country or race he belongs to, or what religion he professes.[62]

In addition to defeating Spain on the battlefield, Gómez had dismantled the semantic framework that allowed blood to link race, nation, and religion. He echoed the Cuban anthropologists who articulated a Darwinist concept of a single human species—albeit with varying degrees of civilization—undivided by differences of blood. But unlike the leaders of the Sociedad Antropológica, Gómez more or less denied the very existence of races, even if he still distinguished Englishmen from swarthy savages. Blanco's proposal drew on centuries of Spanish invocations of blood and race to explain political allegiance to empire. Moving even beyond Martí's "Cuban blood" that united opponents of colonial rule, Gómez invalidated the significance of blood and race to the coherence of any political community.

Conclusion

Michel Foucault referred to a transformation in the late nineteenth century in which a revitalized "thematics of blood" aided the formation of racism, "racism in its modern, 'biologizing' statist form."[63] Setting aside arguments that modern, biologized, statist racism emerged much earlier, and out of the specific experience of Spanish colonialism, we can tell a quite different story from Foucault's that acknowledges Cuba and attends to the open conceptual envelope of blood—but one in a stack that also includes open envelopes of race and political allegiance.[64] Changing deployments of blood discourse, never fully vanishing, took a secondary role in coalescing forms of belonging that necessarily shifted in relation to the transformations occurring in Cuba. If loyalty to Spanish rule appears more as a continuity rather than a change, it helps to recognize the shifting techniques of rule within those persistent politics: the diminished political importance of blood purity, revised vocabularies to express allegiance, and an alliance between new social science and new politicians attuned to evolutionary principles. By the early decades of the twentieth century, still newer links drawn between anthropology and race sought to link physiognomy to criminality and cast African-derived religious practices as atavistic and uncivilized.[65] The idea of eugenics guided scientists and politicians who struggled to reconcile biological difference and national unity.[66] Bodily characteristics still inscribed racial identification indexed political subjectivities, but each of those ideas had received transfusions that gave them a new lease on life.

Notes

1. Cited by Robert L. Paquette in *Sugar Is Made with Blood: The Conspiracy of La Escalera and the Conflict between Empires over Slavery in Cuba* (Middletown, CT: Wesleyan University Press, 1988), 56.
2. *An Address to the People of Great Britain, on the Utility of Refraining from the Use of West India Sugar and Rum*, 5th edn (London: M. Gurney and W. Darton, 1791), 2–3.
3. On the Bois-Caïman ceremony, see Laurent Dubois, *Avengers of the New World: The Story of the Haitian Revolution* (Cambridge, MA: Harvard University Press, 2004), 99–102.
4. "Documentos que se refiere al acta del cabildo celebrado por el Ayuntamiento fecha Stgo. de Cuba 25 de junio de 1811 en que se trató del proyecto presentado a las cortes sobre abolición de la esclavitud," Archivo Nacional de Cuba, Fondo Asuntos Políticos, legajo 213, expediente 81.
5. Alejandra Bronfman, *Measures of Equality: Social Science, Citizenship, and Race in Cuba, 1902–1940* (Chapel Hill: University of North Carolina Press, 2004).
6. Roland Greene, *Five Words: Critical Semantics in the Age of Shakespeare and Cervantes* (University of Chicago Press, 2013), 108–11.
7. Iberian associations of the Curse of Ham with black Africans contributed to linking servitude and stained genealogy.
8. María Elena Martínez, *Genealogical Fictions: Limpieza de Sangre, Religion, and Gender in Colonial Mexico* (Stanford University Press, 2008), 6. Blood was not the only biological marker of social difference in the early modern Iberian world; early settlers explained differences between Europeans and indigenous peoples through their bodies as well as their behavior, especially the humors (the blood, phlegm, black bile, and yellow bile that governed bodily health) and the food that they ate. See Rebecca Earle, *The Body of the Conquistador: Food, Race, and the Colonial Experience in Spanish America, 1492–1700* (Cambridge University Press, 2012).
9. *Oración inaugural que en la apertura del tribunal de la Real Audiencia de la isla de Cuba el 2 de enero de 1812* (Havana: En la oficina nueva de Arazoza y Soler, 1812), 8.
10. Adolfo de Castro (ed.), *Cortes de Cádiz: Complementos de las sesiones verificadas en la isla de León y en Cádiz*, 2 vols (Madrid: Imprenta de Prudencio Pérez de Velasco, 1913), 1: 178.
11. On these distinctions and deliberations, see Tamar Herzog, *Defining Nations: Immigrants and Citizens in Early Modern Spain and Spanish America* (New Haven: Yale University Press, 2003), ch. 7.
12. Verena Martínez-Alier, *Marriage, Class, and Colour in Nineteenth-Century Cuba: A Study of Social Values in a Slave Society* (Cambridge University Press, 1974), 18.
13. José Antonio Saco, *La situación política de Cuba y su remedio* (Paris: Imprenta de E. Thunot y Compañía, 1851), 38.
14. La Escalera (The Ladder) refers to the repression of an alleged antislavery conspiracy between slaves, free people of color, and British abolitionists, which inaugurated a period of harsh repression. See Paquette, *Sugar is Made with Blood*, and Michele Reid Vazquez, *The Year of the Lash: Free People of Color in Cuba and the Nineteenth-Century Atlantic World* (Athens: University of Georgia Press, 2011).

15. Miguel Rodríguez Ferrer, *Los nuevos peligros de Cuba entre sus cinco crisis actuales* (Madrid: Imprenta de Manuel Galiano, 1862), 26.
16. Entry for 9 April 1869, Archivo Provincial de Cienfuegos, Cuba, Fondo Actas Capitulares, Libro 12.
17. Francisco de Camps y Feliú, *Españoles é insurrectos. Recuerdos de la Guerra de Cuba por el Coronel retirado*, 2nd edn (Havana: Imprenta de A. Álvarez y Comp., 1890), 376.
18. Emilio Bacardí y Moreau, *Crónicas de Santiago de Cuba*, 2nd edn, 10 vols (Madrid: Gráficas Breogán, 1973), 4: 397.
19. Rafael Serra, *Ensayos políticos* (New York: Imprenta El Porvenir, 1892), 113.
20. Bacardí y Moreau, *Crónicas de Santiago de Cuba*, 6: 293.
21. Walter Reed's work in Cuba as part of the US Yellow Fever Commission stirred debates about Finlay's credit for the breakthrough. Research about blood certainly preceded Finlay: during the 1832 cholera epidemic, for example, doctors identified anemia and other blood diseases as contributing factors. See Adrián López Denis, "Higiene pública contra higiene privada: cólera, limpieza y poder en La Habana," *Estudios Interdisciplinarios de América Latina y el Caribe* 14 (2002–3), 103–25.
22. Manuel Cañizares y Venegas, *¿Los medios analíticos conocidos para distinguir las manchas de sangre de las que no lo son llevan las exigencias de la ciencia?* (Havana: Imprenta La Antilla, 1880), and Manuel Delfín y Zamora, *Manchas de sangre; estudio químico legal. Tesis leída el día 20 de noviembre en esta Universidad, en el ejercicio para aspirar al grado de doctor en la Facultad de Farmacia* (Havana: Imprenta La Propaganda Literaria, 1890).
23. See, for example, Thomas C. Holt, "'Blood Work': Fables of Racial Identity and Modern Science," in *Race and Blood in the Iberian World*, ed. Max S. Hering Torres, María Elena Martínez, and David Nirenberg (Berlin: Lit Verlag, 2012), 191–203.
24. Carlos Finlay, "El mosquito hipotéticamente considerado como agente de trasmisión de la fiebre amarilla," *Anales de la Academia de ciencias médicas, físicas y naturales de la Habana* 18 (1882), 162.
25. Alejandra Bronfman notes that institutionalization does not mark the "origins of a discipline" but, rather, "it can nonetheless open possibilities for formalized channels of interaction and opportunities for far-flung exchanges among participants." "On Swelling: Slavery, Social Science, and Medicine in the Nineteenth Century," in *Obeah and Other Powers: The Politics of Caribbean Religion and Healing* (Durham, NC: Duke University Press, 2012), 105.
26. Esteban Borrero Echeverría, "La vieja ortodoxia y la ciencia moderna," *Revista de Cuba* 5 (1879), 292.
27. "Qué es la antropología?" *Revista de Cuba* 2 (1877), 159.
28. "Consideraciones generales sobre el estado e importancia de la Antropología en la Isla de Cuba," *Revista de Cuba* 2 (1877), 365.
29. *Actas: Sociedad antropológica de la isla de Cuba*, ed. Manuel Rivero de la Calle (Havana: Comisión Nacional Cubana de la UNESCO, 1966), session of 5 May 1878, 32.
30. Thomas Skidmore, *Black into White: Race and Nationality in Brazilian Thought* (Oxford University Press, 1974), 77. See also Emilia Viotti da Costa, *The Brazilian Empire: Myths and Histories* (University of Chicago Press, 1985), ch. 9.
31. *Actas*, session of 6 May 1883, 156.

32. "Consideraciones generales sobre el estado e importancia de la Antropología en la Isla de Cuba," *Revista de Cuba* 2 (1877), 364. Between 1868 and 1874, anthropological societies appeared in London, New York, Moscow, St Petersburg, Madrid, Manchester, Florence, Berlin, Vienna, Stockholm, and Tbilisi.
33. *Actas*, session of 6 May 1883, 157.
34. On the presumed climatological and racial bases of Spanish colonial authority, see *Actas*, session of 6 November 1887, 192, and Earle, *The Body of the Conquistador*.
35. Pedro M. Pruna and Armando García González, *Darwinismo y sociedad en Cuba: Siglo XIX* (Madrid: Editorial CSIC, 1989), 131.
36. Pedro M. Pruna Goodgall, "Biological Evolutionism in Cuba at the End of the Nineteenth Century," in *The Reception of Darwinism in the Iberian World*, ed. Thomas F. Glick, Miguel Ángel Puig Samper, and Rosaura Ruiz (Dordrecht: Kluwer Academic Publishers, 2001), 54–5.
37. "Origen natural del hombre. Discurso leido en el Liceo Artístico y Literario de Guanabacoa," *Revista de Cuba* 5 (1879), 429.
38. *Actas*, session of 1 March 1891, 221.
39. Francisco Calcagno, *En busca del eslabón* (Havana: Editorial Letras Cubanas, 1983 [1888]), 111.
40. Ibid., 201.
41. A. W. Reyes, "Estudio comparativo de los negros criollos y africanos," *Revista de Cuba* 5 (1879), 155.
42. Armando García González and Consuelo Naranjo Orovio, "Antropología, 'raza' y población en Cuba en el último cuarto del siglo XIX," *Anuario de Estudios Americanos* 55.1 (1998), 273. The *Revista* promoted science, law, literature, and fine arts.
43. On Jovellar, see "Trabajos preparatorios para la constitución de la Sociedad Antropológica de la Isla de Cuba," *Boletín de la sociedad antropológica de la Isla de Cuba* 1 (1879), 4, and "Consideraciones generales sobre el estado e importancia de la Antropología en la Isla de Cuba," *Revista de Cuba* 2 (1877), 365; on Calleja, see *Actas*, session of 5 May 1878, 32.
44. The connections made by Cuban leaders between political life and evolutionary thought should not suggest consensus, on or beyond the island, about the social, political, or biological legacy of Darwin. See Piers J. Hale, *Political Descent: Malthus, Mutualism, and the Politics of Evolution in Victorian England* (University of Chicago Press, 2014).
45. David N. Livingstone employs "geography of reading" to refer to "how scientific proposals are read in different venues and how they are marshaled in particular places for particular projects." *Dealing with Darwin: Place, Politics, and Rhetoric in Religious Engagements with Evolution* (Baltimore: Johns Hopkins University Press, 2014), 1–2. See also Ronald L. Numbers and John Stenhouse (eds), *Disseminating Darwinism: The Role of Place, Race, Religion, and Gender* (Cambridge University Press, 2001).
46. "Conservadores, democratas, y liberales," *Revista de Cuba* 7 (1880), 555.
47. Martínez, *Genealogical Fictions*, 13.
48. On humoralism, see Earle, *The Body of the Conquistador*, ch. 1.
49. "Discurso pronunciado por el Sr. Rodolfo de Lagardère en la noche del 11 de marzo de 1882 en la reinauguración del Casino Español de color de la

Habana," Archivo Histórico Nacional, Madrid, Sección Ultramar, legajo 4884, tomo 8, no. 142.
50. "La evolución psicológica," *Revista de Cuba* 5 (1879), 12, 18, 19.
51. Céspedes's pamphlet is referred to in *Miguel Figueroa, 1851–1893: Discurso leído ... el 6 de julio de 1943, en conmemoración del cincuecentenario de su muerte*, ed. Jorge Mañach, vols 119–21 (Havana: El Siglo XX, 1943), 42.
52. Antonio Maceo to José A. Rodríguez, writing from Kingston, Jamaica, 1 November 1886, in *Antonio Maceo: Ideología política: Cartas y otros documentos*, 2 vols (Havana: Editorial de Ciencias Sociales, 1988), 295.
53. Manuel Sanguily, "Las reformas políticas y el darwinismo," *El Cubano*, 29 October 1887.
54. *Miguel Figueroa, 1851–1893*, 9.
55. "El General Calleja," *La Igualdad*, 2 January 1894.
56. "Al Gobernador General. Un grave error," *El Día*, 5 January 1894.
57. José Martí, "En Cuba," *Patria*, 30 October 1894, in José Martí, *Obras completas*, 27 vols, 2nd edn (Havana: Editorial de Ciencias Sociales, 1975), 4: 320, and "Carta al *New York Herald*," 2 May 1894, *Obras completas*, 4: 155.
58. Joshua Goode, *Impurity of Blood: Defining Race in Spain, 1870–1930* (Baton Rouge: Louisiana State University Press, 2009), 29.
59. "Discurso pronunciado en Zaragoza por D. Segismundo Moret, 19 de Julio de 1897," Carlos O'Donnell y Abreu, *Apuntes del ex-ministro de Estado Duque de Tetuán, para la defensa de la política internacional y gestión diplomática del gobierno Liberal-Conservador*, 2 vols (Madrid: Tip. de R. Péant, 1902), 2: 87.
60. "The Great Weyler Ape" is reproduced in David Sartorius, *Ever Faithful: Race, Loyalty, and the Ends of Empire in Spanish Cuba* (Durham, NC: Duke University Press, 2013), 213.
61. Cited in Louis A. Pérez, Jr, *Cuba in the American Imagination: Metaphor and the Imperial Ethos* (Chapel Hill: University of North Carolina Press, 2008), 87.
62. Josep Conangla i Fontanilles, *Memorias de mi juventud en Cuba: Un soldado del ejército espanol en la guerra separatista (1895–1898)* (Barcelona: Ediciones Peninsula, 1998), 202–3.
63. Cited in Ann Laura Stoler, *Carnal Knowledge and Imperial Power: Race and the Intimate in Colonial Rule* (Berkeley: University of California Press, 2002), 151.
64. See Irene Silverblatt, *Modern Inquisitions: Peru and the Colonial Origins of the Modern World* (Durham, NC: Duke University Press, 2004).
65. See Bronfman, *Measures of Equality*, and Stephan Palmié, *Wizards and Scientists: Explorations in Afro-Cuban Modernity and Tradition* (Durham, NC: Duke University Press, 2002). The work of Cuban criminologist Israel Castellanos provides countless examples of scholarly linkages. See *Contribución al estudio craneométrico del hombre negro delincuente* (Seville: Policlínica Sevillana, 1915) and "La brujería y el nañiguismo desde el punto de vista médico-legal," *Anales de la Academia de Ciencias Médicas, Físicas y Naturales de la Habana* 53 (1916), 267–370. Castellanos also wrote about blood, as a substance rather than an idiom: *La plasmogenia* (Havana: Rambla, Bouza, 1921); *Hematología forense* (Havana: La Propagandista, 1932); and *La sangre en policiología* (Havana: Carasa, 1940).
66. Nancy Leys Stepan, *"The Hour of Eugenics": Race, Gender, and Nation in Latin America* (Ithaca, NY: Cornell University Press, 1991).

Further Reading

Alpers, Paul. *What is Pastoral?* University of Chicago Press, 1996.
Anidjar, Gil. *Blood: A Cultural Critique of Christianity*. New York: Columbia University Press, 2014.
Anonymous. *Genuine Memoirs of the Late Celebrated Jane D****s*. London, 1761.
Appelbaum, Robert. *Literature and Utopian Politics in Seventeenth-Century England*. Cambridge University Press, 2002.
Aquinas, St Thomas. *Summa Theologica*, trans. Fathers of the English Dominican Province. New York: Benziger Bros, 1947.
Bay, Mia. *The White Image in the Black Mind: African-American Ideas on White People, 1830–1925*. Oxford University Press, 2000.
Benassy-Berling, Marie-Cécile. *Humanismo y religión en Sor Juana Inés de la Cruz*, trans. Laura López de Belair. Mexico City: UNAM, 1983.
Bevington, David, and Peter Holbrook, eds. *The Politics of the Stuart Court Masque*. Cambridge University Press, 1998.
Biller, Peter. "Views of Jews from Paris around 1300: Christian or 'Scientific'?," in *Christianity and Judaism: Studies in Church History* 29, ed. D. Wood. Cambridge, MA: Blackwell, 1992. 187–207.
—— "A 'Scientific' View of Jews from Paris around 1300," *Micrologus* 9 (2001), 137–68.
Blight, David W. *Race and Reunion: The Civil War in American Memory*. Cambridge, MA: Harvard University Press, 2001.
Bondio, Mariacarla Gadebusch, ed. *Blood in History and Blood Histories*. Florence: Sismel, Galluzzo, 2005.
Brace, Loring. *Race Is a Four-Letter Word: The Genesis of the Concept*. Oxford University Press, 2005.
Broberg, Gunnar. "Linnaeus's Classifications of Man," in *Linnaeus: The Man and His Work*, ed. Tore Frängsmyr. Berkeley: University of California Press, 1983. 156–94.
Bronfman, Alejandra. "On Swelling: Slavery, Social Science, and Medicine in the Nineteenth Century," in *Obeah and Other Powers: The Politics of Caribbean Religion and Healing*, ed. Diana Paton and Maarit Forde. Durham, NC: Duke University Press, 2012. 1–3–120.
Browne, Thomas. "The Sixth Book: Concerning Sundry Tenets Geographical and Historical," in *Pseudodoxia Epidemica: or, Enquiries into Very Many Received Tenets and Commonly Presumed Truths*. London, 1672.
Burton, R. *Anatomy of Melancholy*, ed. Thomas C. Faulkner, Nicholas K. Kiessling, and Rhonda L. Blair. Oxford University Press, 1989.
Buxó, José Pascual. *Sor Juana Inés de la Cruz: Amor y conocimiento*. Mexico: UNAM, 1996.
Carey, David. *Locke, Shaftesbury, and Hutcheson: Contesting Diversity in the Enlightenment and Beyond*. Cambridge University Press, 2006.
Carretta, Vincent. *Phillis Wheatley: Biography of a Genius in Bondage*. Athens: University of Georgia Press, 2011.

Caulker, Tcho Mbaimba. *The African-British Long Eighteenth Century: An Analysis of African-British Treaties, Colonial Economics, and Anthropological Discourse.* Lanham, MD: Lexington Books, 2009.

Clubb, Louise George. *Italian Drama in Shakespeare's Time.* New Haven and London: Yale University Press, 1989.

Conrad, L. I. et al. *The Western Medical Tradition: 800 BC to AD 1800.* Cambridge University Press, 1995.

Cook, H. J. *The Decline of the Old Medical Regime in Stuart London.* Ithaca, NY: Cornell University Press, 1986.

Cruz, Sor Juana Inés de la. *Obras completas.* Mexico: Fondo de Cultura Económica, 1951.

Dain, B. A. *Hideous Monster of the Mind: American Race Theory in the Early Republic.* Cambridge, MA: Harvard University Press, 2002.

Derrida, Jacques. *Geneses, Genealogies, Genres, and Genius: The Secrets of the Archive,* trans. Beverley Bie Brahic. New York: Columbia University Press, 2003.

Des Chenes, Dennis. *Life's Form: Late Aristotelian Conceptions of the Soul.* Ithaca, NY: Cornell University Press, 2000.

Dominique, Lyndon J. *Imoinda's Shade: Marriage and the African Woman in Eighteenth Century British Literature, 1759–1808.* Columbus: Ohio State University Press, 2012.

Downs, Jim. *Sick from Freedom: African-American Illness and Suffering During the Civil War and Reconstruction.* Oxford University Press, 2012.

Drummond, Ian. "John Duns Scotus on the Passions of the Will," in *Emotion and Cognition in Medieval and Early Modern Philosophy,* ed. Martin Pickavé and Lisa Shapiro. Oxford University Press, 2012. 53–74.

Duras, Claire de. *Ourika,* trans. John Fowles; introd. Joan DeJean and Margaret Waller. New York: MLA, 1995.

Eamon, William. *Science and the Secrets of Nature: Books of Secrets in Medieval and Early Modern Culture.* Princeton University Press, 1994.

Earle, Rebecca. *The Body of the Conquistador: Food, Race, and the Colonial Experience in Spanish America, 1492–1700.* Cambridge University Press, 2012.

Elgersman, Maureen G. *Unyielding Spirits: Black Women and Slavery in Early Canada and Jamaica.* New York: Routledge, 1999.

Eliav-Feldon, Miriam, et al. *The Origins of Racism in the West.* Cambridge University Press, 2009.

Fabian, Ann. *The Skull Collectors: Race, Science, and America's Unburied Dead.* University of Chicago Press, 2010.

Fallon, Stephen M. *Milton among the Philosophers: Poetry and Materialism in Seventeenth-Century England.* Ithaca, NY: Cornell University Press, 2007.

Faust, Drew Gilpin. *This Republic of Suffering: Death and the American Civil War.* New York: Knopf, 2008.

Feerick, Jean. *Strangers in Blood: Relocating Race in the Renaissance.* University of Toronto Press, 2010.

Floyd-Wilson, Mary. *English Ethnicity and Race in Early Modern Drama.* Cambridge University Press, 2003.

Foucault, Michel. *The History of Sexuality: An Introduction,* trans. Robert Hurley. London: Penguin, 1978.

——— *"Society Must Be Defended": Lectures at the Collège de France, 1975–76.* London: Picador, 2003.

García González, Armando, and Consuelo Naranjo Orovio. "Antropologia, 'raza,' y población en Cuba en el último cuarto del siglo XIX," *Anuario de Estudios Americanos* 55.1 (1998), 267–89.

Glick, Thomas F., et al. (eds). *The Reception of Darwinism in the Iberian World*. Dordrecht: Kluwer Academic Publishers, 2001.

Goode, Joshua. *Impurity of Blood: Defining Race in Spain, 1870–1930*. Baton Rouge: Louisiana State University Press, 2009.

Gowland, Angus. *The Worlds of Renaissance Melancholy: Robert Burton in Context*. Cambridge University Press, 2006.

Granada, Fray Luis de. *Introducción el símbolo de la fe*. Madrid: Cátedra, 1989.

Greene, Roland. *Five Words: Critical Semantics in the Age of Shakespeare and Cervantes*. University of Chicago Press, 2013.

Greg, Walter W. *Pastoral Poetry and Pastoral Drama: A Literary Inquiry, with Special Reference to the pre-Restoration Stage in England*. New York: Russell & Russell, 1959.

Hacking, Ian. *The Social Construction of What?* Cambridge, MA and London: Harvard University Press, 1999.

Hankins, James. "Monstrous Melancholy: Ficino and the Physiological Causes of Atheism," in *Laus Platonici philosophi: Marsilio Ficino and his Influence*, ed. S. Clucas, P. Forshaw, and V. Rees. Leiden: Brill, 2011. 25–43.

Harvey, W. *The Anatomical Exercises*, ed. Geoffrey Keynes. New York: Dover, 1995.

Hatfield, Gary. "The Cognitive Faculties," in *The Cambridge History of Seventeenth-Century Philosophy*, ed. Daniel Garber and Michael Ayers, 2 vols. Cambridge University Press, 2008. 2: 955–61.

Heng, G. "The Invention of Race in the European Middle Ages I & II," *Literature Compass* 8 (2011), 258–93.

Henke, Robert. *Pastoral Transformations: Italian Tragicomedy and Shakespeare's Late Plays*. Newark: University of Delaware Press, 1997.

—— "Pastoral as Tragicomedic in Italian and Shakespearean Drama," in *The Italian World of English Renaissance Drama: Cultural Exchange and Intertextuality*, ed. Michele Marrapodi. Newark and London: University of Delaware Press, 1998.

Herring Torres, Max S., María Elena Martínez, and David Nirenberg (eds). *Race and Blood in the Iberian World*. Berlin: Lit Verlag, 2012.

Heuman, Gad J. *Between Black and White: Race, Politics, and the Free Coloreds in Jamaica, 1792–1865*. Westport, CT: Greenwood Press, 1981.

Hill, Christopher. "William Harvey and the Idea of Monarchy," *Past & Present* 27 (1964), 54–72.

Hobbes, T. *Leviathan*, ed. C. B. Macpherson. Harmondsworth: Penguin, 1968.

Hobson, Janelle. *Venus in the Dark: Blackness and Beauty in Popular Culture*. New York: Routledge, 2005.

Holt, Thomas C. "'Blood Work': Fables of Racial Identity and Modern Science," in *Race and Blood in the Iberian World*, ed. Max S. Hering Torres, María Elena Martínez, and David Nirenberg. Berlin: Lit Verlag, 2012. 191–203.

Jarrett, Gene. "'To Refute Mr. Jefferson's Arguments Respecting Us': Thomas Jefferson, David Walker, and the Politics of Early African American Literature," *Early American Literature* 46.2 (2011), 291–318.

Johnson, Willis. "The Myth of Jewish Male Menses," *Journal of Medieval History* 24 (1998), 273–95.

Keevak, Michael. *Becoming Yellow: A Short History of Racial Thinking.* Princeton University Press, 2011.
Kessler, Eckhard. "Psychology: The Intellective Soul," in *The Cambridge History of Renaissance Philosophy*, ed. Quentin Skinner, C. B. Schmitt, Eckhard Kessler, and Jill Kraye. Cambridge University Press, 1988. 485–534.
King, Peter. "Dispassionate Passions," in *Emotion and Cognition in Medieval and Early Modern Philosophy*, ed. Martin Pickavé and Lisa Shapiro. Oxford University Press, 2012. 9–31.
Kleist, Heinrich von. "The Betrothal in Santo Domingo," in *The Marquise of O and Other Stories*, trans. David Luke and Nigel Reeves. New York: Penguin, 1978.
Knuutila, Simo. "Sixteenth-Century Discussions of the Passions of the Will," in *Emotion and Cognition in Medieval and Early Modern Philosophy*, ed. Martin Pickavé and Lisa Shapiro. Oxford University Press, 2012. 116–32.
Knuuttila, Simo, and Juha Sihvola (eds). *Sourcebook for the History of the Philosophy of Mind: Philosophical Psychology from Plato to Kant.* Heidelberg: Springer, 2014.
Koerner, Lisbet. *Linnaeus: Nature and Nation.* Cambridge, MA: Harvard University Press, 1999.
Lagier, Raphaël. *Les races humaines selon Kant.* Paris: Presses Universitaires de France, 2004.
Lampert, L. "Race, Periodicity, and the (Neo-)Middle Ages," *MLQ* 65 (September 2004), 391–421.
Law, J. *The Social Life of Fluids: Blood, Milk, and Water in the Victorian Novel.* Ithaca, NY: Cornell University Press, 2010.
Lettow, Susanne (ed.). *Reproduction, Race, and Gender in Philosophy and the Early Life Sciences.* New York: SUNY Press, 2014.
Levecq, C. *Slavery and Sentiment: The Politics of Feeling in Black Atlantic Antislavery Writing, 1770–1850.* Hanover: University of New Hampshire Press, 2008.
Li Causi, Pietro. *Generare in comune. Teorie e rappresentazioni dell'ibrido nel sapere zoologico dei Greci dei Romani.* Palermo: Palumbo, 2008.
Lindley, David (ed.). *The Court Masque.* Manchester University Press, 1984.
Loomba, Ania. "Race and the Possibilities of Comparative Critique," *New Literary History* 40 (2009), 501–22.
López Beltrán, Carlos. "Hippocratic Bodies, Temperament and Castas in Spanish America (1570–1820)," *Journal of Spanish Cultural Studies* 8 (2007), 253–89.
Lugt, Maaike van der, and Charles de Miramon (eds). *L'hérédité entre Moyen Age et époque moderne.* Florence: SISMEL—Edizioni del Galluzzo, 2008.
Lynch, Deidre Shauna. *The Economy of Character: Novels, Market Culture, and the Business of Inner Meaning.* University of Chicago Press, 1998.
Macready, William. *The Irishman in London; or, The Happy African.* London, 1793.
Manning, S., and T. Ahnert (eds). *Character, Self, and Sociability in the Scottish Enlightenment.* Basingstoke: Palgrave Macmillan, 2011.
Martínez, María Elena. *Genealogical Fictions: Limpieza de sangre, Religion, and Gender in Colonial Mexico.* Stanford University Press, 2008.
Mazzolini, Renato G. "Las Castas: Inter-Racial Crossing and Social Structure (1770–1835)," in *Heredity Produced: At the Crossroads of Biology, Politics and Culture, 1500–1870*, ed. Staffan Müller-Wille and Hans-Jörg Rheinberger. Cambridge, MA: MIT Press, 2007.
McCune Smith, J. *The Works of James McCune Smith: Black Intellectual and Abolitionist*, ed. J. Stauffer. Oxford University Press, 2006.

McMullan, Gordon. *The Politics of Unease in the Plays of John Fletcher.* Amherst: University of Massachusetts Press, 1994.
Medin, Douglas L., and Scott Atran (eds). *Folkbiology.* Cambridge, MA, and London: MIT Press, 1999.
More, Anna. "Cosmopolitanism and Scientific Reason in New Spain: Sigüenza y Góngora and the Dispute over the 1680 Comet," in *Science in the Spanish and Portuguese Empires, 1500–1800*, ed. Daniela Bleichmar et al. Stanford University Press, 2009.
Morgan, Philip D. "British Encounters with Africans and African-Americans, circa 1600–1780," in *Strangers within the Realm: Cultural Margins of the First British Empire*, ed. Bernard Bailyn and Philip D. Morgan. Chapel Hill: University of North Carolina Press, 1991. 157–219.
Naranjo Orovio, Consuelo, and Armando García González. *Racismo e Inmigración en Cuba en el siglo XIX.* Madrid: Doce Calles, 1996.
Nelson, Megan Kate. *Ruin Nation: Destruction and the American Civil War.* Athens: University of Georgia Press, 2012.
Newmyer, Stephen T. *Animals in Greek and Roman Thought: A Sourcebook.* London and New York: Routledge, 2011.
Nirenberg, David. *Communities of Violence: Persecution of Minorities in the Middle Ages.* Princeton University Press, 1996.
—— "Was there Race before Modernity? The Example of 'Jewish' Blood in Late Medieval Spain," in *The Origins of Racism in the West.* Cambridge University Press, 2009. 232–64.
Norbrook, David. *Poetry and Politics in the English Renaissance.* London: Routledge, 1984.
Nussbaum, Felicity A. *The Limits of the Human: Fictions of Anomaly, Race and Gender in the Long Eighteenth Century.* Cambridge University Press, 2003.
Opie, Amelia. *Adeline Mowbray; or, The Mother and Daughter.* Peterborough: Broadview Press, 2009.
Park, Katharine. "The Organic Soul," in *The Cambridge History of Renaissance Philosophy*, ed. Quentin Skinner, C. B. Schmitt, Eckhard Kessler, and Jill Kraye. Cambridge University Press, 1988. 464–84.
Pasnau, Robert. *Thomas Aquinas on Human Nature: A Philosophical Study of Summa Theologiae 1a 75–89.* Cambridge University Press, 2002.
Paster, Gail. "Nervous Tensions: Networks of Blood and Spirit in the Early Modern Body," in *The Body in Parts: Fantasies of Corporeality in Early Modern Europe*, ed. David Hillman and Carla Mazzio. London and New York: Routledge, 1997. 107–25.
Paster, Gail, et al. (eds). *Reading the Early Modern Passions: Essays in the Cultural History of Emotion.* Philadelphia: University of Pennsylvania Press, 2004.
Patterson, Annabel. *Censorship and Interpretation: The Conditions of Writing and Reading in Early Modern England.* Madison: University of Wisconsin Press, 1984.
—— *Pastoral and Ideology.* Berkeley and Los Angeles: University of California Press, 1987.
Pérez-Amador, Adam Alberto. *De finezas y libertad: Acerca de la Carta Atenagórica de Sor Juana Inés de la Cruz y las ideas de Domingo de Báñez.* Mexico: Fondo de Cultura Ecónomica, 2011.
Peterson, Carla. "Untangling Genealogy's Tangled Skeins: Alexander Crummell, James McCune Smith, and Nineteenth-Century Black Literary Traditions," in

A Companion to American Literary Studies, ed. Caroline Levander and Robert Levine. Malden, MA: Blackwell, 2011.

Phillips, Carla Rahn, and William D. Phillips, Jr. *Spain's Golden Fleece: Wool Production and the Wool Trade from the Middle Ages to the Nineteenth Century.* Baltimore and London: Johns Hopkins University Press, 1997.

Prince, Mary. *The History of Mary Prince*, ed. Sara Salih. New York: Penguin, 2001.

Prince, Nancy. *Black Woman's Odyssey Through Russia and Jamaica: The Narrative of Nancy Prince.* New York: Markus Weiner Publishing, 1990.

Pruna, Pedro M., and Armando García González. *Darwinismo y sociedad en Cuba: Siglo XIX.* Madrid: CSIC, 1989.

Resnick, I. "Medieval Roots of the Myth of Jewish Male Menses," *Harvard Theological Review* 252 (2000), 241–63.

—— *Marks of Distinction: Christian Perceptions of Jews in the High Middle Ages.* Washington, DC: Catholic University of America Press, 2012.

Root, Michael. "Race in the Social Sciences," in *Handbook of Philosophy of Science: Philosophy of Anthropology and Sociology.* Amsterdam and Boston: Elsevier/ North Holland (2007), 735–53.

Russell, Nicholas. *Like Engend'ring Like: Heredity and Animal Breeding in Early Modern England.* Cambridge University Press, 2006.

Saillant, J. *Black Puritan, Black Republican: The Life and Thought of Lemuel Haynes, 1753–1833.* Oxford University Press, 2003.

Sanborn, G. "Mother's Milk: Frances Harper and the Circulation of Blood," *English Literary History* 72.3 (Fall 2005), 691–715.

Sawday, J. *The Body Emblazoned: Dissection and the Human Body in Renaissance Culture.* London: Routledge, 1995.

Schmitt, C. B. "The Rise of the Philosophical Textbook," in *The Cambridge History of Renaissance Philosophy*, ed. Quentin Skinner, C. B. Schmitt, Eckhard Kessler, and Jill Kraye. Cambridge University Press, 1988.

Scodel, Joshua. *Excess and the Mean in Early Modern English Literature.* Princeton University Press, 2002.

Sensbach, Jon F. *Rebecca's Revival: Creating Black Christianity in the Atlantic World.* Cambridge, MA: Harvard University Press, 2005.

Shapin, Stephen. *The Scientific Revolution.* University of Chicago Press, 1996.

Shapin, S., and S. Schaffer. *Leviathan and the Air-Pump: Hobbes, Boyle, and the Experimental Life.* Princeton University Press, 1985.

Sharpe, Jenny. *Ghosts of Slavery: A Literary Archaeology of Black Women's Lives.* Minneapolis: University of Minnesota Press, 2003.

Sicroff, Albert. *Los estatutos de limpieza de sangre: Controversias entre los siglos XV y XVIII*, trans. Mauro Amiño. Madrid: Taurus, 1985.

Simonsohn, S. *The Apostolic See and the Jews.* Toronto: Pontifical Institute of Mediaeval Studies, 1991.

Sloan, Phillip R. "The Gaze of Natural History," in I*nventing Human Science: Eighteenth-Century Domains*, ed. Christopher Fox, Roy Porter, and Robert Wokler. Berkeley: University of California Press, 1995. 112–51.

Soriano Vallès, Alejandro. *El Primero sueño de Sor Juana Inés de la Cruz: Bases tomistas.* Mexico: Universidad Nacional Autónoma de México, 2000.

Stauffer, J. *The Black Hearts of Men: Radical Abolitionists and the Transformation of Race.* Cambridge, MA: Harvard University Press, 2002.

Stevens, Peter F., and Steven P. Cullen, "Linnaeus, the Cortex-Medulla Theory, and the Key to His Understanding of Plant Form and Natural Relationships," *Journal of the Arnold Arboretum* 71 (1990), 179–220.
Sullivan Jr, Garrett. *Sleep, Romance, and Human Embodiment: Vitality from Spenser to Milton*. Cambridge University Press, 2012.
Thackeray, William Makepeace. *Vanity Fair*. Oxford University Press, 1983.
Thorndike, Ashley H. "The Pastoral Element in the English Drama before 1605," *Modern Language Notes* 14.4 (1899), 114–23.
Townley, James. *High Life Below Stairs*. London, 1759.
Voegelin, Eric. *Die Rassenidee in der Geistesgeschichte von Ray bis Carus*. Berlin: Junker und Dünnhaupt Verlag, 1933.
Wheeler, Roxann. *The Complexion of Race: Categories of Difference in Eighteenth-Century British Culture*. Philadelphia: University of Pennsylvania Press, 2000.
Wilson, Kathleen. *The Island Race: Englishness, Empire and Gender in the Eighteenth Century*. New York: Routledge, 2002.
Wood, Roger J., and Vítězslav Orel. *Genetic Prehistory in Selective Breeding: A Prelude to Mendel*. Oxford University Press, 2001.
Wright, Thomas. *Circulation: William Harvey's Revolutionary Idea*. London: Chatto & Windus, 2012.
Yerushalmi, Yosef Hayim. "Assimilation and Racial Anti-Semitism: The Iberian and German Models," *Leo Baeck Memorial Lecture* 26. New York: Leo Baeck Institute, 1982.
Yoon, Carol Kaesuk. *Naming Nature: The Clash Between Instinct and Science*. New York and London: W. W. Norton, 2009.

Index

Page numbers in italics refer to illustrations.

Academia de Ciencias Médicas, Fisicas, y Naturales de la Habana (Havana Academy of Medical, Physical, and Natural Sciences)
  shift in focus from genealogy to material substance of blood, 235
  studies as instrumental in racial differentiation, 235
Acts 17:26, and antislavery literature, 157
Adelman, Janet, 119, 120
African blood
  commodification criticized in *Clotel* (Brown), 161, 163
  Declaration of Independence's omission of, 13–14, 152–3, 157–8
  excluded from blood imagery in Civil War era white print culture, 17, 211–12, 217–20
  linked with colonial money by mulatto heiresses, 86–7
  references to in Civil War black print culture, 223–4
  whitening equations for, 53–4
  see also black people
Agricultura (1513) (Gabriel Alonso de Herrera), 49
Agustin, Friar Miguel
  complexion as essence, 47
  dog breeding, 59
  plant hybridization, 48
Albert the Great, 110–11
Alcalá Galan, Mercedes, 28, 30
Alicia Seldon (character in *Edmund and Eleonora*)
  as beneficiary of father's wealth, 97
  exogamous marriage, 97–8
  as figure for social reform, 87, 95–6, 98–101
  progressivism of father, 96
  as reproduction of African mother, 98–100
  similarities to Olivia Fairfield, 86
Alonso de Herrera, Gabriel, 49
Alpers, Paul, 68
Aminta (Tasso), 69
Anatomy of Melancholy (1621) (Robert Burton)
  humoralism in, 15–16, 176–80
  melancholy humor applied to body politic in, 178–80
  utilitarianism in, 179–80
  utopianism in, 180
Anidjar, Gil, 27
animal husbandry
  and folkbiological essentialism, 47–9
  and hereditary race, 3–4
  as origin of whitening equations, 45–6, 58
  and the race concept, 3–4
anthropology, Cuban. *See* Sociedad Antropológica de la Isla de Cuba (Anthropological Society of the Island of Cuba)
Aphorisms (Hippocrates), 110
Appeal, in Four Articles; ... to the Coloured Citizens of the World, but in Particular, and Very Expressly, to Those of the United States of America (1829) (David Walker), 158–9
Aquinas, Thomas, 129, 131–3, 135
  see also Thomist Aristotelianism
Arderne, John, 110
Aristotelianism
  and folkbiological essentialism, 45, 47, 49, 55
  persistence in early modern Iberia, 127
  soul as bound to body in, 132

Aristotelianism, late
  divergence from Thomism, 127,
    132, 138
  and "First Dream" (Sor Juana), 138
  and "intentional species," 139
Ars magna lucis (Athanasius Kircher),
    142
"Athenagoric Letter" (Sor Juana Inés
    de la Cruz), 130
Atran, Scott, 45, 56
Augustine, Saint, 109, 114–15
Avicenna, 110, 111, 112

Barletta, Vincent, 31, 32
Bartholin, Thomas, as possible
    Linnaeus source, 196
Bartlett, Robert, 2
"The Battle of Lexington"
    (Lemuel Haynes), 156
Beckles, Hilary, 91
Behn, Aphra, Oroonoko (1688), 8,
    99–100
Bell, James Madison, "To the
    Cleveland Union-Savers," 223
Benezet, Anthony, Some Historical
    Account of Guinea (1771), 157, 158
Bernard of Gordon, 113
biological essentialism vs
    folkbiological essentialism, 47, 56
Bisset, Robert
  Douglas; or, The Highlander (1800),
    86–7, 90–2, 100
  see also Mrs Dulman
Black, Georgina Dopico, 35
black people (Civil War era)
  and blood imagery in white print
    culture, 17, 211–12, 217–19
  lack of documentation of injuries or
    mortality, 17, 210–12, 220–2, 225
  viewed as labor force, 210–11, 220,
    225
  see also African blood
The Blessed Jew of Marocco (Thomas
    Calvert), 119
Blight, David W., 216–17
blood
  ambivalence of in sympathy-based
    accounts of character, 13
  in Anatomy of Melancholy (Burton),
    177–8
  associated with race in Notes
    (Jefferson), 154
  as conduit of vital spirits in "First
    Dream" (Sor Juana), 137, 138
  excluded as metric by
    anthropologists, 236
  fused with slave trade in Liberty
    (Haynes), 155–8
  "fusion" as nineteenth-century
    American literary concern, 160,
    161
  linked with skin color in
    folkbiological essentialism, 7,
    54–5, 57
  material substance as focus of
    Cuban scientists, 235
  milk as concocted from, 4, 46, 50
  mutability of in Comus (Milton),
    67, 80–1
  mutability of in The Faithful
    Shepherdess (Fletcher), 68, 71–3,
    75
  as repository of elite identity, 65,
    66–7, 69
  universal nature stressed, 157–8,
    159, 163, 245–6
  whitening equations for, 45–6,
    53–4, 56–8, 60
blood as metaphor (Civil War era)
  applied to nation, 17, 216–17,
    223–4
  applied to suffering of white
    soldiers, 17, 212–19
  in black print culture, 223–4
  exclusion of black experience
    in white print culture, 17,
    211–12
  and interpretation of historical
    record, 17, 222
  mythology of war debunked by, 17,
    213–16
  in white print culture, 214–19
blood as metaphor (in nineteenth-
    century Cuba)
  as increasingly unstable signifier,
    18, 230–4
  and Máximo Gómez, 245–6
  as referencing political allegiance,
    231–4, 245–6
  as referencing slavery, 17–18

260 Index

blood circulation
  analogized to money in *Leviathan* (Hobbes), 183, 185–8
  discovered in horses, 51
  humoralism undermined by Harvey's theory of, 172–4
  influence on Hobbes, 15–16, 172, 181
  see also blood movement; Harvey, William
blood equations. *See* whitening equations
bloodlines
  and animal husbandry, 3–4
  as divine in Italian pastorals, 69–70
  divinity challenged in English pastorals, 74–5, 77–8, 81
  as essence, 47–8, 65–7
  as "tainted" in *Don Quixote* (Cervantes), 6, 26–7, 29–30, 35–6
  virtue conveyed through, 5, 9
  see also noble blood; *raza*
blood movement
  in *Anatomy of Melancholy* (Burton), 177–8
  humoral vs mechanistic model, 174
  see also blood circulation
blood purity
  attached to national character in *Notes* (Jefferson), 153–4
  ironic treatment of in "Heads of the Colored People" (Smith), 146–7
  as referencing political allegiance in Spanish America, 231–3
  see also limpieza de sangre doctrine; whitening equations
body. *See* mind/body relationship
body politic
  and Harvey's impact on concept, 15–16, 172–4
  humoralism applied to in *Anatomy of Melancholy* (Burton), 15–16, 178–80
  mechanistic concept of in *Leviathan* (Hobbes), 15–16, 181–8
  and shift in "community" in early modern Iberia, 27–8

Boose, Lynda, 85
Bradley, Richard, 198–9
Bright, Timothy, 10, 21
Broberg, Gunnar, 192, 196
Broca, Paul, 236
Brody, Jennifer DeVere, on blood, 9, 84–6
Browne, Thomas, 121, 204
Buffon, Georges
  and Linnaeus, 196–7
  and races as fluid entities, 194
Burton, Robert, *Anatomy of Melancholy* (1621). *See Anatomy of Melancholy* (1621) (Robert Burton)

Caesarius of Heisterbach, 108
Cain
  as denoting Jews, 115–17, 124n25
  Jewish bleeding disease as "mark of," 107, 108–9
Calgano, Francisco, 239
Calleja, Captain General Emilio, 244
Calvert, Thomas, 119, 120–1
Carew, Thomas, *Coelum Britannicum*, 77
Carey, John, 76
casta
  flexibility of meaning, 46
  and Margrave's classification of humans, 197
casta paintings, 50, 55–6
Castleman, Alfred Lewis, 217
Cervantes, Miguel de. *See Don Quixote* (Miguel de Cervantes Saavedra)
Céspedes, Carlos Manuel de, 243
character
  sympathy in formation of, 13–14, 147–50, 151
  see also national character
Charles I (England)
  and *The Faithful Shepherdess* (John Fletcher), 66, 75–6
  and William Harvey, 174
Chloe Llwhyddwhuydd (character in *The Dramatic History of ... Mrs. Llwhyddwhuydd* (Stevens))
  narrative negation of, 86–7, 89–90
  as racially ambiguous figure, 89–90

rank-elevating strategies, 89–90
"tainted" genealogy, 88–9
Christian theology, and justification of Jewish bleeding disease, 107–9, 112–16, 122
Cisneros (Cardinal), 31
"Civilization" (James McCune Smith), 163
Clement XI (Pope), 58
Clifford, George (Dutch East India Company director), 199
Clotel (1853) (William Brown)
criticism of slave trade, 161, 163
medical practice and blood "fusion" in, 161
as saga of Jefferson's "mixed blood" family, 160–1
Coelum Britannicum (Thomas Carew), 77
Cohen, Jeffrey Jerome, 11
Coker, Daniel, *A Dialogue between a Virginian and an African Minister* (1810), 159, 160
color
applied to both skin and "complexions" of mind by Hume, 149–50
as essence, 7, 48, 54–5
Comentarios reales de los Incas (Garcilaso), 52, 55, 60
Communipaw. *See* Smith, James McCune
community, shift in concept in early modern Iberia, 27–8
Compendio de albeitería (1729) (Fernando de Sande y Lago), 51, 55
complexion
applied to mind by David Hume, 149–50
as essence, 47
as fluid concept, 87–8
and humoralism, 4–5
see also mulatto heiresses; wealthy women of color
The Complexion of Race (Roxann Wheeler), 87
Comus (John Milton)
embodied rank exposed as fiction, 7, 77–9, 81

The Faithful Shepherdess (John Fletcher) as source, 67, 76, 78
historical context, 66, 76–7
mutability of blood in, 67, 80–1
noble blood associated with depravity in, 7–8, 77, 80
as puzzle to critics, 66
theory of immanence challenged by, 77–82
virtue as product of labor in, 8, 75, 78–81
"conceptual envelope" concept and changing paradigms, 230
and racial differences, 236
"The Constitutional Evolution of Cuba" (Céspedes), 243
Cortina, José Antonio, 235–6
Council of Toledo, Seventeenth, on Jewish "perpetual servitude," 116
Covarrubias, Sebastián de, 49
and essences, 7, 46
on paper types, 30
Critica Botanica (Carl Linnaeus)
discussion of human varieties, 195
distinction of species and varieties, 194–6
crossbreeding. *See* hybridization
Cuban independence movement, and Darwinism, 243–6
Curación racional de irracionales (1728) (Francisco García Cabero), 51

Darwin, Charles. *See* Darwinism
Darwinism
and Cuban independence movement, 243–6
influence on Cuban anthropologists, 238–9
political exclusion and inclusion rationalized by, 240–3
as universal paradigm, 241
Davenant, William, appended dialogue to *The Faithful Shepherdess* (John Fletcher), 71, 76
De Circulatione Sanguinis (1649) (William Harvey), 173–5
compared with *De Motu Cordis* (1626), 174–5

Declaration of Independence (draft of Thomas Jefferson)
  compared with *Liberty Fully Extended* (Haynes), 157–8
  emphasis on "English" blood, 14, 151–2
  exemption of slaves and blacks from national character, 13–14, 152–3, 157–8
  referenced in *Clotel* (Brown), 161
  sympathy-based account of character, 13, 148
  see also Jefferson, Thomas; Query XIV of *Notes on the State of Virginia* (Thomas Jefferson)
Defoe, Daniel, "A True-Born Englishman" (1701), 9, 14–15
degeneration equations
  and classification of Indians, 57–8
  for dog breeding, 59–60
  and jump-back (*salta-atrás*) principle, 55
  see also hybridization
De Motu Cordis (1626) (William Harvey), 173
  compared with *De Circulatione Sanguinis* (1649), 174–5
Descartes, René, 12–13, 22nn36–7, 42–3, 172, 181
De Secretis Mulierum, 112–13
Diccionario (Spanish Royal Academy of Language), 47, 49
Diccionario de Autoridades, on "inclinación," 133
diet
  associated with melancholy humor, 111–12
  linked with hemorrhoidal bleeding, 110–11, 113
difference, embodied, historical overview, 1–2
"Discorsi del Poema" (Tasso), 69
Discourse on Method (René Descartes), 172
discrimination, based on physiological vs cultural difference, 1–2
Don Quixote (Miguel de Cervantes Saavedra)
  Alacaná market episode, 25, 30–3, 36
  chivalric ideals as anachronism in, 6–7
  Don Quijote (character) conflated with, 6–7, 33
  first-edition front matter, 39–40
  first-edition title page, *38*
  limpieza de sangre doctrine parodied, 6, 26–7, 29, 33
  as metafictional, 26, 36–7
  metalepsis in, 28, 36–7, 39–40
  metaliterary discourse in, 26
  metamateriality in, 26–7, 39–40
  Sancho Panza (character) as indicator of cultural shift, 7
  sociocultural references of Alcaná episode, 30–3
  tainted "genealogy" of physical text as comment on culture, 26–7, 29–30, 35–6
  text-as-body in, 27, 33, 35–6, 39–40
Douglas; or, The Highlander (1800) (Robert Bisset), 86–7, 90–2, 100
Douglass, Frederick, 219
The Dramatic History of Master Edward, Miss Ann, Mrs. Llwhyddwhuydd, and Others (1743) (George Alexander Stevens), 86–7, 88–90, 94–5

Eclogues (Virgil), 68
Edmund and Eleonora (1797) (Edmund Marshall)
  Ôroonoko (Behn) as precedent, 99–100
  progressivism of slave owner in, 96–7
  as reflection of historical fact, 86
  as vehicle for social reform, 85–7, 95–6, 97–101
  see also Alicia Seldon
Elements of Philosophy, Epistle Dedicatory to (Thomas Hobbes), 181
Elyot, Thomas, *Boke named the Governour* (1531), 5
En busca del eslabón (Looking for the Link) (Francisco Calgano), 239
England
  and enforcement of Jewish badge law, 117–18

idea of Jewish bleeding disease
  discredited in, 118–22
environment
  and folkbiological essentialism,
    46–8, 57–9
  and human diversity in *Systema
    Naturae* (Linnaeus), 16, 192–4,
    207n10
Escritura desatada (Mercedes Alcalá
  Galán), 28
essence
  bloodline as, 47–8, 65
  color as, 7, 54–5
  complexion as, 47
  and environmental factors, 48, 57–9
  and folkbiology, 46–7
  Platonic view of rejected in *Comus*
    (John Milton), 78–80
"essence placeholder," 47, 58, 60
essentialism
  and Linnaeus's concept of human
    diversity, 192, 193, 196, 204–5
  see also folkbiological essentialism
Eurocentrism
  and David Hume, 149–50
  and folkbiological essentialism, 7,
    60
evolution, theories of, influence
  on Cuban science and politics,
    238–46
*Examen de ingenios para las ciencias
  (1574)* (Juan de Huarte), 137

Faerie Queene (Edmund Spenser), 73
The Faithful Shepherdess (John
  Fletcher)
  botanical motifs, 72
  compared with *Faerie Queene*
    (Edmund Spenser), 73
  contrasted with Italian pastorals,
    70–2, 74–5
  divinity through labor in, 71–2,
    74–5
  early performance history, 66, 75–6
  mutability of blood, 71–3
  as puzzle to critics, 66
  rustic figure as signifying gentility,
    73–4
  as satire of noble blood, 67–8, 71–2

  as source for *Comus* (John Milton), 76
  transfer of power in, 74–5
  virtue as result of conscious choice
    in, 72
  and William Davenant's royal
    panegyric, 71, 76
  see also Fletcher, John
Faust, Drew, 215
Feijoo y Montenegro, Benito
  Jerónimo, "Color etiópico" in
  Teatro crítico universal (1778), 58
"feminine mulattaroon"
  disgust of blood in, 85–6
  effect of juxtaposition with "rich
    woman," 85–6, 87
  narrative negation of, 84–5, 86–7
  as trope, 84–5
  see also mulatto heiresses; wealthy
    women of color
Feyerabend, Paul, 205–6
Figueroa, Miguel, 243
Sor Filotea de la Cruz, and Sor Juana's
  "Response," 130
Finlay, Carlos
  association of skin thickness and
    yellow fever infection, 235
  identification of yellow fever vector,
    235
  and shift in focus to material
    substance of blood, 235
"First Dream" ("Primero sueño")
  (c. 1691) (Sor Juana Inés de la
  Cruz)
  lighthouse metaphor, 138
  magic lantern metaphor, 141–2
  mind/body relationship in, 12,
    136–42
  Phaeton metaphor, 141
  philosophical influences on, 12,
    128–9, 136, 137
  pyramid metaphor, 138–40
  see also Sor Juana Inés de la Cruz
Fistula in Ano (John Arderne), 110
Fletcher, John
  and Italian pastoral, 69, 70, 74
  motivation for *Shepherdess*, 66–7
  and tragicomedy, 66, 67
  see also The Faithful Shepherdess
    (John Fletcher)

Floyd-Wilson, Mary, 203–4
folkbiological essentialism
  and Aristotelianism, 45, 47, 49, 55
  vs biological essentialism, 47, 56
  concept of, 47
  as constructionist, 47
  and environmental factors, 48, 57–9
  and Eurocentric ideology, 7, 60
  and hybridization, 45–6, 48–53
  opposed by David Hume, 148–9
  reasoning compared with modern science, 56
  skin color linked with blood in, 7, 54–5, 57
  undifferentiation of animal and plant domains, 46–53, 55–6
  in veterinary and farmers' handbooks, 47–9
folk biology. *See* folkbiological essentialism
Foucault, Michel, 18–19n1, 56, 65, 246
Fourth Lateran Council, 117–18
Frederick Douglass' Paper, 146
freed slaves. *See* black people (Civil War era)
Fuchs, Barbara, 34

Gafford, Tia, "The Reformation of the Mulatto Hero/Heroine," 86
Galán, Alcalá, 32
La Galatea (Miguel de Cervantes), 37
Galenic physiology. *See* humoralism
García Cabero, Francisco, 51
García Conde, Pedro, 48
Garcilaso, the Inca, classification of humans, 52, 55, 60
"The Gaze of Natural History" (Philip R. Sloan), 194
gender
  and "blood," 57
  and Church's criticism of Sor Juana, 12, 129–30, 143n12
  and slave status, 87, 155
genealogy, and tainted "bloodlines" of *Don Quixote* (Cervantes), 6, 26–7, 29–30, 35–6
Genesis 4:15, as prooftext of hereditary Jewish bleeding disease, 107

geohumoralism
  and Carl Linnaeus, 16, 203–4
  see also humoralism
Gómez, Máximo, 245–6
Góngora, Luis de, "Solitudes" (Soledades), 128
Goode, Joshua, 244
Granada, Luis de, *Introduction of the Symbol of Faith* (Introducción del Símbolo de Fe), 12, 136, 137
Greene, Roland, 230
Greer, Margaret, 27
Greg, W. W., 68, 70
Guarini, Giovanni Battista. *See Il pastor fido* (Guarini)
Gumilla, José
  influence of, 53
  and jump-back (*salta-atrás*) principle, 55
  on neophyte classification, 58, 63n59
  whitening equations, 53–4, 56–7, 60

Hacker, J. David, 222
Hall, Prince, *Charge of 1797*, 158
Ham (Noah's son), as figure for hereditary servitude, 115–16
Harper, Francis Ellen Watkins, 223
Harvey, William
  De Circulatione Sanguinis (1649), 173
  De Motu Cordis (1626), 173
  influence on Thomas Hobbes, 15–16, 172, 181
  political opinions, 174–5
  see also blood circulation
Haynes, Lemuel. *See Liberty Further Extended: Or Free Thoughts on the Illegality of Slavery*
"Heads of the Colored People" (James McCune Smith)
  as contesting Enlightenment moral philosophy, 147
  double sense of character in, 147
  ironic treatment of blood purity, 146–7
  "mixed blood" representation of Thomas Jefferson, 146–7

heart
  as mechanical pump in *De Circulatione Sanguinis* (Harvey), 175–6
  as metaphor in *De Motu Cordis* (Harvey), 174–5
  as regulator of humors in *Anatomy of Melancholy* (Burton), 177–8
Hemmings, Sally, 14, 155
hemorrhoidal bleeding
  analogized to menstruation in *Problemata varia anatomica*, 120, 125n51
  diet linked with, 110–11, 113
  early scientific discourses on, 110–12, 113
  Jewish bleeding disease characterized as, 107
  melancholy humor linked with, 110, 111–12, 113
  see also Jewish bleeding disease
Heng, Geraldine, 117–18
hereditary race. *See* race, hereditary
Heredity Produced: At the Crossroads of Biology, Politics, and Culture, 1500–1800 (ed. Steffan Müller-Wille and Hans-Jörg Rheinberger), 16
Herrera, Gabriel Alonso de, 49
Hill, Christopher, 174–5
Hippocrates, 110, 113
La historia medicinal de las cosas que se traen de nuestras Indias Occidentales (1565–74) (Nicolás Monardes), 49
History of Jamaica (1774) (Edward Long), 86
History of Jerusalem (Jacques de Vitry), 108–9
Histriomastix (William Prynne), 76
Hobbes, Thomas
  De Cive (1642), 174
  The Elements of Law (1640), 174
  and Harvey's circulation theory, 15–16, 22n42, 172, 181
  see also Leviathan (1651) (Thomas Hobbes)
horses
  hybridity taxa applied to humans, 52–3

likened to humans, 51–3
  nobility of, 50–2
  as status symbol, 50
Hortus Cliffortianus (Carl Linnaeus), 199, *200*
Huarte, Juan de, *Examen de ingenios para las ciencias (1574)*, 137
Hugh of St Cher, 109
human classification
  ad hoc scheme in *A Philosophical Account of the Works of Nature* (Bradley), 198–9
  by *casta* in *Historia Naturalis Brasiliae* (1648) (Marcgrave), 197
  by horse hybridity taxa in *Comentarios reales de los Incas* (Garcilaso), 52, 55, 60
  by skin color in "About the Various External Appearance of Men" (Vallerius), 197–8
  in *Systema Naturae* (Linnaeus), 191–3, 197–8, 200–4
human diversity
  correlated with four continents by Linnaeus, 191–3, 197, 198
  represented in *casta* paintings, 55–6
  see also human hybridization; racial difference
human hybridization
  analogized to plant/animal hybridization, 49–3, 56
  and *casta* paintings, 55–6
  horse taxa applied to, 52–3
  and jump-back (*salta-atrás*) principle, 55
  whitening equations, 53–4, 56–7, 60
  see also human diversity; racial difference
Humboldt, William von, 163
Hume, David
  and differing "complexions" of mind, 149–50
  distinction between animal and human breeding, 148–9
  and "moral causes" in character formation, 148–9
  "Of National Characters" (1748), 13, 148–9
  and polygenesis, 149–50

humoralism
  in *Anatomy of Melancholy* (Burton), 15–16, 176–80
  and complexion, 4–5
  defined, 9
  in "First Dream" (Sor Juana), 12, 129, 136, 137–8
  ideological importance, 3–4
  in Lagardère's speech on universality of man, 242
  and noble blood, 4–5
  persistence of, 172–4
  and the soul, 9–11
  undermined by Harvey's theory of blood circulation, 172–4
  see also geohumoralism
hybridization and folkbiological essentialism, 45–6, 48–53
  see also degeneration equations; whitening equations

Iberia, early modern
  book trade practices in, 30
  chivalric ideals in, 6–7
  culture reflected in physical "genealogy" of *Don Quixote* (Cervantes), 26–7, 29
  emergence of race in, 26–9
  "New Christians" vs "Old Christians" in, 6
  persistence of Aristotelianism in, 127
  shift in systems of power, 6–7
  status of horse, 50–3
  see also limpieza de sangre doctrine
Impossible Purities (Jennifer DeVere Brody), 86
"inclination"
  as desire for knowledge in the "Response" (Sor Juana), 133–5
  and Thomist "appetites," 131–3
  vs will in "Response" (Sor Juana), 128–9, 130–1
Indians
  classification of, 57–8, 63n59
  whitening equations for, 56–7
Innocent III (Pope), 116–17
Inquisition
  Arabic texts as target of, 31
  and *probanzas de limpieza*, 34–5
  role in classification of Iberians, 27

Introduction of the Symbol of Faith (Introducción del Símbolo de la Fe) (Luis de Granada), 12, 136, 137
Isidore of Seville, 115–16

James I (England)
  and *The Faithful Shepherdess* (John Fletcher), 66
  influence of Descartes on, 181
  and William Harvey, 174
Jarrett, Gene, 159
Jefferson, Thomas
  African American writers' response to, 146–7, 155–3
  depiction of slaves as members of paternal household, 153
  exclusion of black people from national character formation, 13–14, 152–3
  "mixed blood" representations of, 146–7, 160–1
  whitening equations in 1815 letter, 154–5
  see also Declaration of Independence (draft of Thomas Jefferson); *Query XIV of Notes on the State of Virginia* (Thomas Jefferson)
Jewish bleeding disease
  analogized to menstruation, 107, 108, 109, 114, 119, 120
  characterized as hemorrhoidal, 107, 110, 112, 119
  discrediting correlated with disempowerment, 11, 119–22, 126n63
  and embodied difference, 10–11, 107–10, 112–13
  and embodied inferiority, 107–9, 113–14
  and hereditary servitude, 108–10, 119
  historical overview, 11, 107–8
  linked with melancholy humor, 112, 113
  as punishment for crucifixion, 107–9, 113–14
  theological justification for, 107–9, 112–16, 122
  see also hemorrhoidal bleeding

Jewish hereditary servitude
  and Jewish bleeding disease, 108–10
  as justification for political
    subordination, 119
  justified theologically, 11, 107–10,
    114–17, 118–19
Jews
  associated with melancholy humor,
    111–12
  expulsion from Spain, 6
  servitude institutionalized, 116–19
  John 8:24 (New Testament), as
    authority for Jewish servitude,
    114–15
Johnson, Carroll B., 28
Johnson, Christopher, 140
Sor Juana Inés de la Cruz
  "Athenagoric Letter," 130
  reputation of, 127
  see also "First Dream" (Sor Juana);
    "Response" (Sor Juana)
jump-back (*salta-atrás*) principle, 53,
  55–6
Jussieu, Bernard de, 197

Kant, Immanuel, 206
Katz, David, 119
Kessler, Eckhard, 11, 139
Kircher, Athanasius, 128, 142

labor
  divinity through in *The Faithful
    Shepherdess* (Fletcher), 71–2,
    74–5
  virtue as product of in *Comus*
    (Milton), 8, 75, 78–81
Lagardère, Rodolfo de, 241–2, 243
Lateran Council (1513), 132
Leclerc, Georges-Louis, Comte de
  Buffon. *See* Buffon, Georges
Lemnius, Levinus, 4
Lequerica, José María, 232
Leviathan (1651) (Thomas Hobbes)
  blood as money in, 183, 185–8
  concept of voluntary vs involuntary
    motion in, 182–3
  human body as natural machine in,
    181–2
Liber de anima, Philip Melanchthon,
  10

*Liberty Further Extended: Or Free
  Thoughts on the Illegality of Slavery*
  (c. 1776) (Lemuel Haynes)
  blood as metaphor in, 155–8
  compared with Declaration of
    Independence, 157–8
  connotations of "character"
    exploited, 159–60
  critique of Jefferson, 159–60
  scriptural allusions in, 157–8
  shared bloodshed as authorizing
    membership, 156–7
Libro de albeitería (1583) (Franciso de
  la Reina), 51, 54–5
*Libro de los secretos de agricultura, casa
  de campo y pastoril* (1617) (Friar
  Miguel Agustin), 47, 48, 59
Lilium Medicinae (Bernard of Gordon),
  113
limpieza de sangre doctrine
  contemporary studies of, 28–9
  decreasing importance of in 19th
    century Cuba, 231–4
  difference essentialized by, 33
  and emergence of race, 26–9
  establishing proof of, 34
  and the Inquisition, 27, 31, 34–5
  and legal status, 6, 27
  parodied in *Don Quixote*
    (Cervantes), 6, 26–7, 29, 33
  replacement by other linkages of
    purity, 231
  social effects of, 34
  statutes defended by horse-trading
    analogy, 52
  and virtue, 6–8
  see also blood purity; Iberia, early
    modern; whitening equations
Lincoln, Abraham, 220
Linnaeus, Carl
  and concept of race, 16, 191–6, 204–6
  distinction of species and varieties,
    194–6
  and humoralism, 16, 203–4
  interest in differences among
    continents, 199
  later views on human diversity,
    204–5
  legacy of geopolitical stereotypes,
    193, 204–5

Linnaeus, Carl – *continued*
  misconceptions regarding, 16, 192
  and the term "race," 194
  use of geographic grid for accumulation of facts, 199–200
  see also *Systema Naturae* (1735) (Linnaeus); *Systema Naturae* (1738)
Llave de albeytería (1734) (Domingo Royo), 51
Locke, John, 152
Long, Edward, *History of Jamaica* (1774), 86
Loomba, Ania, 118
Lope de Vega, 39

Maceo, Antonio, 243
Manganelli, Kimberly Snider, *Transatlantic Spectacles of Race: The Tragic Mulatta and the Tragic Muse (2012)*, 86
Maravillas de naturaleza (1629) (Manuel Ramírez de Carrión), 48, 49, 50, 54
Marcgrave, Georg, 197
Marlowe, Philip, "Passionate Shepherd," 75
Marshall, Edmund. See *Edmund and Eleonora* (1797) (Edmund Marshall)
Martí, José, 244
Martínez, María Elena, 241
"A Masque Presented at Ludlow Castle" (John Milton). See *Comus*
Mathews, Mrs Charles
  Memoirs of a Scots Heiress (1791), 86–7, 92–4, 94–5, 100
  see also Miranda Vanderparcke (character)
Matthew 27:25 (New Testament)
  as prooftext of hereditary Jewish bleeding disease, 107, 109, 114
  as prooftext of hereditary servitude of Jews, 115–16
Mazzolini, Renato, 206
McHale, Brian, 37
McMullan, Gordon, 75
Medin, Douglas, 45

melancholy humor
  applied to body politic in *Anatomy of Melancholy* (Burton), 178–80
  identified as Jewish complexion, 111–12
  linked with hemorrhoidal bleeding, 110, 111–12, 113
  linked with Jewish bleeding disease, 112, 113
Melanchthon, Philip, *Liber de anima*, 10
Memoirs of a Scots Heiress (1791) (Mrs Charles Mathews), 86–7, 92–4, 94–5, 100
menstruation
  hemorrhoidal bleeding analogized to in *Problemata varia anatomica*, 120, 125n51
  Jewish bleeding disease analogized to, 107, 108, 109, 114, 119, 120
Merchant of Venice (Shakespeare), and Jewish bleeding disease, 11, 119, 120
mestizo
  application of term, 49
  legal status of, 57–8
  whitening equations for, 56–8
Mestre, Antonio, 238
metalepsis
  as fictional device, 37
  and metamateriality in *Don Quixote* (Cervantes), 28, 36–7, 39–40
metamateriality (in *Don Quixote* (Cervantes)), 28, 36–7, 39–40
  and metalepsis, 28, 36–7, 39–40
  and representations of Arab-Iberian culture, 26–7
  and text-as-body, 27, 33, 35, 39
  used to parody *limpieza de sangre*, 6, 26–7, 33
Michie, Elsie B., 84–5
milk
  as concocted blood, 4, 46, 50
  and wet-nurse selection, 4–5, 20n12, 49–50
Mill, John Stuart, 163
Milton, John. See *Comus* (John Milton)
mind/body relationship

in "First Dream" (Sor Juana), 12, 128, 136–42
and humoralism, 9–11
in Neoplatonism, 132
and Thomist Aristotelianism, 132–3
Miramon, Charles de, 4
Miranda Vanderparcke (character in *Memoirs of a Scots Heiress*)
  narrative negation of, 86–7, 94–5, 100
  "respectable" complexion of, 92–3
  "tainted" geneology, 93–5
  "mixed blood," and Jefferson's descendants, 146–7, 160–1
Monardes, Nicolás, 49
"moral causes," and character formation, 148–50
Moret, Segismundo, 244
moriscos
  cultural identity preserved in Arab texts, 31–2
  cultural status reflected in *Don Quixote's* "genealogy," 26–7, 28, 29
  defined, 6, 26
  repression of, 31, 43n16
  see also New Christians; Old Christians
Mrs Dulman (character in *Douglas* (Bisset))
  depiction as unattractive, 90–1, 103n41
  narrative negation of, 86–7, 92
  strategies to transform complexion, 91–2, 100
mudéjars, 6
mulatto heiresses
  African blood linked with African respectability by, 97–101
  as reflection of historical fact, 86
  responses generated by, 96
  transformations of complexion, 98–100
  as vehicle for social reform, 9, 85–7, 95–6, 98–101
  vs wealthy women of color, 8–9, 86–7
  see also Alicia Seldon (character in *Edmund and Eleonora*); complexion; "feminine mulattaroon"; Olivia Fairfield (character in *The Woman of Colour*); "rich woman"

narrative negation, and stabilization of prejudice, 86–7, 89–90, 92, 94–5, 100
nation
  and concept of national family, 14–15
  dynasty supplanted by in Jefferson's writings, 150–1, 154
  metaphors of blood applied to, 16–17, 216–17, 222–4, 231–4, 245–6
  shared bloodshed as authorizing membership within, 14, 156–7
national character
  black people excluded in Jefferson's writings, 13–14, 152–3
  Jefferson's concept of, 13, 155
  shaped by "moral causes" in Hume's political theory, 148–50
  see also character
nationhood
  and Civil War metaphors of blood, 17, 217, 222–4
  and postbellum African American writers, 224
Nebrija, Antonio de, 49
Neoplatonism
  influence on "First Dream" (Sor Juana), 128, 138, 142
  soul as independent of body in, 132
New Christians
  impact of blood purity doctrine on, 34
  status of, 6
  see also moriscos; Old Christians
"The New Pen and Old Graveyards" (James McCune Smith), 162–3
New Testament, as authority for Jewish servitude, 114–15
noble blood
  associated with depravity in *Comus* (Milton), 7–8, 77, 80
  associated with horses, 50–2
  equated with virtue in *Il pastor fido* (Guarini), 69, 71
  and humoralism, 4–5
  and Renaissance ideology, 65–7
  satirized in *Don Quixote* (Cervantes), 6–7, 26–7, 29, 33

noble blood – *continued*
   satirized in *The Faithful Shepherdess* (Fletcher), 67–8, 71–2
   and wet-nurse selection, 4–5
   see also bloodlines; *raza*
Notes on the State of Virginia (Thomas Jefferson). *See* Query IV *of Notes on the State of Virginia* (Thomas Jefferson)
Nussbaum, Felicity, 88

"Of National Characters" (1748) (David Hume), 13, 148–50
The Old Arcadia (Philip Sidney), 68–9
Old Christians
   fear of Arabic texts, 32
   obsession with blood purity, 34
   passing as, 34–5
   status of, 6
   see also moriscos; New Christians
Olivia Fairfield (character in *The Woman of Colour*)
   endogamous marriage, 97
   as figure for social reform, 85, 87, 95, 98–101
   as object of disgust, 95–6
   as reincarnation of African mother, 98, 99–100
   similarities to Alicia Seldon, 86
   as "unportioned," 96–7
Omnes homines, 113–14
On the Fourteenth Query of Thomas Jefferson's Notes on Virginia (James McCune Smith), 163
Oostindie, Gert, 93
El Orinoco ilustrado (1741–45) (José Gumilla), 53–4, 55, 56–7, 58, 60, 63n59
Oroonoko (1688) (Aphra Behn), 8, 99–100
Oudin, César, 49

Palmer, Benjamin Franklin, 217
Pasnau, Robert, 131–2
"Passionate Shepherd" (Christopher Marlowe), 75
The Passions of Mind (1601) (Thomas Wright), 11
Passman, Bert, 93

pastoral
   appropriated as courtly mode, 66, 67, 68–9
   bloodlines as divine in, 69–79
   classical origins, 67, 68
   duality in, 68
   Italian expressions of, 69–70
   Renaissance flowering of, 68
   seen as escapist by modern readers, 66
   used to interrogate ruling ideologies, 66
Il pastor fido (Giovanni Battista Guarini)
   botanical motifs, 70, 72
   contrasted with *The Faithful Shepherdess* (John Fletcher), 70–2, 74
   and divine bloodliness, 69–70
   performance in England, 70
Patterson, Annabel, 67, 74
Paul (in New Testament), as authority for Jewish servitude, 114–15
Paz, Octavio, 141
Philip II (of Spain), Royal Decree of 1566, 31
Poey, Felipe, "The Unity of the Human Species," 238
polygenesis, and David Hume, 149–50
Pomponazzi, Pietro, 9–10, 11, 132
Pragmática, 31, 43n16
prejudice
   corrected by mulatto heiresses, 85–7, 95–6, 98–101
   stabilization by narrative negation, 86–7, 89–90, 92, 94–5, 100
Problemata Aristotelis. *See Problemata varia anatomica*
Problemata varia anatomica
   English publication history, 119
   on hemorrhoidal bleeding, 120, 125n51
   on theological origin of Jewish bleeding disease, 113–14
Pruner, Franz Ignatz (Pruner Bey), 237
Prynne, William, *Histriomastix*, 76
Psalm 78:66, as textproof of Jewish bleeding disease, 108, 113
Pseudodoxia Epidemica (Thomas Browne), 121
Puttenham, George, 66

Index   271

Query XIV of Notes on the State of Virginia (Thomas Jefferson)
  blood as possible source of skin color in, 153, 222n24
  blood purity attached to national character, 153–4
  dynasty supplanted by nation in, 154
  racial mixing avoided by expatriation, 154
  referenced in *Clotel* (Brown), 161
  see also Declaration of Independence (draft of Thomas Jefferson); Jefferson, Thomas

race concept
  and animal husbandry, 3–4
  applied to national groups by Defoe, 9
  and bodily difference, 1–3
  and Cuban anthropologists, 235–9
  as cultural, 2
  emergence in early modern Iberia, 26–9
  as fluid in eighteenth century, 193
  and folkbiological essentialism, 45
  and Georges Buffon, 194
  and Hume's polygenesis, 149–50
  and Linnaeus, 191–6, 204–6
  see also bloodlines; *raza*
race, hereditary. *See* bloodlines
racial difference
  as accidental in *En busca del eslabón* (Calcagno), 239
  and changing paradigms in Cuba, 236
  and Cuban anthropologists, 235–9
  see also human diversity; human hybridization
racial mixing
  approach of European and Cuban anthropologists contrasted, 237–8
  and articulation of "Latin race," 244
  and Jefferson's family, 146–7, 160–1
  legitimized in *Edmund and Eleonora* (1797) (Edmund Marshall), 101
  and US propaganda on Cuban intervention, 245
racism, scientific. *See* scientific racism
Ramírez de Carrión, Manuel

  color as essence, 48, 54
  inference on wet-nurses from animal domain, 49–50
  on whitening of blackness, 54
raza
  duality of meaning, 46
  as essence, 47–8, 55
  as hereditary, 46–8, 60
  see also bloodlines; noble blood
"The Reformation of the Mulatto Hero/Heroine" (Tia Gafford), 86
Reina, Francisco de la, 51, 54–5
religious difference, humoral construction of, 10–11
Reply to Faustus (Saint Augustine), 114–15
Resnick, Ivan, 119, 125n51, 126n63
"The Response of the Poetess to the Very Illustrious Sor Filotea de la Cruz" ("La Respuesta de la Poetisa a la muy ilustre Sor Filotea de la Cruz") (Sor Juana Inés de la Cruz)
  distinction between "inclination" and "will" in, 128–9, 130–1, 133–5
  obedience to God vs others contrasted, 130–1, 133
  "pleasure" as mediator between "will" and "inclination," 134–5
  as response to Church criticism, 129–30, 143n12
  see also "First Dream" (Sor Juana)
Revista de Cuba
  new vocabularies applied to race concept, 236
  publication of anthropological studies, 234–5
Reyes, Agustín, 240
"rich woman"
  and disgust of money, 85–6
  effect of juxtaposition with "feminine mulattaroon," 85–6, 87
  narrative negation of, 84–5, 86–7
  as trope, 84–5
  see also mulatto heiresses
Rodríguez, Juan Carlos, 28
Royal Decree of 1566 (Philip II of Spain), 31
Royo, Domingo, *Llave de albeytería* (1734), 51

Saco, José Antonio, 232–3
Saillant, John, 158
salta-atrás (jump-back) principle, 53, 55–6
Sanborn, Geoffrey, 218
Sancho, Ignatius, 163
Sande y Lago, Fernando de, 51, 55
Sanguily, Manuel, 243
scientific racism
  and biology/culture linkage, 118
  vs folkbiological essentialism, 47, 56
  and Foucault, 18–19n1, 56
  and Linnaeus's medulla-cortex theory, 204
  and naturalized conceptions of difference, 1–2
Servetus, Michael, 171
Shakespeare, and Jewish bleeding disease, 11, 119, 120
Shapiro, James, 119, 122
Sidney, Philip, *The Old Arcadia*, 68–9
Simonsohn, Shlomo, 116
skin color
  and blushing vs flushing, 218
  as first mark of distinction in *Systema Naturae* (Linnaeus), 191–3, 200
  as fixed racial designation in eighteenth-century Britain, 87–8
  linked with blood in folkbiological essentialism, 7, 54–5, 57
  and Linnaeus's geohumoralism, 203–4
  localized by Marcgrave, 197
  Thomas Jefferson on, 153, 222n24
  as variable in Linnaeus's writings, 192–3, 195–6
skin thickness, and yellow fever infection, 235
slavery
  and blood imagery in postbellum black literature, 224
  criticized in *Clotel* (Brown), 161, 163
  fused with blood in *Liberty Further Extended* (Haynes), 155–8
  as hereditary through mother, 57, 87

Sloan, Philip R., "The Gaze of Natural History," 194
Smith, Adam, 187
Smith, James L. (former slave), 211, 224
Smith, James McCune
  concept of blood, 163
  concept of character, 162–3
  criticism of Jefferson, 163
  essay on human brotherhood and the meaning of "Communipaw," 163
  see also "Heads of the Colored People" (McCune Smith); *On the Fourteenth Query of Thomas Jefferson's Notes on Virginia*
Sociedad Antropológica de la Isla de Cuba (Anthropological Society of the Island of Cuba)
  approach to racial mixing, 236–7, 238
  influence on public discourse, 235, 240–1
  political significance of, 240–1
  race concept reformulation, 236
"Solitudes" (Soledades) (Luis de Góngora), 128
Some Historical Account of Guinea (1771), 157
Soriano Vallès, Alejandro, 137, 138–9, 140–1
soul
  attempt to escape body in "First Dream" (Sor Juana), 12, 128, 136–42
  as bound to body in Aristotelianism, 132–3
  Galen's definition, 10
  as independent of body in Neoplatonism, 132
  as manifest in body, 9–11
  as non-corporal in Christianity, 132
Spanish Constitution of 1812
  distinction between indigenous and African men, 231–2
  and shift in signification of blood, 232
Spanish Royal Academy of Language *Diccionario*

on *mestizo*, 49
on *raza*, 47
Spenser, Edmund, 67, 73
Sponsalia Plantarum (1746), 196
Stevens, George Alexander, *The Dramatic History of Master Edward, Miss Ann, Mrs. Llwhyddwhuydd, and Others* (1743), 86–7, 88–90, 94–5
Stevens, Thaddeus, 216–17
Stolke, Verena, 232
Suarez, Francisco, 133, 139
substantive soul
 challenged by Christianity, 132
 in "First Dream" (Sor Juana), 141
 in Thomist Aristotelianism, 132, 139
Systema Naturae (1735) (Carl Linnaeus)
 entry for *Homo sapiens*, 190–2, *192*
 human diversity correlated with four continents, 191–3, 197, 198
 human varieties interpreted as subspecies, 194
 possible anthropological sources, 196–9
 skin color as first mark of distinction, 191–3
Systema Naturae (1758) (Carl Linnaeus)
 annotated page, *202*
 classification as fluid, 201–2
 entry for *Homo sapiens*, 201
 expansion of 1735 classification, 200–4
 geohumoralism in, 16, 203–4
 medulla-cortex theory, 204–5

Tasso, Torquato, and pastoral mode, 69
Taylor, Susie King, 224
Templador veterinario de la furia vulgar en defensa de la facultad veterinaria o medicina de bestias (1727) (Francisco García Cabero), 51
Ten Years' War (Cuba 1868–78), 233–4
Tesoro de la lengua castellana o española (Sebastián de Covarrubias) 1611 dictionary, 7, 30, 46, 49

Thomas of Cantimpré, 109
Thomist Aristotelianism
 debate regarding mind/body relationship, 132–3, 138
 debates on mediation of material stimuli and immaterial soul, 139
 and folkbiology, 47, 51
 and "intentional" vs "sensible" species, 139
 and philosophical framework for "First Dream" (Sor Juana), 128–9, 135
 substantive soul as concept in, 132, 139
 see also Aquinas, Thomas
throwback. *See* jump-back (*salta-atrás*) principle
Transatlantic Spectacles of Race: The Tragic Mulatta and the Tragic Muse (2012) (Kimberly Snider Manganelli), 86
Trésor des deux langues françoise et espagnolle (1607) (César Oudin), 49
"A True Born Englishman" (1701) (Daniel Defoe), 9, 14–15
Truth, Sojourner, 224

"The Unity of the Human Species," Felipe Poey, 238
University of Paris, debate on Jewish bleeding disease, 111–12
utilitarianism, in *Anatomy of Melancholy* (Burton), 179–80
utopianism, in *Anatomy of Melancholy* (Burton), 180

Vallerius, Harald Johannson, 197–8
Verdadera albeytería (1707) (Pedro García Conde), 48
Viera, Antonio (Portuguese Jesuit), 129–30
Virgil, and the pastoral, 67, 68
virtue
 conveyed through bloodlines, 5, 9
 as effect of humoral complexion, 5
 embodied in rustic figures in *The Faithful Shepherdess* (John Fletcher), 72, 74

virtue – *continued*
  equated with nobility in *Il pastor fido* (Guarini), 71
  equated with rank in royal masques, 78
  as essence, 48
  and *limpieza de sangre* doctrine, 6–8
  as product of labor in *Comus* (John Milton), 8, 75, 78–81
Vitry, Jacques de, 108–9, 115

Walker, David, *Appeal in Four Articles; ... to the Coloured Citizens of the World, but in Particular, and very Expressly, to Those of the United States of America* (1829), 158–9
Wealth of Nations (1776) (Adam Smith), 187
wealthy women of color
  vs mulatto heiresses, 8–9, 86–7
  prejudice stabilized by failures to transform complexion, 86–7, 90, 92, 94–5, 100
  types of, 88
  see also Chloe Llwhyddwhuydd (character in *The Dramatic History of ... Mrs. Llwhyddwhuydd* (Stevens)); complexion; Miranda Vanderparcke (character in *Memoirs of a Scots Heiress* (Mathews)); Mrs Dulman (character in *Douglas* (Bisset))
Weyler, Valeriano, 245

Wheatley, Phyllis, 163
Wheeler, Roxann, 87
White, John S., 176
whiteness, as essence, 54–5
whitening equations
  for human hybridization, 53–4, 56–7, 60
  in Jefferson's 1815 letter, 154–5
  origins in animal husbandry, 45–6, 58
  see also degeneration equations; hybridization
"will"
  and "appetite" in Thomistic philosophy, 131–3
  vs "inclination" in "Response" (Sor Juana), 128–9, 130–1
The Woman of Colour (anonymous)
  colonial money used to perpetuate ownership in, 96–7
  Oroonoko (Behn) as precedent, 99–100
  as reflection of historical fact, 86
  treatment of African mother's blood, 98
  as vehicle for social reform, 85–7, 95–6, 98–101
  see also Olivia Fairfield (character)
Wright, Thomas, *The Passions of Minde* (1601), 11

yellow fever infection, and skin thickness, 235

GPSR Compliance
The European Union's (EU) General Product Safety Regulation (GPSR) is a set of rules that requires consumer products to be safe and our obligations to ensure this.

If you have any concerns about our products, you can contact us on

ProductSafety@springernature.com

In case Publisher is established outside the EU, the EU authorized representative is:

Springer Nature Customer Service Center GmbH
Europaplatz 3
69115 Heidelberg, Germany

www.ingramcontent.com/pod-product-compliance
Lightning Source LLC
Chambersburg PA
CBHW071615100426
42873CB00004B/53